Central and Southern Appalachian Sutures: Results of the EDGE Project and Related Studies

Edited by

Lynn Glover, III
Department of Geological Sciences
Virginia Polytechnic Institute and State University
Blacksburg, Virginia 24061-0420

and

Alexander E. Gates
Department of Geology
Rutgers University
Newark, New Jersey 07102

SPECIAL PAPER

314

1997

Published by The Geological Society of America, Inc.
3300 Penrose Place, P.O. Box 9140, Boulder, Colorado 80301

Printed in U.S.A.

GSA Books Science Editor Abhijit Basu

Library of Congress Cataloging-in-Publication Data

Central and southern Appalachian sutures : results of the EDGE Project and related studies / edited by Lynn Glover, III, and Alexander E. Gates.

p. cm. -- (Special paper ; 314)
Includes bibliographical references.
ISBN 0-8137-2314-0
1. Suture zones (Structural geology)--Appalachian Region.
2. Plate tectonics--Appalachian Region. I. Glover, Lynn.
II. Gates, Alexander E., 1957- . III. Series: Special Papers (Geological Society of America) ; 314.
QE511.48.C46 1997
551.8'2'0975--dc21 96-39846
CIP

Cover: Upper plate model of Middle Atlantic states continental crust, eastern Goochland (1 Ga) basement version.

10 9 8 7 6 5 4 3 2 1

Contents

Preface

Thirty-three years ago a seminal paper, "History of Ocean Basins" by Harry Hess (1962), gave a mechanism and a measure of credibility to the early twentieth-century hypothesis of continental drift. That paper led directly to the modern theory of plate tectonics, now recognized as a revolutionary paradigm of Earth processes. In view of the acceptance of the theory, particularly in explaining Mesozoic and Cenozoic tectonics, it is all the more remarkable that today there is no general agreement on the number, age, and location of sutures in the Appalachian orogen. It is a curious phenomenon that although these are critical, first-order questions of ancient plate tectonics, relatively few investigators are directly probing them.

A major factor contributing to the slow progress in defining plate boundaries is a shortage of money to finance sophisticated field mapping studies. Laboratory investigations of age, geochemistry, and metamorphism rely on such maps. Only about 30 percent of the Blue Ridge and Piedmont are covered by modern geological mapping at 1:24,000 scale. Thus specialized studies cannot always be conducted in the optimum localities or for the best reasons.

This volume contributes to the understanding of sutures and plate margins in the central and southern Appalachians. Field mapping is the foundation of most of the studies. Our hope is that these papers will contribute to a quickening of research on these problems, including applications of the splendid cache of geoscience specialties we can now bring to bear on them. Perhaps in the future we may yet know whether the theory of plate tectonics is adequate to explain the evolution of the Appalachian orogen or, more likely, whether Appalachian studies will contribute to modifications of the theory.

We dedicate this volume to the geologists and geophysicists of the U.S. Geological Survey, state geological surveys, and academic institutions who have contributed most of the field studies that form the basis of current Appalachian tectonic models. Continuing efforts to reduce and restructure all of these organizations are being accomplished at the expense of basic research in this and many other fields. This trend does not bode well for the future of tectonics studies in the Appalachians. We may be in for a long dry time, but our intellectual curiosity will eventually replenish this effort. The human condition depends on the idea that "Nothing is so practical as theory."

In the first chapter, Hooper, Hatcher, Troyer, Dawson, and Redmond describe the character of the metavolcanic Avalon terrane and its boundary with the Piedmont in central Georgia. The boundary occurs along the Ocmulgee-Middleton-Lowndesville fault and can be traced northeastward on the basis of changes in stratigraphy, metamorphic grade, and aeromagnetic signature. To the southwest the location of the boundary is problematic, probably because of offset along northwest-trending brittle faults.

Hanley, Chalokwu, and Steltenpohl, in the second chapter, discuss the problem of locating this boundary southwest of the study area of Hooper et al. Field mapping, petrology, and petrochemistry do not reveal contrasts across the boundary as projected into central Georgia and Alabama. If the Carolinian/Avalonian terrane rocks are present in Alabama, they may be concealed beneath the Atlantic Coastal Plain because of offset by left-slip faults.

The third chapter by Baird and Glover, discusses fold nappe emplacement of Carolinian rocks in southern Virginia. These magmatic arc rocks appear to have been deformed in Iapetus during the ca. 620 to 546 Ma Virgilinian orogeny and then to have collided with Laurentia during the ca. 535-Ma Carolinian orogeny. The Middle Ordovician Taconian orogeny may represent continued Carolinian collision or more likely collision of eastern Carolinia with western South America. During the late Paleozoic the North American margin experienced Acadian dextral shear and Alleghanian shear and orthogonal(?) collision.

In the fourth chapter, Gates shows that the Carolinian-Laurentian boundary at Brookneal, Virginia, is a major zone of crustal weakness that was reactivated during every subsequent tectonic event. Not only did these reactivations involve significant fault movement, but they also served to localize both sedimentary deposits and igneous intrusions. Evidence is present for at least Carolinian, Taconian, Ordovician-Silurian, Alleghanian, and Mesozoic activations. Such fundamental boundaries are subordinate to major plate boundaries but act as primary recording devices for the effects of large-scale tectonism on within-plate settings.

For the past 20 years there has been general agreement that the mafic-ultramafic suite in the Blue Ridge represents an ophiolitic mélange emplaced along a suture of Late Proterozoic, Cambrian, or Ordovician age, according to various authors. Wang and Glover examine this question in the fifth chapter of the volume. They find that the principal geochemical tectonic discriminant criteria previously used to identify mid-ocean ridge basalt (MORB) components fail for continental basalts. Additionally, the field relations do not support a MORB origin. Wang and Glover conclude that the mafic-ultramafic suite is a product of magmatism during continental rifting and that there is no suture in the Blue Ridge of Virginia.

In the sixth chapter, Kasselas and Glover present field evidence in the northern Virginia Blue Ridge to show that an unbroken stratigraphic sequence exists from the Grenville basement through the Lynchburg/Fauquier Groups and into the Catoctin Formation of the Late Proterozoic Laurentian margin. Therefore there is no evidence to support a suture as recently drawn through this terrane.

The last chapter, by Glover, Sheridan, Holbrook, Ewing, Talwani, Hawman, and Wang, interprets the "Taconic"/ Carolinian suture and the Laurentian margin through the central Appalachians and offshore. Geophysical and basement sample data under the Atlantic Coastal Plain and seismic reflection interpretations along the continental edge are collated and interpreted. The authors present two new models for the Cambrian and Ordovician tectonic evolution. Then a discussion follows of upper- and lower-plate variants of each model as Mesozoic extension developed new structures that were guided by Paleozoic tectonic heredity. The chapter authors suggest that dating of the basement gneiss under the eastern Chesapeake and Delaware Bay areas might result in the most economical solution to the ambiguity between the two sets of models.

REFERENCES CITED

Hess, H. H., 1962, History of ocean basins, *in* Engle, A. E. J., James, H. L., and Leonard, B. F., eds., Petrologic studies: A volume in honor of A. F. Buddington: New York, Geological Society of America, p. 599–620.

Lynn Glover, III
Alexander E. Gates

Geological Society of America
Special Paper 314
1997

The character of the Avalon terrane and its boundary with the Piedmont terrane in central Georgia

Robert J. Hooper
Conoco Inc., P.O. Box 2197, Houston, Texas 77252-2197
Robert D. Hatcher, Jr.
Department of Geological Sciences, University of Tennessee, Knoxville, Tennessee 37996-1410, and Environmental Sciences Division, Oak Ridge National Laboratory, Box 2008, Oak Ridge, Tennessee 37831
Paulette K. Troyer, Roger J. Dawson, and Cindy G. Redmond
Department of Geology, University of South Florida, Tampa, Florida 33620-5200

ABSTRACT

The Avalon terrane in central Georgia comprises a series of layered gneisses and orthogneisses, deformed multiple times, that have been intruded by plutonic rocks of ultramafic to felsic composition. The boundary between Avalon and the more inboard Piedmont terrane can be located based on changes in stratigraphy, metamorphic grade, and aeromagnetic signature. The boundary lies within a steep local gravity gradient of ~2 mgals/km that separates low values over the Piedmont terrane from relatively high values over the Avalon terrane. Avalonian rocks are in fault contact with the Piedmont terrane throughout central Georgia along the Ocmulgee-Middleton-Lowndesville fault. The Ocmulgee fault in the Berner area is a broad zone (~2 km) of steeply southeast-dipping foliation containing localized syn- and postmetamorphic right-lateral strike-slip shear zones. The Piedmont-Avalon boundary in the Rock Eagle area is not defined by the same pronounced steep zone; instead, the layered-series gneisses dip moderately to the southeast except where modified by small, near-vertical, ductile shear zones. Gravity surveys have helped elucidate the form of gabbroic bodies within the Avalon terrane. The largest body in the Berner Mafic Complex, the Gladesville Gabbro, is not a single pluton as previously interpreted; instead, it is a series of four or five pods of either olivine-bearing or olivine-free gabbro. The Presley's Mill Gabbro, the largest gabbroic pluton in the Rock Eagle Complex, has previously been interpreted to be barely unroofed and to have a large subsurface expression. Gravity data, however, suggest that the pluton is actually quite small; the extensive thermal aureole on the northwestern side of the body is a composite of the gabbro and a neighboring granite. Late (post–thermal peak; Mesozoic?) brittle strike-separation faults pervade the Piedmont and Avalon terranes; these faults cut and offset all ductile fabrics and faults. The Presley's Mill Gabbro may have been disrupted by a left-lateral strike-separation fault. The Avalon terrane in central Georgia is not a mélange; the current distribution of rocks can be attributed in part to their original emplacement and in part to postemplacement tectonic disruption.

Hooper, R. J., Hatcher, R. D., Jr., Troyer, P. K., Dawson, R. J., and Redmond, C. G., 1997, The character of the Avalon terrane and its boundary with the Piedmont terrane in central Georgia, *in* Glover, L., III, and Gates, A. E., eds., Central and Southern Appalachian Sutures: Results of the EDGE Project and Related Studies: Boulder, Colorado, Geological Society of America Special Paper 314.

INTRODUCTION

Avalonian rocks in central Georgia comprise a distinctive suite of metavolcanic and metaplutonic rocks that can be traced continuously from the area of the Berner Mafic Complex eastward to the Siloam Granite. The boundary between the Avalon and the more inboard Piedmont terrane can be located based on changes in stratigraphy, metamorphic grade, and aeromagnetic signature. The boundary lies within a steep gravity gradient that can be traced through the southern Appalachians for more than 1,500 km.

In this chapter we review stratigraphic, metamorphic, structural, and gravity data from the Avalon terrane in central Georgia (Fig. 1). We use the term "Avalon terrane" as originally defined by Williams and Hatcher (1983). The area described herein lies within the Juliette terrane of Horton et al. (1991). The character of the Piedmont and Avalon terranes in the Berner area has been described previously (Hooper, 1986; Hooper and Hatcher, 1988, 1989a, b). In the Berner area, Piedmont terrane rocks, carried in the hanging-wall of the Box Ankle fault, are separated from arc-affinity rocks of the Berner Mafic Complex by the Ocmulgee fault (Fig. 1). The distinctive suite of metaigneous rocks that comprise the Berner Mafic Complex have been traced 50 km to the east directly into the Rock Eagle Complex (Fig. 1). The Rock Eagle Complex is named after the Rock Eagle Lake quadrangle. The northern boundary of the Rock Eagle Complex is the Middleton-Lowndesville fault zone that separates the complex from rocks of the Piedmont terrane. The Rock Eagle Complex can be traced eastward at least as far as the Siloam granite, where many of the units defined by Vincent (1984) are likely correlative.

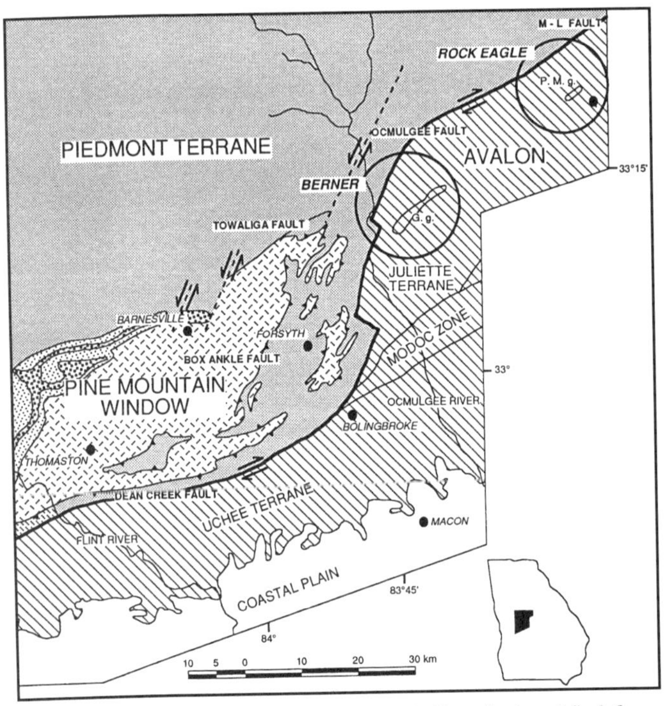

Figure 1. Tectonic sketch map of central Georgia (modified from Hooper, 1990; and West et al., 1995). The Ocmulgee–Middleton-Lowndesville fault and the Dean Creek fault separate the Avalon terrane from the more inboard Piedmont terrane. The Berner and Rock Eagle areas (circled) lie within a subdivision of the Avalon terrane named the Juliette terrane (see Horton et al., 1990; and West et al., 1995). M-L fault, Middleton-Lowndesville fault; G.g., Gladesville Gabbro; P.M.g., Presley's Mill Gabbro.

STRATIGRAPHY

The most diagnostic way of separating the Piedmont and Avalon terranes is by the abrupt contrast in stratigraphy: paragneiss, amphibolite, and schist of the Piedmont are in fault contact with metaigneous rocks of the Avalon terrane.

Piedmont terrane

The Piedmont adjacent to the Avalon terrane throughout central Georgia contains high-grade paragneiss and schist with subordinate metaigneous rocks. The dominant rock type is heterogeneous biotite gneiss (Fig. 2A). This biotite gneiss is locally amphibole bearing and in places grades into amphibole gneiss. Less common schistose layers are found throughout. Felsic orthogneisses occur on a variety of scales from thin sheets (~1 m) to plutons with outcrop areas of several square kilometers. Amphibolite occurs both as meter-scale boudins within biotite gneiss and as more extensive layers.

Minor rock types identified in the Berner area include calc-silicate–rich quartzite, found mostly as small boudins (~1 m), and a variety of pegmatites that occur both as dikes (~1 m) and as boudins of variable size up to several tens of meters long. Most pegmatites are of a simple type (quartz-feldspar-mica) but locally contain garnet and tourmaline. Graphitic quartzite layers have been found at several localities.

Avalon terrane

The Rock Eagle and Berner Mafic Complexes comprise a series of multiply deformed layered gneisses and orthogneisses that have been intruded by plutonic rocks of ultramafic to felsic composition. Although the basic makeup of the two complexes is similar, there are important differences. The distinctive felsic layered gneiss unit adjacent to the Ocmulgee fault and the extensive areas of layered mafic gneisses that characterize the Berner area have not been identified in the Rock Eagle area. The Rock Eagle Complex is dominated by layered intermixed gneisses of similar appearance to those in the Berner area. Although both areas contain gabbroic and ultramafic plutonic rocks, the Rock Eagle area has many more felsic plutons and also a distinctive augen-gneiss unit.

Layered gneiss series. The layered gneiss series is divisible into felsic, mafic, and intermixed members. The "intermixed" gneisses are the same as those previously termed "intermediate" (Hooper, 1986). The divisions are based on the volume percentage of felsic or mafic material present at any given locality and are selected as mappable units with fairly consistent internal stratigraphy.

Figure 2. Field photographs of representative rock types from the Piedmont and layered series in the Avalon terrane. A, Coarse-grained biotite gneiss from the Piedmont terrane. (Scale: U.S. quarter). (From Hooper and Hatcher, 1989a). B, Layered felsic gneiss from the Berner Mafic Complex. (Scale: hammer head = 16 cm) (From Hooper and Hatcher, 1989a). C, Layered felsic gneiss from the Berner Mafic Complex. (Scale: hammer head = 16 cm). D, Layered intermixed gneiss from the Berner Mafic Complex (from Hooper and Hatcher, 1989a). E, Layered intermixed gneiss from the Berner Mafic Complex. (Scale: hammer head = 16 cm). F, Contact zone between layered intermixed gneisses of the Rock Eagle Complex (left) and a felsic pluton (right). The outcrop is cut by numerous small displacement faults that offset the contact between the layered gneisses and the felsic pluton. (Scale: U.S. quarter).

The most conspicuous feature of the layered series rocks is the ubiquitous presence of compositional layering, reflecting either the juxtaposition of several different rock types (Figs. 2B, C, D, and E) or minor variations within a single layer. Layering occurs from centimeter to meter scale. Layering based on the juxtaposition of differing rock types is interpreted to represent transposed primary layering. Layering reflecting minor variations in the volume percentages of phases within an individual layer is interpreted to result from either original local variations or metamorphic segregation. The presence in hand specimen of relict lapilli and phenocrysts of both clinopyroxene and plagioclase and the presence in thin section of relict pyroxenes that commonly retain the form and internal oxide-needle fabric typical of unaltered pyroxenes suggest that the layered gneisses are of metaigneous origin. The layered character, where relatively thin units (~10 m) can be traced from tens to several hundred times their width, implies that they may represent a series of metavolcanic rocks.

Layered felsic gneisses. This important marker unit in the Berner Mafic Complex is found adjacent to the Ocmulgee fault; this unit has not been identified in the Rock Eagle area. The unit comprises a series of well-layered quartz-bearing felsic gneisses with subordinate volumes of more intermediate to mafic rocks that occur as layers and boudins of amphibolite (Figs. 2B and C). The dominant felsic gneiss lithology is a layered quartz, two-feldspar, two-mica gneiss that is locally garnetiferous. The gneiss is locally schistose where the mica content is sufficiently high. The rocks are predominantly dacitic and superficially resemble the more felsic lower-grade metarhyolite and rhyodacite assemblages of the Carolina slate belt. Small plutons (plugs?) of granitic to granodioritic composition locally form pavement outcrops.

Layered intermixed gneisses. The bulk of this unit is layered biotite-rich and/or hornblende-rich gneiss. Layering is formed either by the juxtaposition of layers with differing proportions of mafic phases or by local intralayer mafic-phase variation (Figs. 2D, E, and F). Equal proportions of more felsic gneiss are locally interlayered. Subordinate volumes of mafic gneiss are interlayered throughout the unit. Felsic gneisses are composed predominantly of plagioclase and quartz with varying proportions of biotite and opaque minerals. K-feldspar, amphibole, epidote, garnet, chlorite, and zircon all occur in varying parageneses. Mafic gneisses have the same mineral assemblages as the amphibolites.

Layered mafic gneisses. The layered mafic gneisses occur as amphibolites and amphibole gneisses with localized epidote-rich zones. Subordinate volumes of felsic and biotite gneiss are interlayered throughout. Layering is neither as well developed nor as visually obvious as in the felsic or intermixed units. Layering is predominantly defined by interlayer mafic-phase variation, although, at least locally, it is enhanced by grain-size variation. A differentiated compositional layering based on selective intralayer segregation of felsic and mafic phases occurs locally. All samples contain plagioclase and amphibole as their major constituents; quartz is a ubiquitous accessory. Epidote group minerals, sphene, opaque minerals, and biotite occur in variable parageneses. Epidote group minerals are locally major constituents and comprise over 40 percent of some small centimeter-scale shear zones. Layers up to 1 m thick of almost pure epidosite occur locally. A lapilli-bearing amphibole gneiss unit occurs within the layered mafic gneisses. The unit comprises a medium-grained amphibole gneiss containing highly strained ellipsoidal lapilli. The unit is highly distinctive and can be traced continuously along strike for at least 7 km (20,000 feet).

Metaplutonic series. A series of plutonic rocks of ultramafic to felsic composition intruded the layered-series gneisses. Several mafic and felsic plutonic rocks contain well-developed penetrative foliations that are concordant with local trends in the layered-series gneisses; intrusion must therefore have predated deformation. Other bodies, however, are comparatively free of penetrative deformation, possessing clearly definable thermal aureoles; they clearly postdate deformation. No systematic relationship has been observed between structural age and pluton composition.

Augen gneiss unit. A polydeformed augen gneiss is found within the layered gneisses of the Rock Eagle Complex. This rock type has not been identified within the Berner area. The rock comprises large, oval-shaped, K-spar megacrysts encapsulated in an anastomosing matrix of predominantly biotite, quartz, and mafic phases. The anastomosing character of the foliation imparts two distinct foliations to the rock.

Felsic unit. Felsic plutonic rocks include a series of deformed tonalitic to granitic rocks that are demonstrably intrusive into the layered-series gneisses. The rocks are dominantly quartz, two-feldspar, two-mica gneisses that are locally garnetiferous and locally amphibole-bearing and that contain one or, in some cases, two foliations. Both complexes contain several post-metamorphic plutons. A large postmetamorphic granitic pluton with an extensive contact aureole occurs in the northeast corner of the Rock Eagle Complex.

Mafic unit. A series of mafic metaplutonic rocks is intrusive into the layered series gneisses. Rocks range in composition from olivine gabbro to less mafic two-pyroxene gabbro. Some plutons contain strong foliations; others are essentially free of deformation, with clearly defined contact aureoles. Metaultramafic rocks occur as layers within some bodies.

Gladesville Gabbro. The Gladesville Gabbro is the largest mafic pluton in the Berner Mafic Complex. The gabbro has been regarded in most previous studies (e.g. Matthews, 1967) as a long linear body with a maximum long dimension of 12 km and an average width of 1.5 km. The body has been typically drawn with a bulbous northeastend up to 5 km wide, giving the body a distinctive "fish-shaped profile" clearly recognizable on all regional-scale maps (e.g., Pickering, 1976). The Gladesville Gabbro occupies a distinctive topographic depression, the margins of which give a very approximate indication of the margins of the pluton. Four distinct varieties of Gladesville Gabbro (types 1 to 4 of Hooper, 1986) can be identified in the field. They belong to two basic types: (1) medium-grained, olivine-bearing, two-pyroxene gabbro, with gray/blue plagioclase (bladed and interstitial) and mafic phases segregated into islands; and (2) coarse-grained, olivine-free, two-pyroxene gabbro with dark-purple lath-shaped plagioclase. A coarse-grained amphibole gabbro, with dark-purple lath-shaped plagioclase, and a fine-grained, granular, two-pyroxene gabbro are found in association with olivine-free gabbro. Olivine gabbro has not been found associated with olivine-free gabbro.

Presley's Mill Gabbro. One large gabbroic pluton, the Presley's Mill Gabbro, occurs in the Rock Eagle area. This gabbro, most recently described by Myers (1968), is poorly exposed. The dominant rock type is two-pyroxene gabbro. Unlike the Gladesville Gabbro, olivine-bearing gabbro has not been identified in association with the Presley's Mill Gabbro. No ultramafic layers have been noted. The Presley's Mill Gabbro is surrounded by a thermal aureole that contains high-grade orthopyroxene-bearing hornfels at the contact. The aureole width varies from approximately 1 km at the southern margin to more than 5 km to the north.

Ultramafic unit. Ultramafic rocks within the Berner and Rock Eagle Complexes occur in pods and layers within layered-series gneisses. All ultramafic rocks are highly altered and penetratively deformed. Many rocks have a schistose appearance defined by a strong dimensional preferred orientation of platy minerals. Differential weathering of relict pyroxenes in rocks that are less penetratively deformed imparts a granular, blocky appearance to the rock. The long dimensions of ultramafic bodies, as well as their fabric orientations, are always concordant with local trends. Pods and layers invariably display a preferred long dimension parallel with local foliation trends. Mineral assemblages and modes are highly variable. The primary mineral assemblage of most rocks has been altered to one that now includes amphibole (predominantly actinolitic-hornblende), serpentine group minerals and/or chlorite group minerals, opaque minerals, and minor amounts of talc and anthophyllite. Pseudomorphs of primary pyroxene and olivine are recognizable in areas that do not contain a well-developed penetrative fabric. Pseudomorphs after pyroxene still retain the form and internal oxide-needle fabric typical of unaltered phases; pseudomorphs of olivine often retain oxide-filled fracture networks.

METAMORPHISM

Piedmont terrane

Rocks of the Piedmont terrane juxtaposed against both the Berner Mafic Complex and the Rock Eagle Complex exhibit thermal equilibrium under at least upper amphibolite facies conditions indicated by the presence of sillimanite in appropriate alumina-rich rocks. Sillimanite occurs either as massive blades or as fibers (or packets of fibers) within biotite gneiss and schist. Kyanite has been observed in thin section from several localities. Garnet-bearing amphibolites, found locally in pods within the biotite gneisses, attest to high ambient pressures during metamorphism.

Avalon terrane

Berner Mafic Complex and Rock Eagle Complex rocks exhibit thermal equilibrium under more moderate metamorphic conditions than Piedmont terrane rocks. Determination of the regional metamorphic grade of the complex is, however, problematic in the absence of pelitic rocks. Metamorphic grade has to be determined from mafic and felsic orthogneisses, rocks that are not as sensitive to pressure-temperature variation as pelitic rocks. The highest-grade index minerals found in felsic gneisses are biotite and garnet. Amphibolitic rocks contain actinolitic hornblende (determined optically) and are all garnet free. The mineral assemblages from both felsic and mafic metavolcanic rocks thus loosely constrain the maximum metamorphic conditions reached in both complexes to at, or just below, the greenschist facies–amphibolite facies boundary.

Contact metamorphism. Low-pressure contact metamorphic rocks occur in the Berner Mafic Complex and Rock Eagle Complex, associated with both gabbroic and granitic intrusions. Contact rocks are found only in isolated outcrops, although it is inferred that aureoles form continuous shells around their plutonic cores. The exact form and dimensions of individual aureoles are difficult to assess—low-grade hornfelsic rocks are difficult to distinguish from amphibolites. Mineral assemblages from thermally altered rocks surrounding gabbroic bodies indicate conditions as high as pyroxene-hornfels facies. The thermal aureoles surrounding both the Gladesville and Presley's Mill Gabbros contain high-grade contact rocks characterized by the presence of orthopyroxene-bearing hornfels assemblages. Aureoles surrounding felsic plutons, although much wider than those surrounding the gabbroic plutons, contain lower-grade amphibole-garnet bearing hornfels.

AEROMAGNETIC DATA

The mapped trace of the Ocmulgee-Middleton-Lowndesville fault zone is broadly coincident with a distinctive change in regional aeromagnetic signature (see, for example, the map of Zietz et al., 1980). Low magnetic susceptibilities over the Piedmont and Inner Piedmont contrast sharply with high susceptibilities and high frequencies observed over the Berner Mafic Complex and Rock Eagle Complex. The contrast can be directly related to differences in rock types as well as to the depth to the Blue Ridge–Piedmont megathrust contact, which is greater beneath the Avalon terrane and less beneath the Piedmont terrane (Hatcher and Zietz, 1980). Magnetic lows are associated with paragneiss of the Piedmont, whereas the high volume of mafic material within the Avalon terrane produces magnetic highs. The gabbro bodies, such as the Gladesville and Presley's Mill Gabbros, are easily identified by their high-frequency response. High-frequency anomalies within the Avalon terrane that do not correspond with the location of known gabbro bodies are probably related to bodies in the subsurface.

GRAVITY DATA

Gravity data from the Berner and Rock Eagle areas (Figs. 3A and 4A) were acquired with two aims: to investigate the response of the Piedmont-Avalon boundary and to investigate the response of the metaigneous complexes, in particular the Gladesville and Presley's Mill Gabbros. Data over the igneous bodies were collected with a spacing of 200 m; data to define the regional gradient were acquired with a spacing of 1,000 m. Details of the data acquisition and processing parameters can be found in Dawson (1990) and Redmond (1990).

Regional gradient

The Avalon-Piedmont boundary lies within a steep regional gravity gradient. The gradient links a paired gravity anomaly that extends for ~1700 km throughout the central and southern

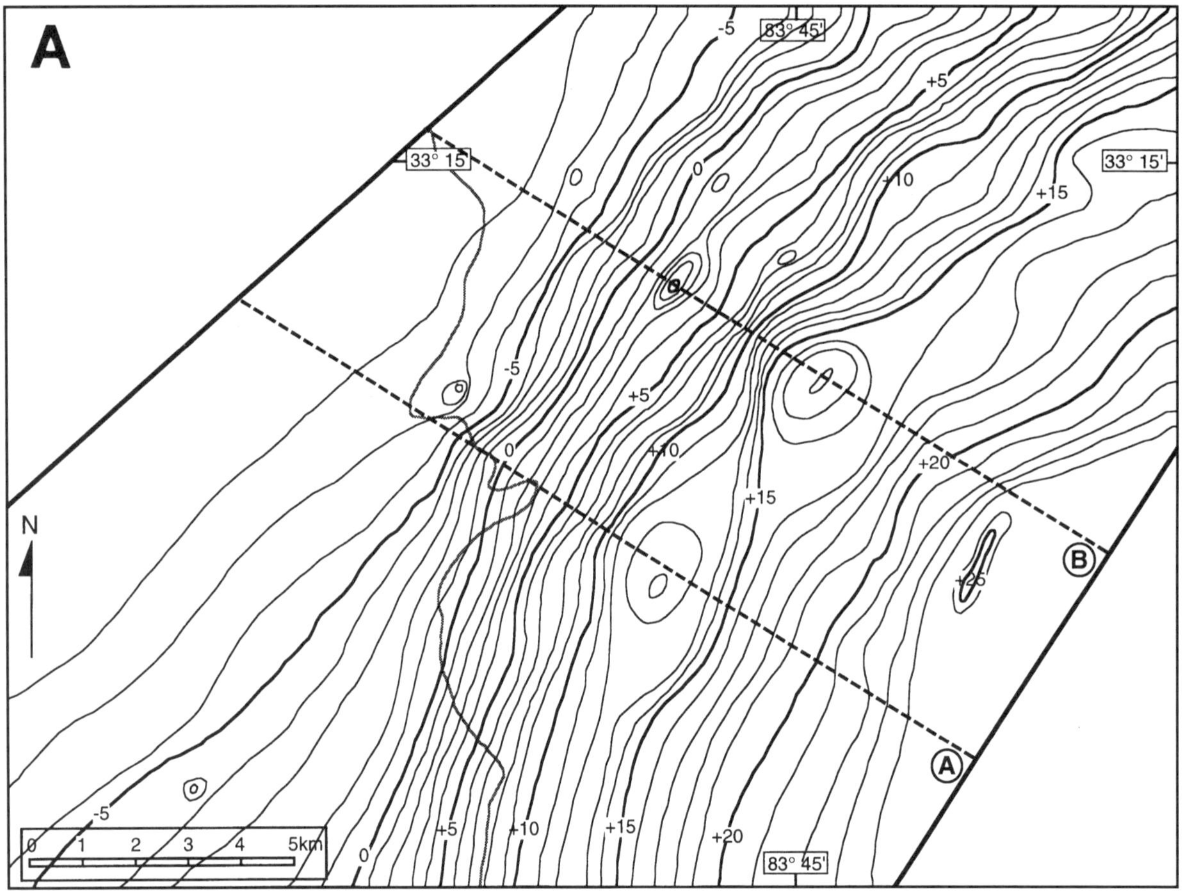

Figure 3 (on this and facing page). A, Bouguer gravity anomaly map of the Berner Mafic Complex (modified from Redmond, 1990). Note the pronounced northwest to southeast gradient from ~ –10 mgals to ~+20 mgals. The center of the anomaly map is dominated by two elliptical enclosed gravity highs with values of 14 and 18 mgal, respectively. Northeast of the highs is a break where the gradient flattens over a distance of 1.5 km between the 10- and 12-mgal contours. The enclosed gravity highs and flattened gradient coincide with the trace of the Gladesville Gabbro. An unnamed mafic body shown to the southeast of the Gladesville Gabbro of the geologic map of Georgia (Pickering, 1976) may be represented by the enclosed high to the southeast of the Gladesville Gabbro. The anomaly pattern is not compatible with a pluton having dimensions of ~12 km by 1.5 km, with a bulbous northeast end, ~5 km wide, as previously interpreted (Matthews, 1967; Hooper, 1986). The Gladesville Gabbro is most likely a composite of several plutons (Fig. 3C) that may have been displaced relative to each other by late brittle faults. The locations of profiles A and B from Figure 3B are marked. B, Profiles and residual profiles (A and B) across the Gladesville Gabbro. The profile locations are indicated on Figure 3A. The average residual anomaly for the pluton, with local gradients removed, is only 4 to 6 mgals, with a corresponding maximum anomaly width of no more than 4 km. C, Geological sketch map of the Berner Mafic Complex (modified from Redmond, 1990). Primary gravity contours from Figure 3A are overlain. The locations of profiles A and B from Figure 3B are indicated.

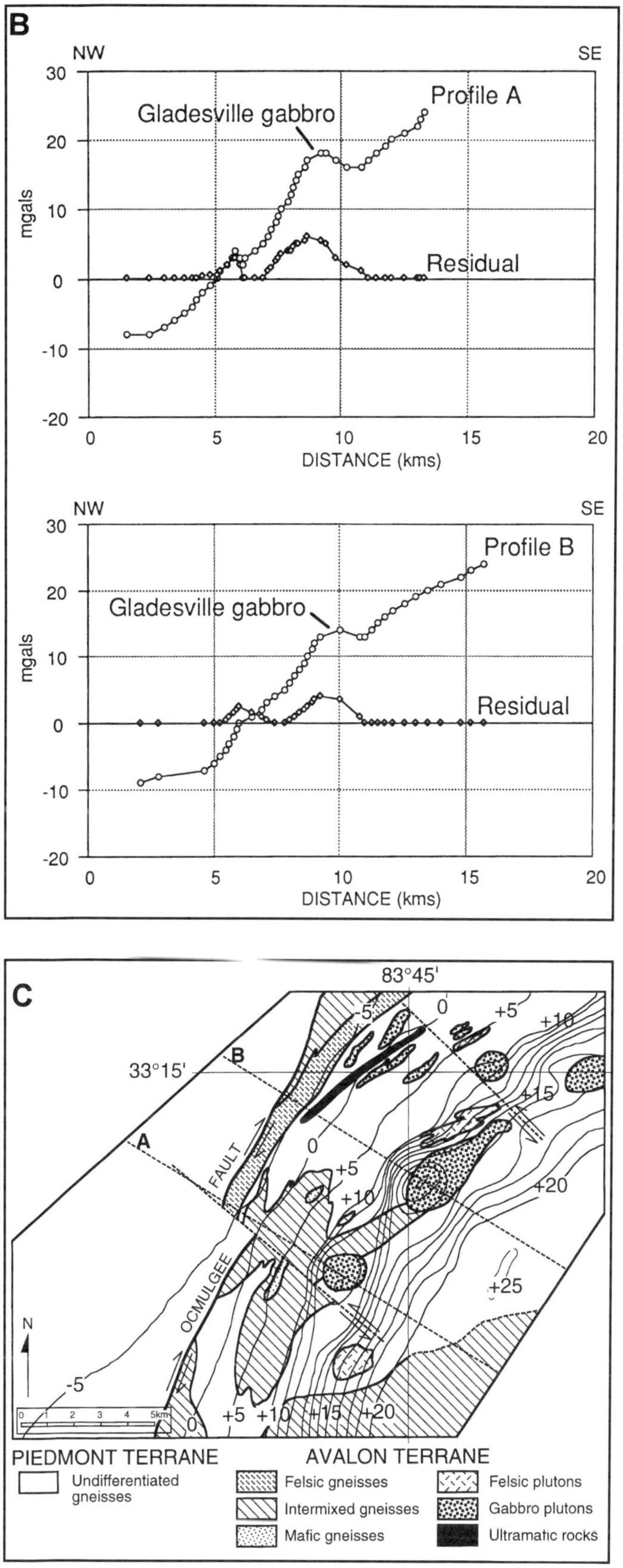

Appalachians. The significance of these anomalies was discussed by Hatcher and Zietz (1980) and Thomas (1983). The gravity gradient in central Georgia has a width of more than 100 km and a change of ~90 mgal from a low of −70 mgals over the Piedmont to a high of more than 20 mgals over the Avalon terrane. The regional gradient is related to a fundamental change in the deep crust. One of the first studies of the regional gradient was by Obaoye (1979), who ran several detailed gravity traverses in eastern Georgia, one of which crossed the Piedmont-Avalon boundary. He concluded that an abrupt change in crustal character was not apparent in the data and considered the crust to be transitional. Thomas (1983), however, concluded that the regional gradient was related to a major collisional suture that juxtaposed thick continental crust underlying the Piedmont terrane against thin oceanic crust underlying the Avalon terrane.

Local gradient

The regional Appalachian gravity gradient (~1 mgal/km) is not the only background gravity element in the Berner and Rock Eagle areas. Data from both the Berner and Rock Eagle areas are dominated by a local background gradient (~1.7 mgal/km in the Berner area and ~2.0 mgal/km in the Rock Eagle area) that is significantly higher than the regional Applachian gravity gradient. The local gradient separates low gravity values of the Piedmont from positive gravity values of the Avalon terrane. The origin of the gradient is most likely the contrast in density across the Piedmont-Avalon boundary caused by the juxtaposition of paragneiss and schist of the Piedmont with the higher-density mafic rocks of the Berner and Rock Eagle Complexes. Modeling of the profiles (Dawson, 1990; Redmond, 1990) supports the observation from the COCORP data of Nelson et al. (1985) that the Piedmont-Avalon boundary dips moderately to the southeast.

Pronounced gravity highs are commonly associated with Piedmont gabbros. Some appear as round features (e.g., the Buffalo gabbro, South Carolina) others as more elongate features (e.g., the Dutchmans Creek gabbro, South Carolina) (Popenoe and Bell, 1975). In contrast, the Gladesville and Presley's Mill Gabbros of the Avalon terrane do not cause pronounced gravity highs. Rather, they appear as undulations or slope changes that perturb an otherwise fairly uniform local gravity gradient of ~2 mgals/km (Figs. 3B and 4B). The local gradient must therefore be quantified and removed (see Dawson, 1990; Redmond, 1990) before interpreting the residual gravity anomalies associated with the plutons (Figs. 3B and 4B).

Berner area

The most prominent feature of the bouguer anomaly map of the Berner area is the pronounced gradient from ~ −10 mgals to ~+20 mgals (Fig. 3A). The center of the anomaly map is dominated by two elliptical enclosed gravity highs with values of 14 and 18 mgal, respectively. Northeast of the highs is a

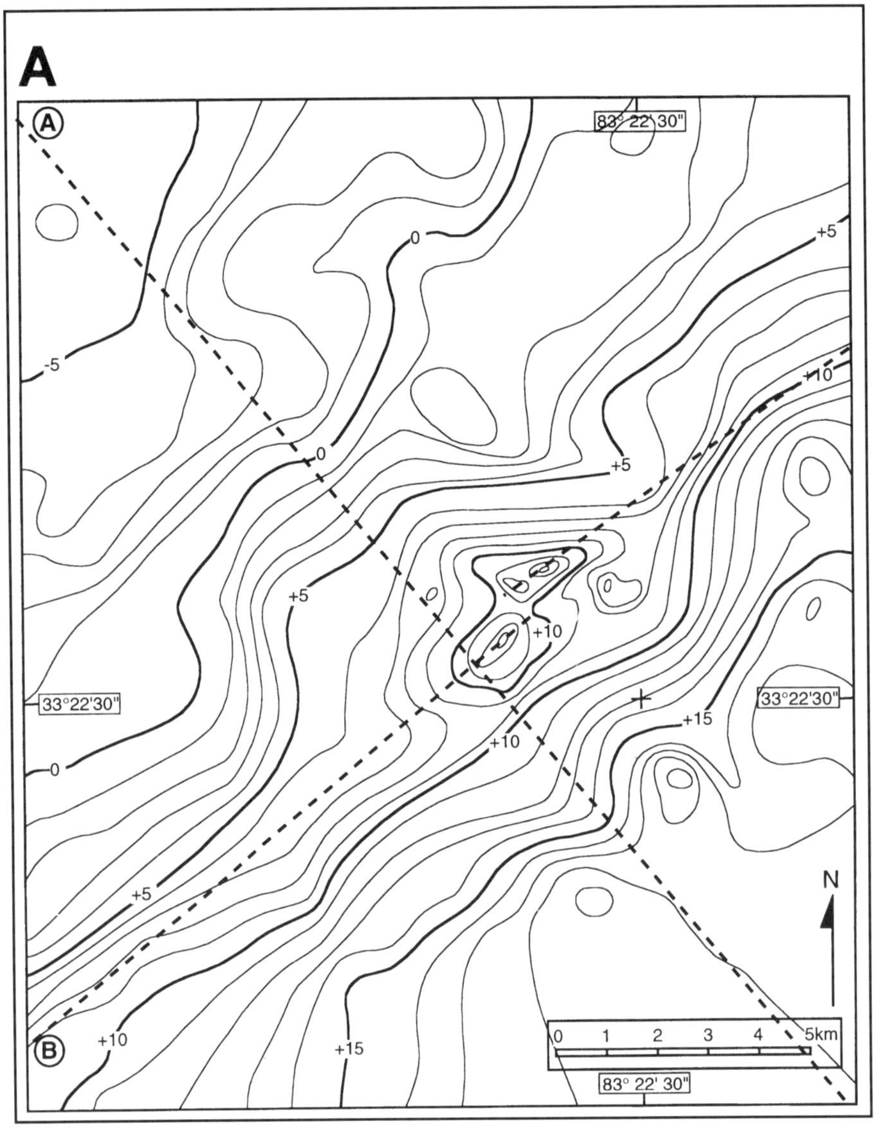

Figure 4 (on this and facing page). A, Bouguer gravity anomaly map of the Rock Eagle Complex (modified from Dawson, 1990). The lowest gravity anomaly, –6 mgals, is located in the northwest corner of the map and corresponds to rocks of the Piedmont terrane. Anomaly values higher than –2 mgals correspond to rocks of the Avalon terrane. The trace of the Middleton-Lowndesville fault (Fig. 4C) roughly follows the –2-mgal contour. The flat area in the northeastern corner of the anomaly map is coincident with the location of a large granitic body. The closed circular gravity high in the center of the anomaly map corresponds to the location of the Presley's Mill Gabbro (Fig. 4C). The double-humped pinched character of the anomaly could be caused by faulting or by two closely spaced bodies. The locations of profiles A and B from Figure 4B are marked. B, Profiles and residual profiles (A and B) across the Presley's Mill Gabbro. The profile locations are indicated on Figure 4A. The pluton has an average residual anomaly with local gradients removed of 6 or 7 mgals. The maximum anomaly width is up to 10 km. The asymmetrical character of the residual anomaly of the pluton (Profile A) may be related to a shallow northerly dip of the pluton. The double-humped residual expression along the body (Profile B) could be caused by faulting or by two closely spaced bodies. C, Geological sketch map of the Rock Eagle Complex (modified from Dawson, 1990). Primary gravity contours from Figure 4A are overlain. The locations of profiles A and B from Figure 4B are indicated.

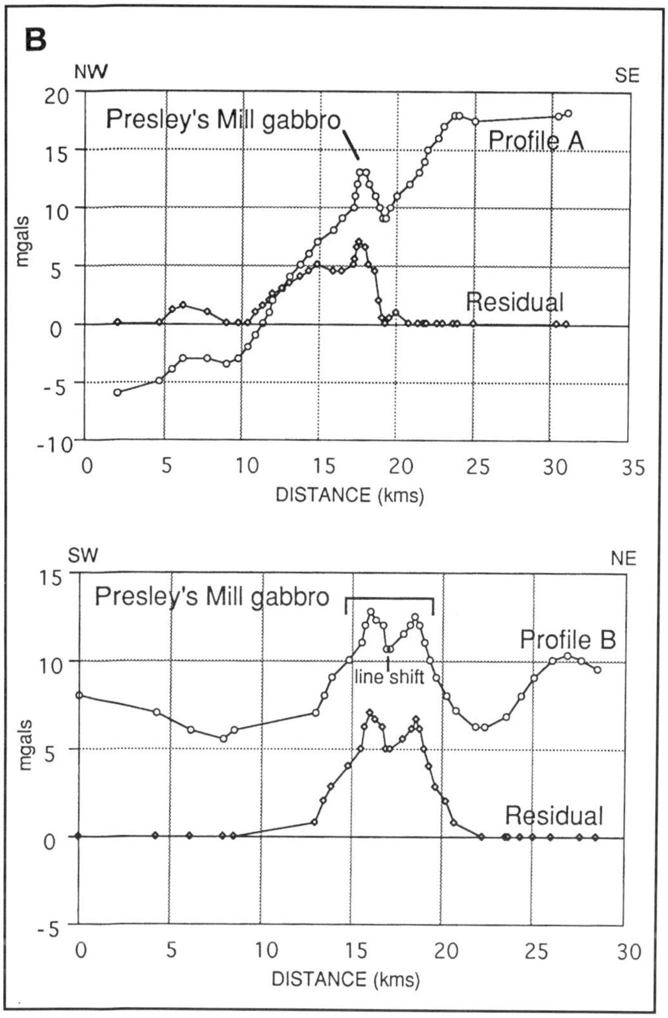

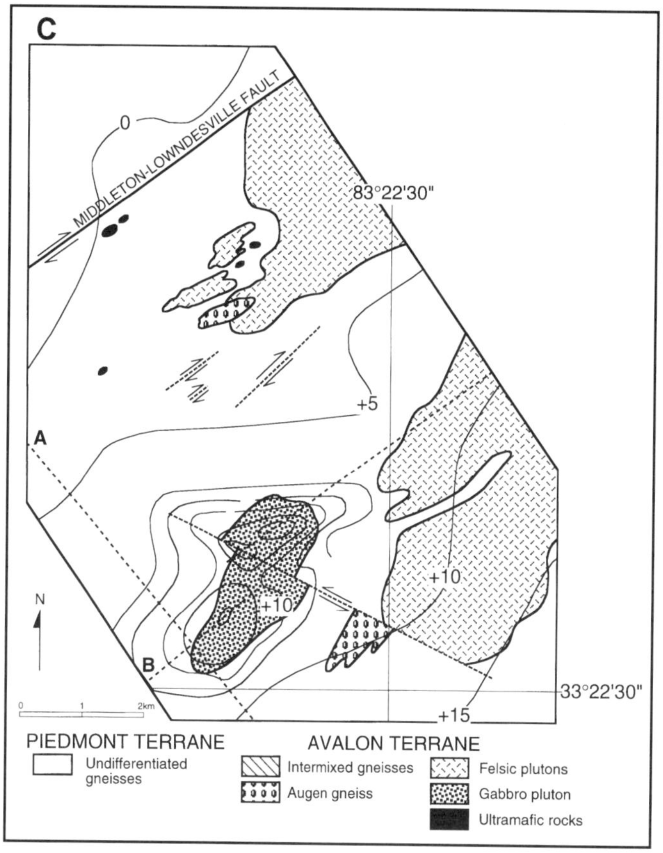

break where the gradient flattens over a distance of 1.5 km between the 10 and 12 mgal contours. The enclosed gravity highs and flattened gradient coincide with the trace of the Gladesville Gabbro (Fig. 3C).

The data station spacing for the Berner survey is 200 m. The spacing was adequate to resolve the local gradient and the form of the gabbroic plutons but could not resolve the numerous small granitic and mafic/ultramafic bodies identified on the local map of Hooper (1986). These small bodies can only be identified where an individual station reading was taken directly on top of one of them; the station reading then appears as an anomalous value. The unnamed mafic body shown to the southeast of the Gladesville Gabbro of the geologic map of Georgia (Pickering, 1976) is not evident in the gravity data acquired during this study (Fig. 3A). The most likely explanations are that the density contrast between body and country rock was too small or that the body is actually smaller than portrayed on the map.

Gladesville Gabbro. The Gladesville Gabbro was interpreted by Matthews (1967) as a single 23,000-ft thick, post-metamorphic layered basic intrusion of Skaergaard/Stillwater type. The pluton was mapped as having dimensions of ~12 km × 1.5 km and was drawn with a bulbous northeast end, ~5 km wide, giving the body a distinctive "fish-shaped" profile. This interpretation was carried through to the gravity survey of Carpenter and Hughes (1970), which assumed the Gladesville Gabbro to be a large single pluton. Mapping by Hooper (1986), despite defining two quite distinct varieties of gabbro within the Gladesville Gabbro (olivine-bearing and olivine-free), again showed the pluton to be a single body.

A single gabbro body of the size proposed for the Gladesville Gabbro (viz., a layered basic intrusion, 12 km × 1.5 km) should produce a large, distinct gravity anomaly. The average residual anomaly for the pluton, with local gradients removed, however, is only 4 to 6 mgals, with a corresponding maximum anomaly width of no more than 4 km (Fig. 3B). The gravity pattern determined during this study does not support the interpretation of a single, large pluton. We interpret the gravity response to indicate that the Gladesville Gabbro is most likely a composite of several plutons (Fig. 3C). The data also suggest that the gabbro pods may have been displaced by late brittle faults. The characteristic "fish-tail" at the northeast end of the Gladesville Gabbro is most likely related to a Mesozoic? left-lateral strike-separation fault. Similarly, the abrupt southwestern termination of the gabbro is also likely to have been caused by faulting. The Juliette Gabbro, located along strike to the southeast of the Gladesville Gabbro, may be a continuation of the Gladesville Gabbro that has been offset along a left-lateral strike-separation fault.

Rock Eagle area

Gravity anomaly contours from the Rock Eagle area generally trend northeast (Fig. 4A). The lowest gravity anomaly, –6 mgals, is located in the northwest corner of the map and corresponds to rocks of the Piedmont terrane. Anomaly values higher than –2 mgals correspond to rocks of the Avalon terrane. The trace of the Middleton-Lowndesville fault (Fig. 4C) roughly follows the –2-mgal contour (Fig. 4A). The contrast of low-density rocks of the Piedmont terrane in fault contact with higher-density rocks of the Avalon terrane is responsible for the steep local gradient. The highest anomalies, between 18 and 19 mgals, form a plateau in the southeastern corner of the map (Fig. 4A). A closed circular gravity high in the center of the anomaly map corresponds to the location of the Presley's Mill Gabbro (Fig. 4C). The flat area in the northeastern corner of the anomaly map is coincident with the location of a large granitic body. The granitic bodies have a lower bulk density than the layered orthogneisses of the Rock Eagle Complex and thus appear on the anomaly map as lows. The closely spaced gravity contours are representative of the local gravity gradient.

Presley's Mill Gabbro. The Presley's Mill Gabbro is composed of two-pyroxene gabbro. This gabbro is denser than the surrounding rocks of the Rock Eagle Complex and produces a gravity anomaly that is 7 mgals higher than the background (Fig. 4C). The most striking feature of the data from the Presley's Mill Gabbro is the pronounced offset in the gravity contours (Fig. 4A, C), expressed as a double hump on Figure 4B, Profile B, that trends along the body (Fig. 4B). Several possible explanations exist for the pinched circular shape of the anomaly. The anomaly could represent two or more closely spaced bodies or a single body offset along a fault trending northwest-southeast. Late brittle faults trending N60W have been documented throughout the Piedmont and Avalon terranes and have been mapped in the Rock Eagle Complex. Fault displacement is considered the most likely explanation for the pinched anomaly.

Contact metamorphic rocks have been found at several locations around the Presley's Mill Gabbro, and a contact aureole is inferred to surround the entire pluton. The width of the hornfelsic zone on the southern side of the pluton is less than 1 km, but increases to several kilometers wide on the northern side of the body. The increased width of the hornfelsic zone could be attributed to the gabbro margin's being steep on the southern side but plunging gently below the surface to the north. This interpretation has support from the asymmetrical residual anomaly of the pluton (Fig. 4B), which is much flatter on the northern side of the body. Part of the increased width of the hornfelsic zone to the north could also be related to interference with the aureole surrounding a large felsic pluton to the northeast of the gabbro (Fig. 4C).

STRUCTURE

The structures present in the Berner area have been described by Hooper (1986) and Hooper and Hatcher (1989b). Some points are repeated below for comparison to the Rock Eagle Complex.

The major structural feature affecting the Berner Mafic Complex is the Ocmulgee fault, interpreted to be the Piedmont-Avalon boundary (Hooper, 1990). The fault comprises a broad zone (~2 km) of steeply southeast-dipping foliation containing localized syn- and postmetamorphic right-lateral strike-slip shear zones. The Piedmont-Avalon boundary in the Rock Eagle area is not defined by the same pronounced steep zone as is the case in the Berner area. The layered-series rocks in the Rock Eagle area dip moderately to the southeast except where modified by small, near-vertical, ductile shear zones.

The Towaliga fault, generally regarded as framing the northern side of the Pine Mountain window, has previously been correlated with the Middleton-Lowndesville fault zone in eastern Georgia (see, for example, the geological map of Georgia by Pickering, 1976). This association is probably incorrect. Our mapping does not trace the Towaliga fault zone across the Ocmulgee River. Instead, the Middleton-Lowndesville fault zone correlates, at least by stratigraphic position, with the Ocmulgee fault.

A regionally extensive brittle fault system has recently been identified in the Piedmont of central Georgia. The fault system has been interpreted to offset major Paleozoic ductile faults, basement, and cover stratigraphy within the Pine Mountain window and units in the Piedmont and Avalon terranes (Hooper and Hatcher, 1989b; Hooper, 1990). The effects of the system are clearly visible on aeromagnetic maps (e.g., that of Zietz et al., 1980). The Ocmulgee fault and units in the Berner Mafic Complex are offset by a series of northwest-southeast–trending left-lateral strike-separation faults. The faults can also be resolved in the gravity data. These faults are brittle and must have developed subsequent to all motion on the Ocmulgee fault. Similar fault sets pervade the Rock Eagle Complex (a small-scale example can be seen in Fig. 2F) and cause the offset observed in the gravity data over the Presley's Mill Gabbro.

DISCUSSION

Is the Avalon terrane a mélange?

Higgins et al. (1984) suggested that the area of the Avalon terrane referred to herein as the Berner Mafic Complex was part of a mélange complex and that the mafic and ultramafic rocks represent fragments of Iapetan oceanic crust. The arguments proposed by Higgins et al. (1984) to favor a mélange origin include the lack of thermal aureoles surrounding mafic bodies and the attitude of igneous layering that in places is reported to be steeply inclined (Matthews, 1967). Although we do not exclude the possibility that mélanges could be present in the Avalonian arc complexes, these particular arguments are not supported by our field data from the Berner Mafic Complex and Rock Eagle Complex. Hooper (1986) demonstrated that a coherent internal stratigraphy can be defined within the Berner

Mafic Complex; similar stratigraphic relations were defined in the Rock Eagle Complex by Troyer (1991). Thin layers (~10 m) of distinctive amphibolite can be traced for as much as 7 km. Several mafic bodies—including the largest bodies in both complexes, the Gladesville and Presley's Mill Gabbros—have clearly definable thermal aureoles that contain mineral assemblages of the hornblende and pyroxene-hornfels facies. Igneous layering was not noted in the field or in a drill core from the northern end of the Gladesville Gabbro. Several outcrops where igneous layering was noted in the Gladesville Gabbro by Matthews (1967) are not within the gabbro but are in layered rocks of the Berner Mafic Complex. The Gladesville Gabbro is in fact a composite pluton comprising several separate intrusions, each with its own crystallization history.

We attribute the current distribution of mafic and ultramafic bodies within both complexes to result in part from their original emplacement and in part from postemplacement tectonic disruption following the schemes originally suggested by Hatcher et al. (1984) and Hooper and Hatcher (1989a). The postemplacement tectonic disruption has been further compounded in both the Berner Mafic Complex and Rock Eagle Complex by the effects of late, brittle strike-separation faults.

Projecting the Avalon-Piedmont boundary into western Georgia and Alabama

The distinctive magnetic boundary that corresponds to the Avalon-Piedmont boundary in central Georgia has been used to project the fault outside the mapped area. The Avalon-Piedmont boundary in central Georgia can be traced, on magnetic character, to the northeast and into the Carolinas; projections to the southwest into western Georgia and eastern Alabama are less certain.

Hooper and Hatcher (1990), projecting the Avalon-Piedmont boundary to the west from central Georgia on the aeromagnetic map of Zietz and others (1980), suggested that the boundary may pass just to the north of Columbus, Georgia. Recent mapping by Hanley et al. (this volume) has shown that this speculation is not correct. Avalonian rocks similar to those we have described from central Georgia are not exposed in the vicinity of Columbus, Georgia (Chalokwu, 1991; Hanley et al., this volume). One complicating factor may be a series of large strike-separation faults that cause the boundary to step beneath the Coastal Plain overlap (Hooper and Hatcher, 1989b; Hooper, 1990). An explanation provided by Hanley et al. (this volume) is that the postulated left-lateral step-off in the Avalon-Piedmont boundary toward the Coastal Plain overlap (Hooper and Hatcher, 1989b; Hooper, 1990) is larger than previously envisaged and that the Avalon-Piedmont boundary in the vicinity of Columbus, Georgia, is in fact beneath the Coastal Plain.

An alternative explanation is provided by the preliminary terrane map of Horton et al. (1991) and the work of West et al. (1995). The Avalon terrane as originally defined by Williams and Hatcher (1983) has, in central Georgia, been subdivided in the preliminary map of Horton et al. (1991) into several smaller terranes (e.g., Carolina, Juliette, and Uchee). The area we described here would lie within the Juliette terrane of Horton et al. (1991) (Fig. 1). West (1994) and West et al. (1995, Fig. 2) show that the Juliette terrane terminates to the south of our study area against the Modoc fault zone (Fig. 1). The Modoc fault in central Georgia is shown by West et al. (1995) to separate the Juliette terrane to the north from the Uchee terrane to the south. The Uchee terrane then continues south of the Pine Mountain window into western Georgia and Alabama (Fig. 1). Rocks in the Columbus area, according to the scheme of West et al. (1995), would belong to the Uchee terrane and thus may be distinct from the rocks described herein from central Georgia, which would belong to the Juliette terrane (Fig. 1). This terrane difference may also explain why the geochemical signature of amphibolitic rocks near Columbus is different from that of the Berner Mafic Complex (Chalokwu, 1991; Hanley et al., this volume).

Relations to surrounding terranes

The Pine Mountain window is framed at its eastern end and separated from the overlying Piedmont terrane by the Box Ankle fault zone (Fig. 5). The Piedmont terrane in central Georgia is separated from Avalonian rocks of the Berner Mafic Complex by the Ocmulgee fault (Fig. 5). West (1994), West et al. (1994), and West et al. (1995) have suggested that the Box Ankle fault zone consists of a shear zone that is 10 to 20 km in outcrop width and bounds the Pine Mountain window to the west and the outboard terranes (our Avalon) to the east. West et al. (1995, Fig. 2) also include large areas of undeformed basement rocks of the Pine Mountain window within the Box Ankle fault zone. Our data from the eastern end of the Pine Mountain window do not support most of the elements of the hypothesis of West et al. (1995), particularly with reference to the geometry of the Box Ankle fault zone. While the Piedmont at the east end of the Pine Mountain window is highly deformed, we can demonstrate that the Box Ankle fault is a much narrower, more concentrated zone of ductile deformation than that proposed by West et al. (1995). Strain gradients can be mapped from the basement into the Piedmont terrane rocks (Fig. 5). The most prominent gradient occurs in the Woodland Gneiss below the Box Ankle fault (Fig. 5) where the Woodland can be traced from an undeformed granitoid, containing euhedral K-spar megacrysts up to 10 cm long in the Pine Mountain window, to annealed augen-gneiss with augen up to 3 cm near the contact, to a medium- to coarse-grained two-mica L-tectonite mylonite gneiss or porphyroclastic mylonite with round (or ellipsoidal) composite porphyroclasts up to 15 cm in diameter in the Box Ankle fault zone, all occurring at sillimanite-grade conditions. This strain gradient and shear zone in the basement rocks ranges up to 1 km thick. Deformation is concentrated in a narrower zone a few tens of meters thick in the Piedmont terrane rocks above the Box Ankle fault. The shear zone present along the Box Ankle fault is tightly folded along the fault surface and

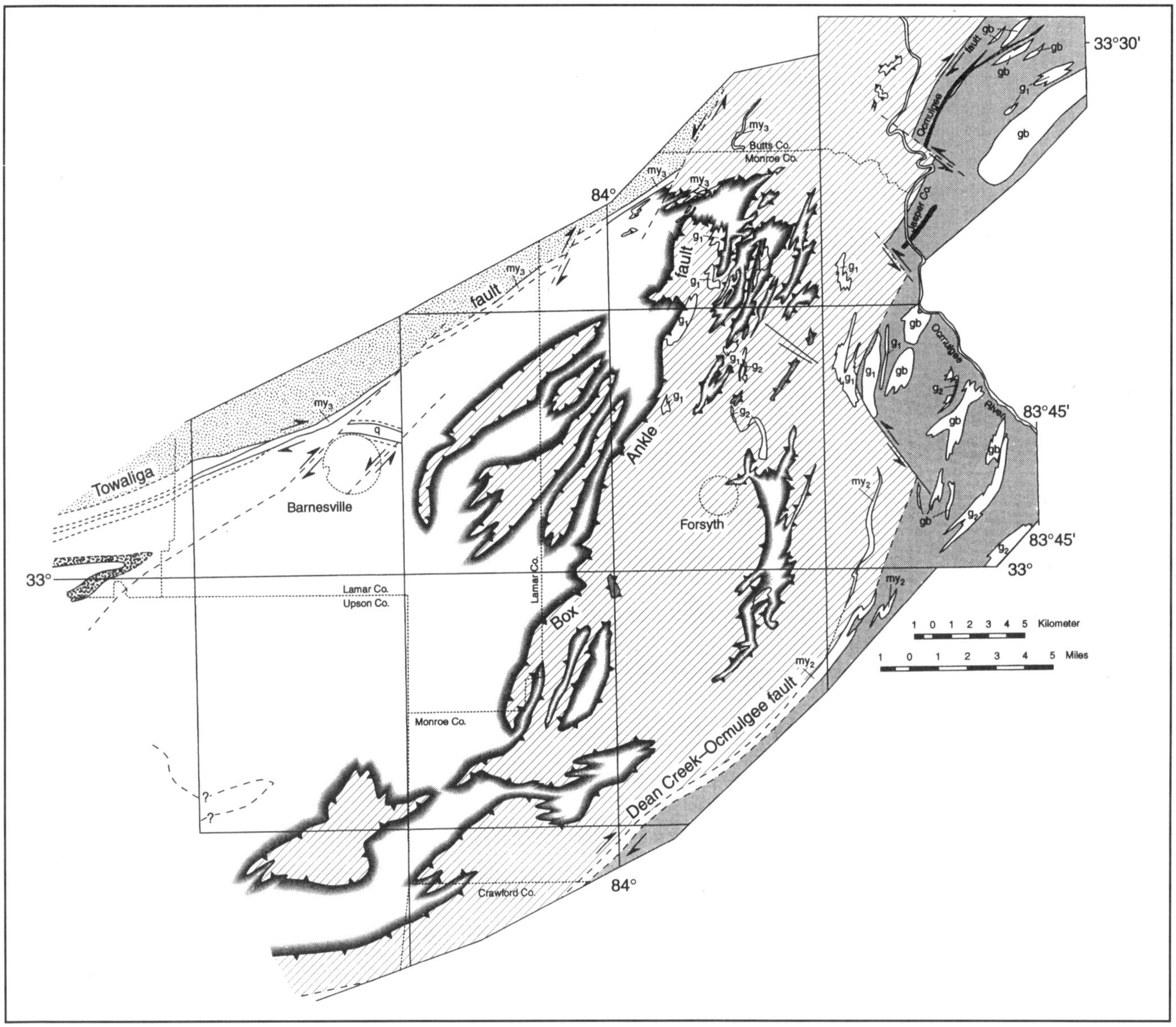

Figure 5. Tectonic map of the eastern end of the Pine Mountain window and adjacent Piedmont and Avalon terranes. Piedmont rocks (diagonal line pattern), carried in the hanging wall of the Box Ankle fault, are thrust over basement rocks of the Pine Mountain window (white). The Box Ankle fault is a narrow zone of ductile deformation preferentially developed in the the basement rocks of the Pine Mountain window (Woodland gneiss). The shear zone thickness in the basement is up to 1 km (gradient pattern along Box Ankle fault); deformation is concentrated in a narrower zone a few tens of meters thick in the Piedmont terrane rocks above the Box Ankle fault. The shear zone present along the Box Ankle fault is tightly folded along the fault surface and exposed around a series of windows through the Piedmont terrane and as klippes in the Pine Mountain window. Inner Piedmont rocks north of the Towaliga fault = fine stipple pattern; Inner Piedmont rocks in the hanging wall of the Box Ankle fault = diagonal line pattern; basement rocks of the Pine Mountain window = white; cover rocks of the Pine Mountain window—heavy stipple pattern = Hollis Quartzite, q = quartzite; Avalon terrane rocks southeast of the Dean Creek–Ocmulgee fault = grey stipple pattern; my = mylonite (2 = mylonites of Dean Creek–Ocmulgee fault system; 3 = mylonites of Towaliga fault system); g = syntectonic granites with 1 or 2 foliations; gb = gabbro.

exposed around a series of windows through the Piedmont terrane and as klippes in the Pine Mountain window (Fig. 5). Our mapping indicates that the basement rocks occur in the hinges of near-isoclinal antiforms, exposing the Box Ankle fault along the contact with the Piedmont terrane rocks (Fig. 5). This contrasts with the interpretation of West et al. (1995), who propose that basement rocks occur as horse blocks in a large shear zone.

The entire Inner Piedmont in the Carolinas and northern Georgia has recently been interpreted to consist of a mid-Paleozoic crustal-scale shear zone (Davis et al., 1991; Davis, 1993). The same sillimanite-grade migmatitic metagraywacke (biotite gneiss)-amphibolite-pelitic schist assemblage and mylonitic fabrics are present in the Piedmont terrane between the Box Ankle and Ocmulgee faults. We regard these rocks as being part of the same crustal shear zone as previously identified in the Carolinas. Odom et al. (1982) reported a ~360-Ma Rb/Sr whole-rock metamorphic age for these rocks. This older mylonitic assemblage is clearly overprinted by the mylonitic assemblage described above associated with the Box Ankle, Ocmulgee, and related faults. Student and Sinha (1992) derived age determinations of ~303 Ma and ~330 Ma, respectively, for the Box Ankle and Ocmulgee faults. These ages reflect Alleghanian deformation and the fact that these faults must have been reactivated during the Alleghanian. The Box Ankle fault may therefore be an abandoned segment of the Blue Ridge Piedmont master detachment but not the final detachment in its present configuration, because of the requirement for balancing sections in this part of the southern Appalachians (see, for example, sections A-A′ in Hatcher et al., 1990).

The Dean Creek fault, exposed along Po Biddy Road south of Thomaston, Georgia, is a late, ductile fault that separates Avalon (Uchee) terrane rocks to the south from Piedmont terrane rocks to the north. This east-west–trending fault zone records dextral motion and retrogrades amphibolite facies assemblages on both sides of it to lower greenschist (chlorite grade) facies assemblages. The Dean Creek fault is mappable as far east as Bolingbroke, Georgia, and at least as far west as south of Thomaston, Georgia (Fig. 1). The Dean Creek fault is not a local zone of retrogression but a major fault that, at least by trend association, probably joins the Modoc fault zone near Bolingbroke. The precise relationships among the Dean Creek fault, the Modoc fault zone, and the Ocmulgee-Middleton-Lowndesville fault are not resolved at this time. The Goat Rock fault, however, as originally identified at Goat Rock Dam near Columbus, Georgia, does not continue into our study area in central Georgia.

CONCLUSIONS

1. The Avalon terrane in central Georgia comprises a distinctive series of metaigneous rocks that can be traced from the Ocmulgee River at least as far east as the Siloam granite.

2. Rocks of the Piedmont terrane are in fault contact with rocks of the Avalon terrane throughout central Georgia along the Ocmulgee-Middleton-Lowndesville fault. The Towaliga fault terminates and does not join the Middleton-Lowndesville fault as has been previously shown.

3. The Avalon-Piedmont boundary can be located based on changes in stratigraphy, metamorphic grade, and aeromagnetic signature.

4. The Avalon-Piedmont boundary lies within a steep local gravity gradient of ~2 mgals/km that separates low values over the Piedmont from relatively high values over the Avalon terrane. This local gradient is superimposed on a broader ~1 mgal/km regional Appalachian gradient.

5. Gravity surveys over the Gladesville and Presley's Mill Gabbros shed light on the form of these rock bodies. The Gladesville Gabbro is not a single pluton; instead it is a series of pods, some with different compositions. The Presley's Mill Gabbro is a small pluton that is disrupted by a left-lateral strike-separation fault. The expansive thermal aureole on the northeast side of the Presley's Mill Gabbro is composite and is not due to shallow gabbro.

6. Offset gravity anomalies in both the Gladesville Gabbro and Presley's Mill Gabbro may be caused by brittle strike-separation faults. These faults are late (post–thermal peak, Mesozoic?) and cut and offset all ductile fabrics and faults.

ACKNOWLEDGMENTS

The work presented here was supported over the years by a variety of sources, including National Science Foundation Grant EAR-7911802 (to RDH) and the Donors of the Petroleum Research Fund, administered by the American Chemical Society (RJH). Mark Stewart (University of South Florida) provided invaluable assistance in processing and interpreting our gravity data and helped guide the theses of Roger Dawson and Cindy Gaddis Redmond. The clarity and focus of the manuscript was greatly improved by the thoughtful comments of the reviewers, Thomas B. Hanley and Donald T. Secor. The patience of volume editor Lynn Glover, III, is sincerely appreciated.

REFERENCES CITED

Carpenter, R. H., and Hughes, T. C., 1970, A geochemical and geophysical survey of the Gladesville norite, Jasper County, Georgia: Georgia Department of Mines, Mining, and Geology Information Circular 37, 7 p.

Chalokwu, C. I., 1991, Ocmulgee fault: The Piedmont-Avalon terrane boundary in central Georgia: Comment: Geology, v. 19, p. 543–544.

Davis, T. L., 1993, Lithostratigraphy, structure, and metamorphism of a crystalline thrust terrane, western Inner Piedmont, North Carolina (Ph.D. thesis): Knoxville, University of Tennessee, 245 p.

Davis, T. L., Hatcher, R. D., Jr., Liu, A., and Tabor, J. R., 1991, Southern Appalachian western Inner Piedmont: A progressive crustal-scale shear zone: Geological Society of America Abstracts with Programs, v. 23, p. A138.

Dawson, R. J., 1990, Implications of the gravity potential field surrounding the Presley Mill gabbro, Putnam County, Georgia [M.S. thesis]: Tampa, University of South Florida, 85 p.

Hatcher, R. D., Jr., and Zietz, I., 1980, Tectonic implications of regional aeromagnetic and gravity data from the southern Appalachians, *in* Wones, D. R., ed., The Caledonides in the USA: Blacksburg, Virginia Polytechnic Institute Department of Geological Sciences Memoir 2, p. 235–244.

Hatcher, R. D., Jr., Hooper, R. J., Petty, S. M., and Willis, J. D., 1984, Structure and chemical petrology of three southern Appalachian mafic-ultramafic complexes and their bearing upon the tectonics of emplacement and origin of Appalachian ultramafic bodies: American Journal of Science, v. 284, p. 484–506.

Hatcher, R. D., Jr., Osberg, P. H., Drake, A. A., Robinson, P., and Thomas, W. A., 1990, Tectonic map of the U.S. Appalachians, *in* Hatcher, R. D., Jr., Thomas, W. A., and Viele, G. W., eds., The Appalachian-Ouchita Orogen in the United States: Boulder, Colorado, Geological Society of America, The Geology of North America, v. F-2, Plate 1.

Higgins, M. W., Atkins, R. L., and Crawford, T. J., and Crawford, R. F., III, 1984, A brief excursion through two crystalline stacks that comprise most of the crystalline terrane of Georgia and Alabama: Georgia Geological Society, 19th Annual Field Trip, Guidebook, 67 p.

Higgins, M. W., Crawford, R. F., III, Atkins, R. L., and Crawford, T. J., 1989, The Macon Complex: an ancient accretionary complex in the southern Appalachians, *in* Horton, J. W., Jr., and Rast, N., eds., Mélanges and olistostromes of the U.S. Appalachians: Geological Society of America Special Paper 228, p. 229–246.

Hooper, R. J., 1986, Geologic studies at the east end of the Pine Mountain window and adjacent Piedmont, central Georgia [Ph.D. thesis]: Columbia, University of South Carolina, 322 p.

Hooper, R. J., and Hatcher, R. D., Jr., 1990, The Ocmulgee fault: The Piedmont-Avalon terrane boundary in central Georgia: Geology, v. 18, p. 708–711.

Hooper, R. J., and Hatcher, R. D., 1988, The Pine Mountain terrane, a complex window in the Georgia and Alabama Piedmont: Evidence from the eastern termination: Geology, v. 16, p. 307–310.

Hooper, R. J., and Hatcher, R. D., 1989a, The origin of ultramafic rocks from the Berner Mafic Complex, central Georgia, *in* Mittwede, S. K., and Stoddard, E. F., eds., Ultramafic rocks of the Appalachian Piedmont: Geological Society of America Special Paper 231, p. 87–92.

Hooper, R. J., and Hatcher, R. D., 1989b, The geology of the east end of the Pine Mountain window and adjacent Piedmont, central Georgia: Geological Society of America Southeastern Section, 1989 Meeting, Guidebook: Atlanta, Georgia Geological Society, 37 p.

Hooper, R. J., 1990, Effects of late, Mesozoic?, brittle faults on Paleozoic configurations: Central Georgia Piedmont: Geological Society of America Abstracts with Programs, v. 14, p. 579.

Horton, J. W., Jr., Drake, A. A., Jr., Rankin, D. W., and Dallmeyer, R. D., 1991, Preliminary tectonostratigraphic terrane map of the central and southern Appalachians: U.S. Geological Survey Miscellaneous Investigations Series, Map I-2 163, scale 1:2,000,000.

Matthews, V., III, 1967, Geology and petrology of the pegmatite district in southeastern Jasper County, Georgia [M.S. thesis]: Athens, University of Georgia, 68 p.

Myers, C. W., 1968, Geology of the Presley's Mill Gabbro, Putnam County, Georgia [M.S. thesis]: Athens, University of Georgia, 67 p.

Nelson, K. D., and 8 others, 1985, New COCORP profiling in the southeastern United States. Part 1: Late Paleozoic suture and the Mesozoic rift basin: Geology, v. 13, p. 714–718.

Obaoye, M. O., 1979, Interpretation of detailed gravity traverses across northeastern Georgia [M.S. thesis]: Athens, University of Georgia, 63 p.

Odom, A. L., Hatcher, R. D., Jr., and Hooper, R. J., 1982, A premetamorphic tectonic boundary between contrasting Appalachian basements, southern Georgia Piedmont: Geological Society of America Abstracts with Programs, v. 14, p. 579.

Pickering, S. M., Jr., 1976, Geologic map of Georgia: Georgia Geological Survey, scale 1:500,000.

Popenoe, P., and Bell, H., III, 1975, Simple Bouguer gravity map of part of the Carolina slate belt including the Haile and Brewer mine areas, north-central South Carolina: U.S. Geological Survey, Map GP-904, scale 1:24,000.

Redmond, C. G., 1990, Gravity investigation of the Gladesville Gabbro and surrounding area, Berner Mafic Complex, Georgia [M.S. thesis]: Tampa, University of South Florida, 80 p.

Student, J. J., and Sinha, A. K., 1992, Carboniferous U-Pb ages of zircons from the Box Ankle and Ocmulgee faults, central Georgia: Implications for accretionary models: Geological Society of America Abstracts with Programs, v. 24, p. 69.

Thomas, M. D., 1983, Tectonic significance of paired gravity anomalies in the southern and central Appalachians, *in* Hatcher, R. D., Jr., Williams, H., and Zietz, I., eds., Contributions to the tectonics and geophysics of mountain chains: Geological Society of America Memoir 158, p. 113–124.

Troyer, P. K., 1991, Geology of the Rock Eagle Complex, central Georgia [M.S. thesis]: Tampa, University of South Florida, 59 p.

Vincent, H. R., 1984, Geologic map of the Siloam Granite and vicinity, eastern Georgia: Georgia Geological Survey, scale 1:100,000.

West, T. E., Jr., 1994, Field evidence that the Appalachian décollement is exposed in central Georgia at the east end of the Pine Mountain terrane [M.S. thesis]: Columbia, University of South Carolina, 26 p.

West, T. E., Jr., Secor, D. T., Jr., and Maher, H. D., Jr., 1994, Is the Appalachian décollement exposed at the east end of the Pine Mountain terrane in central Georgia? New field evidence: Geological Society of America Abstracts with Programs, v. 26, p. 69.

West, T. E. Jr., Secor, D. T., Jr., Pray, J. R., Boland, I. B., and Maher, H. D., Jr., 1995, Is the Appalachian décollement exposed at the east end of the Pine Mountain terrane in central Georgia? New field evidence: Geology, v. 23, p.

Williams, H., and Hatcher, R. D. Jr., 1983, Appalachian suspect terranes, *in* Hatcher, R. D., Jr., Williams, H., and Zeitz, I., eds., Contributions to the tectonics and geophysics of mountain chains: Geological Society of America Memoir 158, p. 33–53.

Zietz, I., Riggle, F. E., and Gilbert, F. P., 1980, Aeromagnetic map of Georgia: U.S. Geological Survey Geophysical Investigations Map GP-941, scale 1:1,000,000.

Manuscript Accepted by the Society April 30, 1996

Printed in U.S.A.

Geological Society of America
Special Paper 314
1997

Constraints on the location of the Carolina/Avalon terrane boundary in the southernmost exposed Appalachians, western Georgia and eastern Alabama

Thomas B. Hanley
Department of Chemistry and Geology, Columbus State University, 4225 University Avenue, Columbus, Georgia 31907-5645
Christopher I. Chalokwu and Mark G. Steltenpohl
Department of Geology, Auburn University, Auburn, Alabama 36849

ABSTRACT

The presence of rocks of the Carolina and Avalon terranes in the southernmost Appalachian Piedmont was suggested first on the basis of late Precambrian to early Cambrian U-Pb zircon ages, followed by a description of the Ocmulgee fault in Georgia as the boundary with the Piedmont. The Piedmont/Avalon terrane boundary was projected through the Uchee belt of western Georgia and Alabama as a moderate- to steep-dipping fault zone with a relatively simple linear trace (Ocmulgee–Dean Creek faults). Although we do not doubt the existence of the Piedmont/Avalon boundary in central Georgia, our work in Alabama and western Georgia does not support the presence of a boundary resembling that described by other workers, for the following reasons. 1. Field mapping does not indicate a distinct break in lithologies. 2. Thermobarometric data indicate that metabasites on either side of the proposed boundary experienced similar amphibolite facies conditions. 3. Petrochemistry of metabasites from westernmost Georgia does not contrast across the proposed boundary but is sharply different from the Berner mafic complex of the Avalon terrane. 4. Structural form lines indicate that the dominant structure southeast of the Goat Rock fault zone in this area is a broad synform, the Lake Oliver synform, and do not support the presence of a moderate- to steep-dipping planar zone of foliation.

Existing data in Alabama imply that if rocks of the Carolina and Avalon terrane are present, they may be locally preserved as strongly overprinted slivers or infolds. Alternatively, the terrane boundary may be concealed beneath the Coastal Plain onlap in this area, offset by left-slip faults.

INTRODUCTION

In western Georgia and eastern Alabama, the Appalachian Piedmont consists of the Inner Piedmont and the Pine Mountain (PMW) and Uchee belts, which are bounded by major mylonite zones (Fig. 1). In central Georgia, the Piedmont is reported to consist of rocks of the Inner Piedmont and Avalon terrane, which are separated by the Ocmulgee fault, a 2-km-wide syn– and post–thermal peak right-slip shear zone (Hooper and Hatcher, 1989; Hooper et al., 1991). Because the PMW tapers to the east, the Uchee belt is the apparent on-strike correlative of parts of the Piedmont and the Avalon terrane. Higgins et al. (1988) interpret the Uchee belt to be part of what they call the Macon Mélange complex, separated from the PMW by the Auchumpkee fault. The latter coincides with what others have called the Goat Rock fault. Hooper and Hatcher (1990) later reported the Ocmulgee/Dean Creek fault to be the boundary between the Piedmont and Avalon terrane. The placement of the

Hanley, T. B., Chalokwu, C. I., and Steltenpohl, M. G., 1997, Constraints on the location of the Carolina/Avalon terrane boundary in the southernmost exposed Appalachians, western Georgia and eastern Alabama, *in* Glover, L., III, and Gates, A. E., eds., Central and Southern Appalachian Sutures: Results of the EDGE Project and Related Studies: Boulder, Colorado, Geological Society of America Special Paper 314.

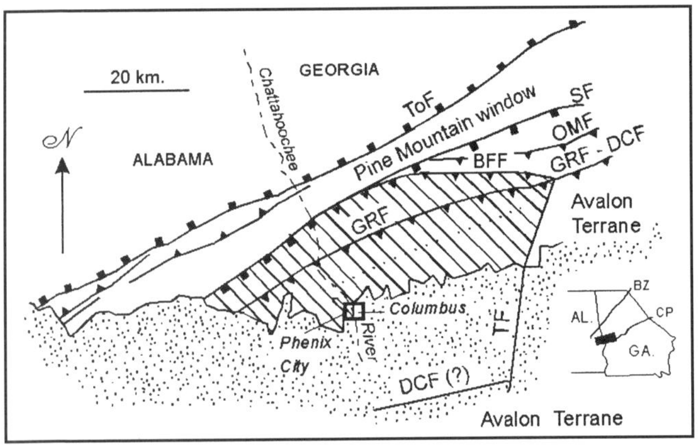

Figure 1. Generalized geologic map of west-central Georgia and east-central Alabama showing the Inner Piedmont, Pine Mountain window, Uchee belt (diagonal line pattern) and a series of faults: Brevard zone (BZ), Towaliga fault (ToF), Shiloh fault (SF), Goat Rock fault (GRF), Oak Mountain fault (OMF), Bartletts Ferry fault (BFF), Talbotton fault (TF), and Dean Creek fault (DCF). Modified from Hanley and Redwine (1986). Coastal Plain is stippled (CP on inset). The dotted line in the Uchee belt shows the location of the Dean Creek fault and possible terrane boundary suggested by Hooper and Hatcher (1989).

Avalon boundary by Hooper and Hatcher (1990) does not correspond to the Auchumpkee fault in the present study area.

Steltenpohl et al. (1991) and Chalokwu (1991) have pointed out disagreements about where the boundary between these terranes lies in this problematic area. Maps presented by Hooper and Hatcher (1989, Fig. 1; 1990, Fig. 1) project the Avalon terrane boundary through the western Uchee belt north of the Coastal Plain, though they point out that this projection is poorly defined and that the boundary may be buried (Hooper and Hatcher, 1990, 1991).

The purpose of this chapter is to present lithostratigraphic, structural, geochemical, thermobarometric, and geophysical considerations that indicate that the terrane boundary is offset beneath the Coastal Plain along left-slip faults recognized by Hooper and Hatcher (1990). We also suggest a magnetic anomaly that may represent the buried mafic rocks of the Avalon terrane.

We use the term "terrane" in this chapter to mean fault-bounded rocks characterized by a geologic history different from that of adjacent terranes (e.g., Jones et al., 1983). The criteria we applied to rocks of this area to determine if they belong to different terranes are those described by Zen (1988).

GENERAL GEOLOGIC CONTEXT

The Uchee belt contains a multiply deformed and metamorphosed suite of plagioclase and K-feldspar gneisses, migmatites, amphibolites, schists, minor quartzites, calc-silicate lenses, and cataclastic and mylonitic rocks.

Northwest of the Uchee belt are rocks of the Pine Mountain and Inner Piedmont belts (Fig. 1). The northern boundary of the Inner Piedmont is the Brevard zone. The Towaliga fault separates the Inner Piedmont from the Pine Mountain belt, and the Bartletts Ferry–Goat Rock fault system separates the latter from the Uchee belt. These features are briefly described below.

The Inner Piedmont is a thin allochthonous thrust sheet (Cook et al., 1979) consisting of polydeformed metasediments and metaigneous rocks (Higgins and Atkins, 1981; Sears et al., 1981) that apparently formed the ancient edge of Laurentia, the North American continent of Cook and Oliver (1981). Intruded by granitoids during early to middle Paleozoic time, the Inner Piedmont was later regionally metamorphosed to amphibolite-facies conditions. These units locally were retrograded under greenschist-facies conditions during late Paleozoic ductile to brittle deformation (Neilson et al., 1990a, b). The Brevard zone, which forms the northern boundary of the Inner Piedmont, is a complex polydeformed zone (see Bobyarchick et al., 1988, for a summary) and may be a splay off the southern Appalachian master décollement (Cook et al., 1980).

A metasedimentary sequence consisting of metamorphosed quartz arenites, schists, and rare marbles (Sears et al., 1981) is exposed in the PMW. These units were deposited nonconformably upon a charnockitic basement of Grenville age (Schamel and Bauer, 1980). The unconformity is tectonically modified in Alabama (Steltenpohl et al., 1990). At the eastern end of the PMW, the basement is in thrust contact with overlying Piedmont lithologies (Hatcher et al., 1988; Hooper and Hatcher, 1988, 1989). Early metamorphism of the cover was possibly Devonian (Steltenpohl and Moore, 1988). Later structures are characterized by ductile deformation zones and brittle faults that penetrate the basement (Neilson et al., 1990a, b).

Steltenpohl (1988) reports that the Towaliga, Bartletts Ferry, and Goat Rock mylonite zones in Alabama are parts of the right-slip eastern Piedmont fault system (Hatcher et al., 1977). Hatcher et al. (1988) interpret fault zones flanking the PMW as a thrust system that was reactivated by later dextral motion. Based on Rb-Sr whole-rock and whole-rock mineral isochrons, Goldberg and Steltenpohl (1990) and Steltenpohl et al. (1990, 1992; Steltenpohl and Kunk, 1993) interpret the right-slip movement along these mylonite zones to have occurred during the late Carboniferous.

The Ocmulgee fault is described as a series of pre- to synmetamorphic ductile shear zones that contain mineral assemblages comparable to the assemblages in the surrounding metamorphic rocks (Hooper and Hatcher, 1988). Foliations are steep to vertical, and the sense of shear is right-slip. Hooper and Hatcher (1990) report that the Ocmulgee fault is well defined along the Ocmulgee River. To the southwest, the Ocmulgee fault is reportedly overprinted by a post–thermal peak right-slip mylonite zone that Hooper and Hatcher (1990) call the Dean Creek fault. The Dean Creek fault follows the trace of the Ocmulgee fault, and Hooper and Hatcher (1990) suggest it might be considered a retrograde phase of that structure. Hooper and Hatcher (1989) also report that the Dean Creek–Ocmulgee fault system is offset in a left lateral sense by the Piedmont brittle fault system, a set of north-northeast trending late-stage faults.

LITHOLOGIC CONTEXT

The Uchee belt is complexly deformed and is the southernmost exposed portion of the Appalachian Piedmont. The belt should feature prominently in unraveling the tectonic evolution of this region (Fig. 2). Unfortunately, the Uchee belt is incompletely mapped. Most of the published data are in field guides based largely on reconnaissance work (e.g., Schamel et al., 1980; Kish et al., 1985; Raymond et al., 1988; Hanley, 1987). A consequence is that many nomenclatural problems exist. It is therefore necessary to describe the most up-to-date results of work on lithologies of this area. Following a brief review, these lithologies are described in detail.

Review

Bentley and Neathery (1970), based on fieldwork in Lee and Russell counties, Alabama, and reconnaissance work in Harris and Muscogee counties, Georgia, subdivided rocks of the Uchee belt into two major units: the Moffits Mill Schist (MMS) and the Uchee Complex. These authors subdivided the latter complex into the Phenix City Gneiss (PCG), Standing Boy Quartz Diorite, and Mulberry Gneiss; they recognized only the MMS and the PCG in Lee County, Alabama.

Along the southern margin of the Goat Rock fault zone is the Motts Gneiss, a strongly lineated mylonitic granitoidal gneiss, which occurs as sill-like bodies. Bentley and Neathery

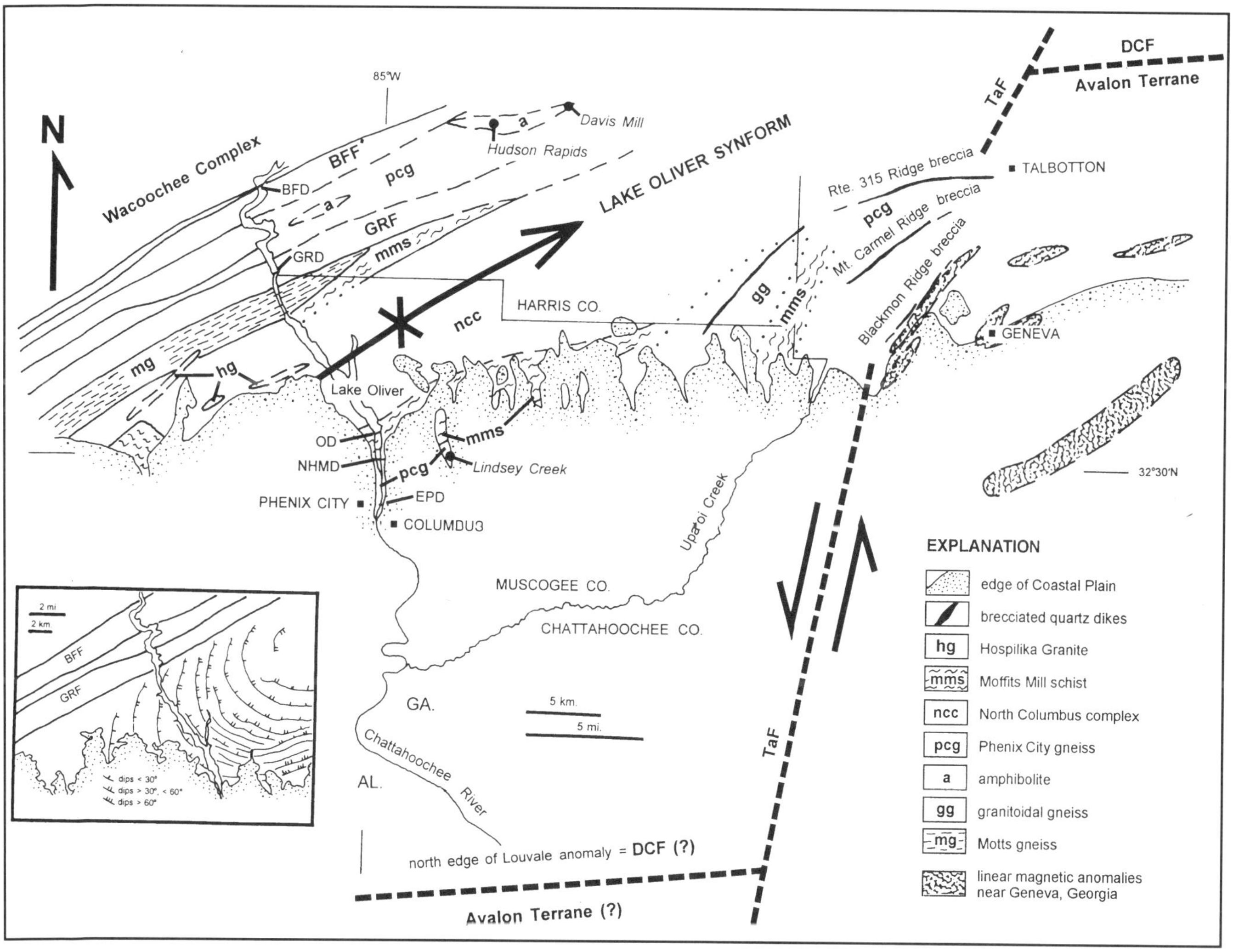

Figure 2. Uchee belt geology of western Georgia and eastern Alabama, modified from Hanley (1987), Szabo et al. (1988), and Hooper and Hatcher (1989). Inset map shows form lines defining the Lake Oliver synform (McRae et al., 1990). The Talbotton fault (TaF) is suggested to offset the Avalon terrane boundary by left-slip movement. Location of DCF is from Hooper and Hatcher (1989). Mottled patterns in the eastern portion of the map are positive linear magnetic anomalies from Zietz et al. (1980) that help determine the location of the Talbotton fault. Dams are identified: Eagle and Phenix Dam (EPD); North Highland Mills Dam (NHMD); Oliver Dam (OD); Goat Rock Dam (GRD); Bartletts Ferry Dam (BFD). Phenix City, Alabama, and Columbus, Georgia, are shown. Compare with Figure 7.

(1970) consider this a transitional unit of the Uchee belt gneisses involved in intense crystal-plastic deformation along the Goat Rock fault zone.

The PCG, characterized by amphibolite bodies in a hornblende-epidote gneiss, dominates the southern part of the Uchee belt. On the basis of geochemistry, Stow et al. (1984) correlate the amphibolites of the PCG with the Hudson Rapids Amphibolite, between the Goat Rock and Bartletts Ferry fault zones (Fig. 2). Probable late Precambrian to Early Cambrian U-Pb zircon ages from those amphibolites suggest a correlation with the Carolina terrane (Russell, 1985). Although the units between the Goat Rock and Bartletts Ferry fault zones are poorly mapped, the geologic map of Georgia (Georgia Geological Survey, 1976) also indicates hornblende gneisses and amphibolite bodies in that area. The "type" Standing Boy Quartz Diorite, a unit described by Bentley and Neathery (1970), is named for outcrops in Standing Boy Creek in northern Columbus, Georgia.

Two units that Bentley and Neathery (1970) described in Alabama but do not mention as occurring in Georgia are the MMS and the Hospilika Granite. These authors interpret the MMS to be a sequence of metamorphosed graywackes and quartz arenites found in the western part of Lee County, Alabama. The Hospilika Granite is an epidote, muscovite quartz diorite to granodiorite, which occurs in small plutons. These granites are considered to be posttectonic, as they cut the lineation of the Goat Rock fault zone (Bentley and Neathery, 1970).

New field and laboratory studies require revision of the terminology and units described by Bentley and Neathery (1970). The "Phenix City Gneiss" is retained, but the "Mulberry Gneiss" and "Standing Boy Quartz Diorite" are combined into the "North Columbus Complex" (NCC) to more appropriately describe the location and accommodate the diverse nature of these units. Between the PCG and NCC in Georgia is a suite of strained augen schists and gneisses, some of which resemble the MMS. This suite has been informally called the Bypass Metamorphic complex by Hanley (1987). We refer to the Bypass Metamorphic complex as the Moffits Mill Schist in this chapter. The use of the term "Uchee Complex" for the diverse gneisses in the Uchee belt is discouraged because of the presence of the MMS which Bentley and Neathery (1970) interpreted as metasedimentary but which we contend may be a zone of distributed strain.

Phenix City Gneiss

The PCG is a layered, coarse-grained, hornblende gneiss with amphibolite lenses and layers and both concordant and discordant pegmatite veins and pods (Schamel et al., 1980). The gneiss has a granoblastic texture and contains varying amounts of andesine, quartz, hornblende, biotite, and minor apatite and opaques. K-feldspar is sparse to absent in the gneiss, though K-feldspar and muscovite are abundant in the pegmatites. The amphibolite bodies and calc-silicate lenses that occur within the PCG are described separately below.

The felsic portion of the gneiss varies from dioritic and quartz dioritic to granodioritic. Its composition, layering, and association with amphibolites and calc-silicates suggest that the unit is a regionally metamorphosed succession of basaltic to dacitic volcanics and/or volcanogenic sediments.

Masses of calc-silicate varying from ovoid pods up to 0.5 m in diameter through thin lenses up to 5 cm thick to tabular bodies at least 3 m thick are found in the feldspathic portions of the PCG as well as in its amphibolite bodies. The larger bodies are known only through construction exposures of waxy green saprolite, though fresh calc-silicate pods have been found in creeks and at construction sites. Similar lithologies are present in the amphibolites between the Bartletts Ferry Fault and Goat Rock fault as well as in the MMS, the NCC, and the mylonites of the Goat Rock fault and Bartletts Ferry fault. Fresh pods contain diopside, actinolite, garnet, plagioclase, scapolite, titanite, epidote, and zoisite.

Tabular amphibolite bodies within the PCG are exposed in creeks and road cuts and along the Chattahoochee River in both the southern and the northern part of the Uchee belt. Structures, textures, and mineralogy as well as whole-rock and mineral chemistry indicate that the amphibolites are ortho-amphibolites (refer to Figs. 3, 4, and 5; Chalokwu and Hanley, 1990). The amphibolites include coarse-grained garnetiferous and nongarnetiferous varieties that vary from massive to thinly layered. Green calc-silicate pods 3 to 5 cm thick, consisting of plagioclase, diopside, quartz, garnet, hornblende, calcite, and titanite, are also present.

The amphibolites are dominated by varying amounts of hornblende and plagioclase with quartz, diopside, epidote, calcite, ilmenite, titanite, and rutile as accessory minerals. Thin, leucocratic lenses occur sporadically in the amphibolites. Thicker green lenses up to 4 cm wide contain plagioclase, clinopyroxene, epidote, and scapolite. This suite is interpreted to be a sequence of volcanic rocks or volcanogenic sediments with some carbonate component. Similarity of some of the amphibolite lithologies, especially those with calc-silicate masses, argues for the correlation of the PCG from south to north across the Uchee belt.

North Columbus Complex

The term "North Columbus Complex" is recommended rather than "Standing Boy Quartz Diorite" and "Mulberry Gneiss" because of the diversity of lithologies encountered in this unit. The NCC consists of a number of lithologies complexly interrelated through migmatization, intrusion, and deformation. The NCC includes massive quartz dioritic gneiss and folded stromatic migmatite gneiss cut by quartz dioritic leucosome as well as foliated augen gneisses. The major minerals in this rock are plagioclase, hornblende, biotite, and quartz. K-feldspar, where present, usually occurs as augens in the leucocratic layers of the migmatite or in pegmatoidal lenses. Amphibolitic, diopsidic, and epidotized calc-silicates and calcareous enclaves are locally common in the gray gneisses and migmatites. Foliated K-feldspar granite sheets (more than 20 m thick) are concordant to fold limbs.

Most NCC exposures contain at least a few small diopside-bearing calc-silicate bodies. These vary from angular masses with gneissosity clearly cut by foliated dioritic material, through disconnected distorted "snakes" to ovoid pods. Dark masses of gneiss and amphibolite commonly show internal layering cut by flow in the enveloping gray gneiss. These amphibolite and calc-silicate enclaves are interpreted to be PCG xenoliths in the NCC gneisses and migmatites.

Moffits Mill Schist

A lithologically diverse and structurally complex unit, the MMS separates the PCG from the NCC. It is recognized as a well-foliated biotite muscovite schist with plagioclase and some microcline augens 2 cm or more in long dimension. Garnet and sillimanite are locally present. Additional lithologies are foliated ilmenite quartzites, graphitic schists, and gneisses. Pods and lensoidal masses of amphibolite, calc-silicate, and pegmatite are locally common. Within the MMS are lineated granitoidal gneiss bodies.

The MMS is covered by Coastal Plain sedimentary rocks from Moffits Mill in Alabama to the Chattahoochee River, where it reemerges and extends northeast through the Uchee belt. The unit is highly susceptible to weathering and is rarely observed except in excavations.

Summary

PCG amphibolites on the north side of the Uchee belt are both garnetiferous and nongarnetiferous depending upon bulk composition, and they contain minor calc-silicate layers and pods. Amphibolites from the south side, though not garnetiferous, also contain minor leucocratic lenses and calc-silicate pods. The NCC contains migmatitically and tectonically disrupted amphibolite and calc-silicate masses similar to lithologies found in the PCG. Similarly, the MMS contains amphibolite and calc-silicate masses similar to lithologies in the PCG. Sharing these lithologies links the NCC and the PCG since the time of formation of the NCC. Masses of schist are not found within the NCC.

STRUCTURE

The overall structure of the Uchee belt is a broad, shallow east-plunging synform, the Lake Oliver synform (LOS) (Hanley, 1986; McRae et al., 1990). The PCG defines the limbs of the synform, and the NCC occupies the core. Between these units is the MMS. The synformal nature of the LOS is indicated by the form-line map (see insert to Fig. 2), based on dominant foliation attitudes in the Uchee belt (McRae et al., 1990). Foliation in the northwestern part of the belt dips toward the southeast and that in the southeastern parts of the belt dips toward the northwest. There are no breaks of foliation trends within the LOS that would indicate a steeply dipping major fault in the position suggested by Hooper and Hatcher (1988).

AMPHIBOLITE GEOCHEMISTRY

Bulk-rock compositions plotted on chemical discrimination diagrams (Figs. 3 and 4) indicate that the protoliths of the PCG amphibolites are igneous. If Avalonian rocks are exposed in westernmost Georgia, Lindsey Creek amphibolites located on the south limb of the LOS should be petrochemically similar to amphibolites from the Berner mafic complex (Hatcher et al., 1984). The great variability in the geochemistry of the PCG amphibolites suggests diverse tectonic settings (Chalokwu and Hanley, 1990) and sharply contrasts with the relatively simple low-K island-arc heritage indicated for the amphibolite protoliths from the Berner complex (Hatcher et al., 1984). Amphibolites from Lindsey Creek have compositions that more closely resemble calc-alkaline basalts from continental margin arcs than do Berner complex rocks (Fig. 5).

The variability in chemistry of PCG amphibolites and absence in western Georgia and eastern Alabama of rocks from a simple arc system similar to Berner complex rocks precludes the presence of Avalonian rocks in this area.

Wang and Glover (1992) demonstrate that mafic rocks from rift environments vary considerably in their geochemistry and cannot be distinguished from mafic rocks of other tectonic environments when plotted on a variety of discriminant diagrams. Thus they call into question the discriminant diagram approach as sole indicator of tectonic environment for igneous rocks. The contrasting degree of geochemical variability of Uchee belt amphibolites with the Berner complex amphibolites, however, indicates a lack of correlation regardless of paleotectonic interpretation and is consistant with other lines of evidence.

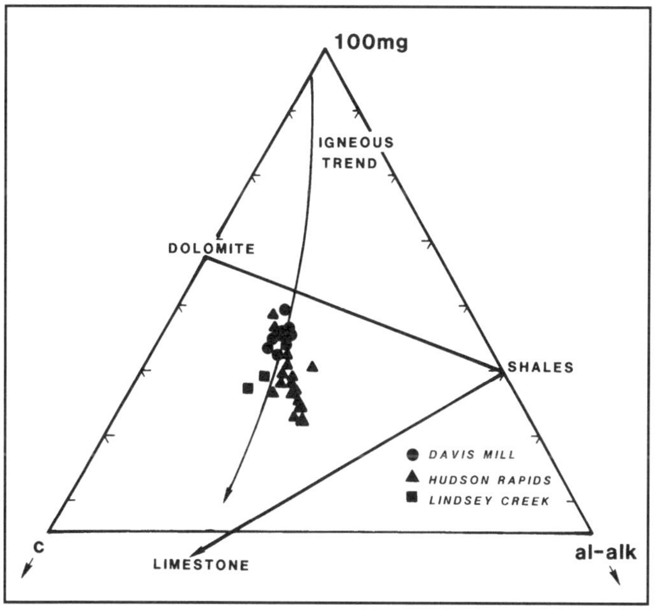

Figure 3. Distinction between ortho- and para-amphibolites based on Niggli c-mg-(al-alk) variation (after Leake, 1964). All Phenix City Gneiss amphibolites plot on or close to the igneous trend curve.

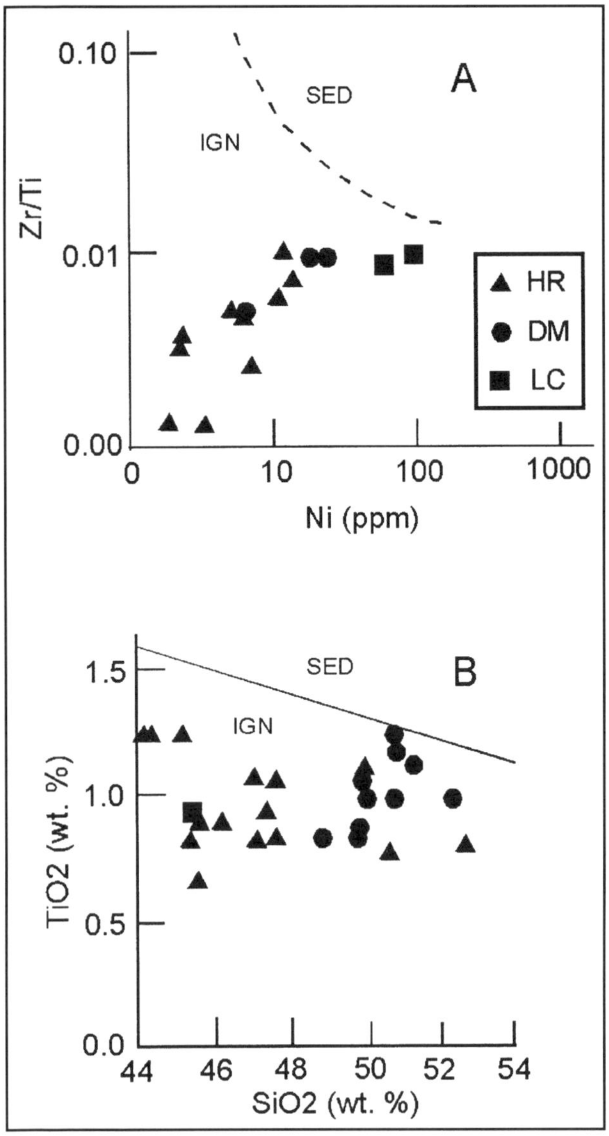

Figure 4. Discrimination plots for separating ortho- from para-amphibolites based on (A) Zr/Ti versus Ni variation and (B) TiO_2 versus SiO_2. HR = Hudson Rapids; DM = Davis Mill; LC = Lindsey Creek. Igneous and sedimentary fields after Winchester and Max (1982) and Tarney (1977).

P-T CONDITIONS FOR THE AMPHIBOLITES

Chalokwu (1989) and Chalokwu and Kuehner (1992) explored the pressure-temperature (P-T) conditions for the metamorphism of some of the amphibolites of the PCG on both limbs of the LOS. These authors conclude that metamorphic grade varies from upper amphibolite facies (garnet zone) to retrograde epidote-amphibolite facies. The Davis Mill amphibolites were more deeply buried or metamorphosed (at a depth of 36 km) than Hudson Rapids (at 30 km) and Lindsey Creek counterparts (at 20 km). Although the Davis Mill and Lindsey Creek amphibolites, located on opposite limbs of the LOS (Fig. 2), were metamorphosed at different depths, both amphibolites followed a similar uplift path after the thermal peak was reached in each locality (Chalokwu and Kuehner, 1992). This implies that there is no significant contrast in postpeak metamorphic history between these rocks. The similarity and overlap of metamorphic conditions (Fig. 6) for amphibolites on the north and south side of the Uchee belt lead us to conclude that there is no major terrane boundary between them.

Mineral assemblages in MMS and PCG are also consistent with equilibration at amphibolite facies conditions, indicating no significant contrast in metamorphic conditions across the Uchee belt in western Georgia or eastern Alabama.

GEOPHYSICS

Hooper and Hatcher (1990) used the trends of magnetic and Bouguer gravity anomalies to project the trace of the Avalonian boundary through the Uchee belt. Figure 7, reduced from the 1:500,000 map of Zietz et al. (1980), shows the Charleston magnetic terrane and Altamaha magnetic anomaly trough. The broad moderate intensity zone on the north side of the Altamaha magnetic anomaly trough separates the magnetically defined terranes on the south from the Piedmont terranes on the north (Higgins and Zietz, 1983). In the Columbus area, Higgins and Zietz (1983) draw the boundary along a well-defined linear anomaly pattern that crosses the Chattahoochee River at Oliver Dam. Hatcher et al. (1988) and Hooper and Hatcher (1989) appear to have used this anomaly to project the Dean Creek fault into western Georgia.

South of Talbotton, several small positive anomalies break away to the southwest along a N30°E trend (Figs. 2 and 7). This trend is parallel to an unnamed brittle fault that passes west of Talbotton, which we refer to as the Talbotton fault. A regional map by Hooper and Hatcher (1989, Fig. 1) shows the Talbotton fault offsetting the Dean Creek (Avalon terrane boundary) fault with a left lateral sense. West of Geneva, ridges held up by brecciated quartz dikes are also oriented N30°E; mylonites in the vicinity have a N20°E strike. The Talbotton fault is one of a set of right-stepping brittle faults recognized by Hooper and Hatcher (1989).

Southeast of Columbus, Georgia, a string of intense magnetic highs, referred to herein as the Omaha–Warner Robins magnetic trend, curves southwest from the north side of the west-tapering Charleston magnetic terrane into the Altamaha magnetic anomaly trough and terminates in the Louvale anomaly. A positive gravity anomaly also follows this trend and is compatible with the interpretation that the Omaha–Warner Robins trend represents mafic rocks concentrated along a terrane boundary. The Omaha–Warner Robins magnetic trend appears offset locally along a set of northwest-oriented linear patterns.

The Berner mafic complex, one of the defining units of the Avalon terrane, produces a strong magnetic anomaly and is centered on a strong Bouguer gravity anomaly (Higgins and Zietz, 1983). The shape of the Berner complex magnetic anom-

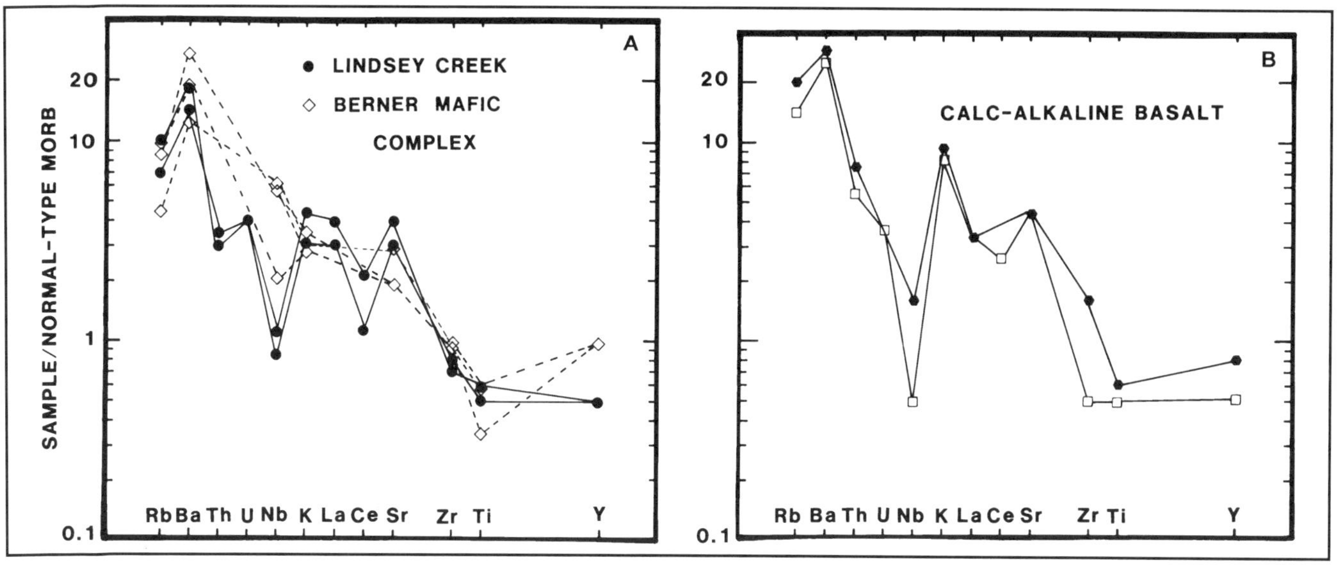

Figure 5. Spidergrams comparing incompatible elements in (A) amphibolites from the Lindsey Creek and the Berner mafic complex (Hatcher et al., 1984) with (B) the compositions of primitive calc-alkaline basalt (open symbol) from Sun (1980) and average calc-alkaline basalt from continental margin arcs (closed symbol) given in Condie (1985). All compositions have been normalized to a normal-type midocean ridge basalt.

aly is similar to that of the Louvale anomaly, though the Berner complex anomaly is smaller than the Louvale anomaly and of slightly lower intensity. Both consist of a double anomaly that converges to the west like an arrow; additionally, both seem to have tails that extend to the east-northeast, consisting of intense linear highs separated from the "head" by a broader but less intense saddle. A lack of a strong gravity or magnetic signature for the PCG amphibolites along both the northern and southern limb of the LOS also indicates that the PCG amphibolites do not correlate with the Berner mafic complex.

To summarize the geophysical data, the best candidate south of Columbus, Georgia, for the Avalon terrane is the material responsible for the Louvale magnetic anomaly. This material was offset to its present location by strike-slip faulting. The size, intensity, and distribution of magnetic anomalies do not seem to match the Piedmont. The Omaha–Warner Robins magnetic trend may reflect the tectonically thinned and laterally offset termination of the Avalon terrane.

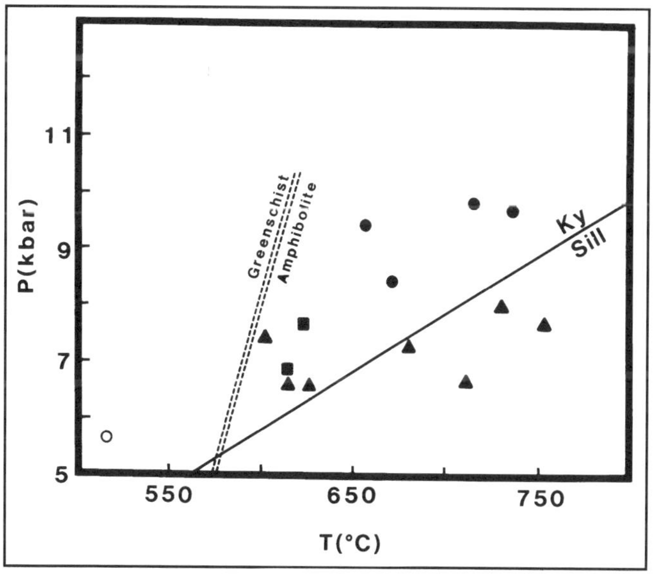

Figure 6. Summary of pressure-temperature data for amphibolites from the Uchee belt. Solid circles = Davis Mill samples, triangles = Hudson Mill Rapids, squares = Lindsey Creek, and open circle = Davis Mill sample with epidote + quartz symplectite. Note that all samples equilibrated at amphibolite-facies conditions except for a sample from Davis Mill with reaction textures indicating retrogression to epidote-amphibolite conditions (see Chalokwu and Kuehner, 1992, for details of the calculations and the thermobarometers used).

SUMMARY AND CONCLUSIONS

The structure of the Uchee belt is a broad synform with no obvious structural data to support the presence of a throughgoing, steeply dipping terrane boundary. Amphibolites repeated on both limbs of the LOS are different in their chemistry and implied tectonic setting, but the grade of metamorphism is not significantly different. These amphibolites differ in compositional variability and implied tectonic setting from Berner mafic complex amphibolites (Hooper and Hatcher, 1988, 1990). Geophysical data indicate that mafic rocks of the Avalon

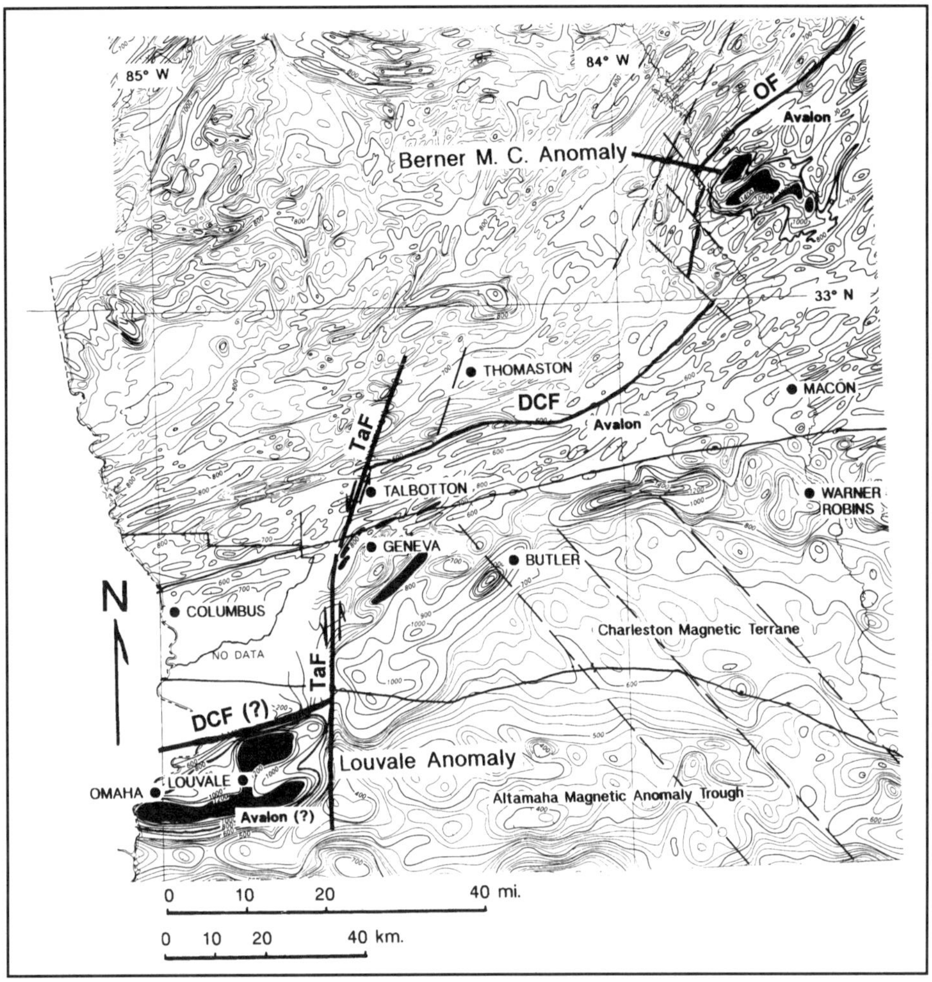

Figure 7. Magnetic anomaly map of western Georgia reduced from the original 1:500,000 state map (Zietz et al., 1980). Contour interval is 100 gammas. The boundaries of the Charleston magnetic terrane and the Altamaha magnetic anomaly trough of Higgins and Zietz (1983) are shown. Magnetic highs near Geneva are darkened. The Dean Creek fault (DCF) and the Ocmulgee fault (OF) bound the north side of the Avalon terrane as shown by Hooper and Hatcher (1989). The DCF is offset by the Talbotton fault (TaF). The arc of strong positive anomalies is the Omaha–Warner Robins trend. The Louvale anomaly occurs at its western end. Note northwest-trending linear patterns in the Charleston magnetic terrane. These may be faults parallel to the northwest-trending faults (Hooper and Hatcher, 1989) that offset the Dean Creek–Ocmulgee fault southwest of the Berner mafic complex anomaly.

terrane may be buried south of Columbus, Georgia, and that a late left-slip fault displaced this terrane to the south beneath the Coastal Plain. The present data are most easily accommodated by a model in which the Avalon terrane boundary passes beneath Coastal Plain sediments somewhere in western Georgia but east of exposures of the Uchee belt.

ACKNOWLEDGMENTS

Acknowledgment is made to the donors of the Petroleum Research Fund, administered by the American Chemical Society, for support of this research (ACS-PRF 23762-GB2 to M. G. Steltenpohl) and to The Columbus College Foundation for support of T. B. Hanley.

REFERENCES CITED

Bentley, R. D., and Neathery, T. L., 1970, Geology of the Brevard Fault Zone, and related rocks of the Inner Piedmont of Alabama: Alabama Geological Society, 8th Annual Field Trip, Guidebook, 119 p.

Bobyarchick, A. R., Edelman, S. H., and Horton, J. W., Jr., 1988, The role of dextral strike-slip in the displacement history of the Brevard Zone, *in* Secor, D. T., Jr., ed., Southeastern geological excursions: Geological Society of America, 1988 Annual Meeting, Field Trip Guidebook, p. 53–104.

Chalokwu, C. I., 1989, Epidote-amphibolite to amphibolite facies transition in the southern Appalachian Piedmont: P-T conditions across the garnet and calc-silicates isograds: Geology, v. 17, p. 491–494.

Chalokwu, C. I., 1991, Ocmulgee fault: The Piedmont-Avalon terrane boundary in central Georgia: Comment: Geology, v. 19, p. 543–544.

Chalokwu, C. I., and Hanley, T. B., 1990, Geochemistry, petrogenesis, and tectonic setting of amphibolites from the southernmost exposure of the Appalachian Piedmont: Journal of Geology, v. 98, p. 725–738.

Chalokwu, C. I., and Kuehner, S. M., 1992, Mineral chemistry and thermobarometry of a southern Appalachian amphibolite with epidote + quartz symplectite: American Mineralogist, v. 77, p. 624–638.

Condie, K. C., 1985, Secular variation in the composition of basalts: An introduction to mantle evolution: Journal of Petrology, v. 26, p. 545–563.

Cook, F. A., and Oliver, J. E., 1981, The late Pre-cambrian–early Paleozoic continental edge in the Appalachian orogen: American Journal of Science, v. 281, p. 993–1008.

Cook, F. A., Albaugh, D. S., Brown, L. D., Kaufman, S., Oliver, J. E., and Hatcher, R. D., Jr., 1979, Thin-skinned tectonics in the crystalline southern Appalachians: COCORP seismic reflection profiling of the Blue Ridge and Piedmont: Geology, v. 7, p. 563–567.

Cook, F. A., Brown, L. D., and Oliver, J. E., 1980, The Southern Appalachians and the growth of continents: Scientific American, v. 243, p. 156–168.

Georgia Geological Survey, 1976, Geologic map of Georgia: Georgia Geologic Survey Map SM-3, scale 1:500,000.

Goldberg, S. A., and Steltenpohl, M. G., 1990, Timing and characteristics of Paleozoic deformation and metamorphism in the Alabama Inner Piedmont: American Journal of Science, v. 290, p. 1169–1200.

Hanley, T. B., 1986, Petrography and structural geology of Uchee belt rocks in Columbus, Georgia, and Phenix City, Alabama, *in* Neathery, T. L., ed., Geological Society of America Centennial Field Guide—Southeastern Section, v. 6, p. 297-300.

Hanley, T. B., 1987, The geology of the Uchee belt, Columbus, Georgia and vicinity, *in* Frazier, William J., and Hanley, T. B., eds., Geology of the Fall Line: A field guide to structure and petrology of the Uchee belt and facies stratigraphy of the Eutaw Formation in southwestern Georgia and adjacent Alabama: Georgia Geological Society, 22nd Annual Field Trip, p. A1–A14.

Hanley, T. B., and Redwine, J. C., 1986, The Bartletts Ferry and Goat Rock fault zones north of Columbus, Georgia, *in* Neathery, T. L., ed., Geological Society of America Centennial Field Guide—Southeastern Section, v. 6, p. 291–296.

Hatcher, R. D., Jr., Howell, D. E., and Talwani, P., 1977, Eastern Piedmont fault system: Speculation on its extent: Geology, v. 5, p. 636–640.

Hatcher, R. D., Jr., Hooper, R. J., Petty, S. M., and Willis, J. D., 1984, Structure and chemical petrology of three southern Appalachian mafic-ultramafic complexes and their bearing upon the tectonics of emplacement and origin of Appalachian ultramafic bodies: American Journal of Science, v. 284, p. 484–506.

Hatcher, R. D., Jr., Hooper, R. J., McConnell, K. I., Heyn, T., and Costello, J. O., 1988, Geometric and time relations of thrusts in the crystalline southern Appalachians, *in* Mitra, G., and Wojtal, S., eds., Geometry and mechanisms of Appalachian thrusting, with special reference to the Appalachians: Geological Society of America Special Paper 222, p. 185–196.

Higgins, M. W., and Atkins, R. L., 1981, The stratigraphy of the Piedmont southeast of the Brevard Zone in the Atlanta, Georgia, area: *in* Wigley, P. B., ed., Latest thinking on the stratigraphy of selected areas in Georgia: Atlanta, Georgia Geologic Survey Information Circular 540A, p. 3–40.

Higgins, M. W., and Zietz, I., 1983, Geologic interpretation of geophysical maps of the pre-Cretaceous "basement" beneath the Coastal Plain of the Southeastern United States, *in* Hatcher, R. D., Jr., Williams, H., and Zietz, I., eds., Contributions to the tectonics and geophysics of mountain chains: Geological Society of America Memoir 158, 223 p.

Higgins, M. W., Atkins, R. L., Crawford, T. J., Crawford, R. F., III, Brooks, R., and Cook, R. B., 1988, The stratigraphy, tectonostratigraphy, and evolution of the southernmost part of the Appalachian Orogen: U.S. Geological Survey Professional Paper 1475, 173 p.

Hooper, R. J., and Hatcher, R. D., Jr., 1988, The Pine Mountain terrane, a complex window in the Georgia and Alabama Piedmont: Evidence from the eastern termination: Geology, v. 16, p. 307–310.

Hooper, R. J., and Hatcher, R. D., Jr., 1989, The geology of the east end of the Pine Mountain and adjacent Piedmont, central Georgia: Geological Society of America Southeastern Section Field Trip Guide: Atlanta, Georgia Geological Society, 37 p.

Hooper, R. J., and Hatcher, R. D., Jr., 1990, Ocmulgee fault: The Piedmont-Avalon terrane boundary in central Georgia: Geology, v. 18, p. 708–711.

Hooper, R. J., and Hatcher, R. D., Jr., 1991, Ocmulgee fault: The Piedmont-Avalon terrane boundary in central Georgia: Comment: Geology, v. 19, p. 544.

Hooper, R. J., Hatcher, R. D., Jr., and Stowell, H., 1991, The Avalon-Piedmont terrane boundary, central Georgia: Geological Society of America Abstracts with Programs, v. 23, p. 46.

Jones, D. L., Howell, D. G., Coney, P. J., and Monger, J. W. H., 1983, Recognition, character, and analysis of tectonostratigraphic terranes in western North America: Journal of Geological Education, v. 31, p. 295–303.

Kish, S. A., Hanley, T. B., and Schamel, S., 1985, Geology of the Southwestern Piedmont of Georgia: Geological Society of America, 98th Annual Meeting, Guidebook, 49 p.

Leake, B. E., 1964, The chemical distinction between ortho- and para-amphibolites: Journal of Petrology, v. 5, p. 238–254.

McRae, M., Steltenpohl, M. G., and Hanley, T. B., 1990, Structural investigation of the Lake Oliver Synform, Uchee belt, Alabama and southwest Georgia: Geological Society of America Abstracts with Programs, v. 22, p. 25.

Neilson, M. J., Goldberg, S. A., and Steltenpohl, M. G., 1990a, Geologic research in the Inner Piedmont and its borders: Progress 1970–1990, *in* Neilson, M. J., Steltenpohl, M. G., Bittner, E. I., and Goldberg, S. A., eds., Tectonometamorphic evolution of Alabama's Inner Piedmont and its borders: Alabama Geological Society, 27th Annual Field Trip, Guidebook, p. 1–11.

Neilson, M. J., Steltenpohl, M. G., Bittner, E. I., and Goldberg, S. A., 1990b, Tectonometamorphic development of the southernmost Inner Piedmont terrane and its borders, Alabama, *in* Neilson, M. J., Steltenpohl, M. G., Bittner, E. I., and Goldberg, S. A., eds., Tectonometamorphic evolution of Alabama's Inner Piedmont and its borders: Alabama Geological Society, 27th Annual Field Trip, Guidebook, p. 13–42.

Raymond, D. E., Osborne, W. E., Copeland, C. W., and Neathery, T. L., 1988, Alabama stratigraphy: Alabama Geological Survey Circular 140, 97 p.

Russell, G. S., 1985, Reconnaissance geochronological investigations in the Phenix City gneiss and Bartletts Ferry mylonite zone, *in* Kish, S. A., Hanley, T. B., and Schamel, S., Geology of the Southwestern Piedmont of Georgia: Geological Society of America, 98th Annual Meeting, Guidebook, p. 9–11.

Schamel, S., and Bauer, D., 1980, Remobilized Grenville basement in the Pine Mountain window, *in* Wones, D. R., ed., The Caledonides in the U.S.A.: Blacksburg, Virginia Polytechnic Institute and State University, Department of Geological Sciences Memoir 25, p. 313–316.

Schamel, S., Hanley, T. B., and Sears, J. W., 1980, Geology of the Pine Mountain Window and adjacent terranes in the Piedmont Province of Alabama and Georgia: Geological Society of America—Southeastern Section, 29th Annual Meeting, Guidebook, 69 p.

Sears, J. W., Cook, R. B., and Brown, D. E., 1981, Tectonic evolution of the western part of the Pine Mountain window and the adjacent inner Piedmont province, *in* Sears, J. W., ed., Contrasts in tectonic style between the Piedmont Terrane and the Pine Mountain window: Alabama Geological Society, 18th Annual Field Trip, p. 1–13.

Steltenpohl, M. G., 1988, Kinematics of the Towaliga, Bartletts Ferry, and Goat Rock fault zones, Alabama: The late Paleozoic dextral shear system in the southernmost Appalachians: Geology, v. 16, p. 852–855.

Steltenpohl, M. G., and Kunk, M. J., 1993, $^{40}Ar/^{39}Ar$ thermochronology and Alleghanian development of the southernmost Appalachian Piedmont, Alabama and southwest Georgia: Geological Society of America Bulletin, v. 195, p. 819–833.

Steltenpohl, M. G., and Moore, W. B., 1988, Metamorphism in Alabama: Geological Survey of Alabama Circular 138, 29 p.

Steltenpohl, M. G., Goldberg, S. A., Kunk, M. J., and McRae, M., 1990, Structural and age relations from the Goat Rock Fault zone, Alabama and

Georgia: Evidence for Alleghanian development: Geological Society of America Abstracts with Programs, v. 22, p. 233.

Steltenpohl, M. G., Chalokwu, C. I., McRae, M., Hanley, T. B., and Kunk, M. J., 1991, Are Carolina/Avalon terrane rocks exposed in Alabama?: Geological Society of America Abstracts with Programs, v. 23, p. 134.

Steltenpohl, M. G., Goldberg, S. A., Hanley, T. B., and Kunk, M. J., 1992, Alleghanian development of the Goat Rock fault zone, southernmost Appalachians: Compatability with the master décollement: Geology, v. 20, p. 845–848.

Stow, S. H., Neilson, M. J., and Neathery, T. L., 1984, Petrography, geochemistry and tectonic significance of amphibolites of the Alabama Piedmont: American Journal of Science, v. 284, p. 416–436.

Sun, S.-S., 1980, Lead isotopic study of young volcanic rocks from mid-ocean ridges, ocean island and island arcs: Royal Society (London) Philosophical Transactions, v. A297, p. 409–445.

Szabo, M. W., Osborne, W. E., Copeland, C. W., Jr., and Neathery, T. L., eds., 1988, Geologic map of Alabama: Geological Survey of Alabama, scale 1:250,000.

Tarney, J., 1977, Petrology, mineralogy and geochemistry of the Falkland plateau basement rocks, Site 330, Deep Sea Drilling Project, *in* Wise, S. W., Jr., ed., Initial report of the Deep Sea Drilling Project, Volume 36: Washington, D.C., U.S. Government Printing Office, p. 893–921.

Wang, P., and Glover, L., III, 1992, A tectonics test of the most commonly used geochemical discriminant diagrams and patterns: Earth-Science Reviews, v. 33, p. 111–131.

Winchester, J. A., and Max, M. D., 1982, The geochemistry and origins of the Precambrian rocks of the Rossclare Complex, SE Ireland: Journal of the Geological Society of London, v. 139, p. 309–319.

Zen E-an, 1988, Evidence for accreted terrane and the effect of metamorphism: American Journal of Science, v. 288-A, p. 1–15.

Zietz, I., Riegle, F. E., and Gilbert, T. P., 1980, Aeromagnetic map of Georgia: U.S. Geological Survey Geophysical Investigations Map GP-940, scale 1:500,000.

Manuscript Accepted by the Society April 30, 1996

Printed in U.S.A.

Geological Society of America
Special Paper 314
1997

Tectonics of the west flank of Carolina terrane in south-central Virginia: Fold nappe emplacement followed by Carboniferous orogen-parallel ductile shear

Robert A. Baird* and Lynn Glover, III
Orogenic Studies Laboratory, Department of Geological Sciences, Virginia Polytechnic Institute and State University, Blacksburg, Virginia 24061-0420

ABSTRACT

Five Latest Proterozoic and Paleozoic structural events occurred during four orogenies in the Charlotte metamorphic belt (CMB) and Carolina slate belt (CSB) of central Virginia. The ca. 546- to 620-Ma Virgilinian orogeny (D1) is recorded by premetamorphic folds in both belts.

The ca. 535-Ma Carolinian orogeny (D2), commonly manifested by intrafolial folds in S3, is the result of collision of the Carolinia magmatic arc terrane with Laurentia. The Carolinian orogeny may be broadly coeval with the Virgilinian orogeny or coeval with an early stage in the protracted development of S3, the Taconian(?) orogeny regional pervasive foliation.

The Taconian orogeny (D3) formed a pervasive regional foliation that affected both belts and is correlated with classic Taconian (or continued Carolinian) collision between Carolinia and Laurentia in the Middle to Late Ordovician. The primary manifestation of D3 is a large westward-verging fold nappe with subhorizontal axial-planar S3 schistosity that characterizes the Virginia CMB. The combination of upright stratigraphy and a reversed metamorphic gradient in the adjacent Laurentian rocks indicates that Carolinian nappes were emplaced over Laurentian rocks during D3. Geothermo-barometry indicates high geothermal gradients indicative of an active magmatic arc regime accompanying regional metamorphism.

The fourth orogenic event resulted in an extensive D4 right lateral shear zone along the eastern margin of North America, possibly recording transform slip of Carolinia along the continental margin during the late Paleozoic Acadian and Alleghanian orogenies. The CMB of Virginia is bounded by such shear zones on both the northwest and southeast. Localization of the northwest (Brookneal) zone was controlled by the Taconic suture, whereas the southeast (Clover) shear zone is associated with the steeper part of a D3 structural flexure. This flexure is a change from subhorizontal to moderately dipping compositional layering and foliations and involves a greenschist/amphibolite retrograde zone that may have acted as a reaction softening front. An east-west principal shortening direction was imposed on the area during

*Present address: 9215 Broad Meadows Road, Glen Allen, Virginia 23060.

Baird, R. A., and Glover, L., III, 1997, Tectonics of the west flank of Carolina terrane in south-central Virginia: Fold nappe emplacement followed by Carboniferous orogen-parallel ductile shear, *in* Glover, L., III, and Gates, A. E., eds., Central and Southern Appalachian Sutures: Results of the EDGE Project and Related Studies: Boulder, Colorado, Geological Society of America Special Paper 314.

D4. Pressures point to crustal thickening between D3 and D4; this, together with decreasing temperatures, suggests a reduction in magmatic activity.

The fifth and last event to affect the region, late Alleghanian D5, formed low-amplitude regional-scale folds. The northeast trend of these structures is consistent with a northwest-southeast principal shortening direction, implying reorientation of the shortening direction between D4 and final D5 docking of Gondwana and North America.

INTRODUCTION

Two generations have passed since the introduction of sea-floor spreading (Hess, 1962) and the concept of plate tectonics (Morgan, 1968), yet there is little agreement on the age and location of sutures in the Piedmont and Blue Ridge (e.g., Hatcher et al., 1989; Horton et al., 1989; and Wang and Glover, this volume). Lack of understanding of these first-order plate tectonic elements stimulated the work presented in this and the other chapters of this volume.

This chapter presents new structural and metamorphic data gathered across the suture between the Laurentian and Carolinian (Secor et al., 1983) terranes in south-central Virginia. These data are then integrated with adjacent studies to provide a tectonic synthesis of the region and an indication of the age of the suture. In a more speculative vein we also attempt to provide a rough analysis of Paleozoic regional stress directions experienced by the Piedmont and integrate these into a history of collisional and strike-slip orogenies. The study area includes portions of the northern Charlotte metamorphic belt (CMB) and Carolina slate belt (CSB) of the central and southern Appalachian orogen (Fig. 1).

STRATIGRAPHY AND REGIONAL STRUCTURAL RELATIONS

Principal references on the CMB in Virginia are Tobisch and Glover (1971), Henika (1977), and Baird (1991). The CMB is a series of layered amphibolite facies gneisses and schists of volcanic and rare sedimentary protolith that have been locally intruded by granitic and gabbroic rocks. To the southeast, the layered gneisses grade into greenschist facies metavolcanic and metaepiclastic rocks of the CSB. The oldest rocks recognized in the study area are Charlotte belt Late Proterozoic and intermediate metavolcanic rocks (southwest part of the area shown in Fig. 2) and rare pelitic schists that have been concordantly intruded by granite. These rocks are succeeded to the northeast by metavolcanic felsic and mafic gneisses and metaepiclastic schists of the central CMB (sequence northwest of the granitic gneiss unit of Figure 2). Southeast of the granitic gneiss unit lies a portion of CMB rocks that grade into greenschist facies rocks of the CSB. CSB rocks in this area consist of the felsic metavolcanic Hyco Formation, the metaepiclastic Aaron Formation, and the mafic metavolcanic/metaepiclastic Virgilina

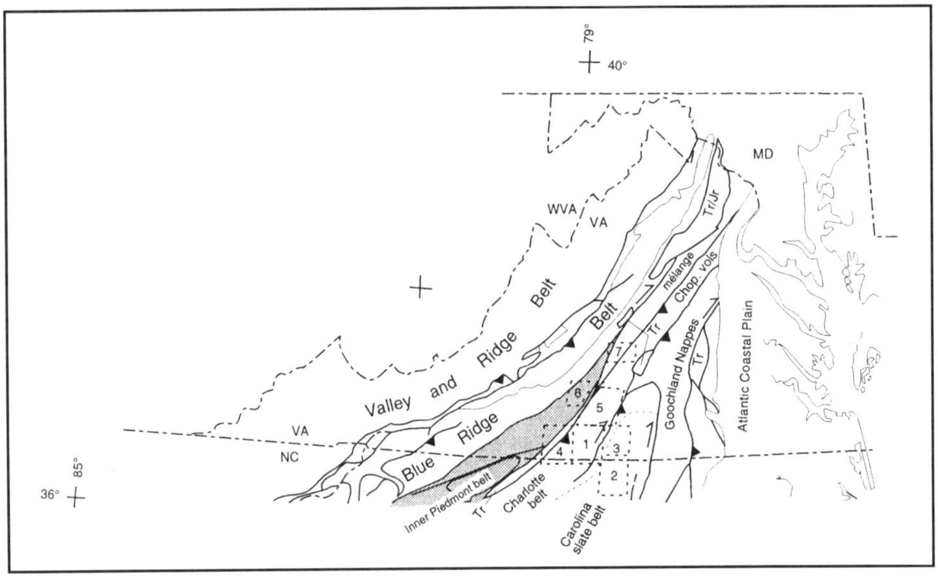

Figure 1. Geologic belts of the central Appalachian orogen with locations of geologic maps cited in the text. Triassic outcrops are marked by Tr. 1, Tobisch and Glover, 1971; 2, Glover and Sinha, 1973; 3, Kreisa, 1980; 4, Henika, 1977; 5, this study (see Fig. 2); 6, Gates, 1981, 1987, 1988; 7, Marr, 1981. Tr/Jr = Triassic/Jurassic rocks; Chop. vols = Late Proterozoic Chopawamsic volcanic rocks; mélange = Late Proterozoic to Cambrian sedimentary and tectonic mélange; shading = Inner Piedmont belt.

Formation (Laney, 1917; Glover and Sinha, 1973; Harris and Glover, 1985, 1988).

Much of the Charlotte belt (Tobisch and Glover, 1971) in the region shown in Figure 1 was called the Milton belt by Horton et al. (1989, 1991). Horton and Zullo (1991) suggested that the Milton belt be incorporated into a larger Charlotte metamorphic belt terrane and the name Charlotte metamorphic belt be applied to the entire belt from Virginia to Georgia. Baird (1991) recommended retaining the name Charlotte metamorphic belt and dropping Milton terrane. Currently Hibbard (1991), Hibbard et al. (1994), and Samson et al. (1992) propose that there may be a Late Precambrian or older initial interaction in the deformation zone between the Milton and CSB terranes and that the deformation zone was reactivated in the Alleghanian. The implication that the boundary between the two belts may be a suture is a very different interpretation than the more gradational stratigraphic and polydeformational relations between the belts as described in Tobisch and Glover (1971) and this chapter.

STRUCTURE AND COMPOSITIONAL LAYERING IN THE STUDY AREA

Definition of one dynamic and four dynamothermal events in the CMB and adjacent CSB is based on cross-cutting relations of the structures they produced (Table 1). The earliest event, the Virgilinian deformation recorded by Glover and Sinha (1973) just to the south (Fig. 1), was not identified in the study area, but its F1 fold axis in the southeast part of the area shown in Figure 2 was projected from work to the south (Tobisch and Glover, 1971; Glover and Sinha, 1973; Kreisa, 1980).

The earliest event in the study area is D1 (Table 1); however, the event that produced the pervasive regional foliation (S3) is D3. S3 is present throughout the Charlotte and Carolina slate belts. Local occurrences of intrafolial folds (F2) within S3 suggest the D2 event or an early stage in a protracted D3 event. D2 and D3 fabrics were in turn overprinted locally by dextral shear zones (D4) on the northwest and southeast margins of the CMB. All fabrics were subsequently affected by a warping event, D5.

The orientations of the various structural elements differ from area to area (Fig. 2), and these differences form the basis for subdividing the region into several structural domains and subdomains. Domain I is characterized by a moderate to steeply southeast-dipping S3 foliation (Fig. 3) overprinted by the D4 Clover shear zone with shear bands parallel to S3 (Baird, 1988). Domain II consists of a subhorizontal S3 foliation folded about regional-scale, open F5 fold axes in IIa but having an easterly dip in IIb. Throughout Domain III, S3 is subhorizontal and F5 folding is very subtle. Domains IIIb and IIIc contain minor variations in F3 folding. Domain IV is characterized by pronounced folding of S3 by F5 and exhibits a strong overprint by the D4 Brookneal shear zone (Gates et al., 1986), more so in IVa than in IVb.

The rest of this section centers on description of the fabrics and structures of five deformations and four metamorphisms. In the Discussion, these events are compared and related to those in adjacent areas.

S0 original layering

Previous workers recognized primary volcano-sedimentary features in the CSB (Laney, 1917; Glover and Sinha, 1973; Harris and Glover, 1988); however, original features and layering are rare in the portion of the CSB examined here because of later recrystallization and shearing. No primary volcano-sedimentary features were recognized in the CMB, although compositional layering reminiscent of relict bedding is present locally.

D1 deformation

In the CSB, in the southeastern part of the area shown in Figure 2, an F1 synclinal fold occurs (refer to Fig. 6). This is the continuation of a structure mapped and better seen in the region immediately to the south (Kreisa, 1980; Glover and Sinha, 1973). As in the region to the south, no S1 foliation or M1 metamorphic mineral assemblage was seen in the study area. The ca. 546- to 620-Ma Virgilinian deformation (Glover and Sinha, 1973) that formed F1 apparently occurred in rocks of this area that were too shallow to become regionally metamorphosed. Undoubtedly at some level, somewhere, metamorphic foliation was generated by this event and may eventually be found. In order to avoid the problem of having S1 associated with D2, in the following we keep the same numerical hierarchy in each event, that is, the first foliation identified is S2 associated with D2 and M2.

D2 deformation

In the CMB, an early foliation, S2, is present locally that is oblique to, and steeper than, the regional foliation S3 (Table 1). Relict S2 is also found refolded by isoclinal folds formed during the later regional event, D3. S2 also is present in garnets (Fig. 4) where the long axis of the garnet is semiconcordant with S3, but the garnet contains an oblique alignment of quartz grains, probably representing the earlier S2 foliation. The garnet shown in Figure 4 appears to have a secondary rim overgrowth on a rounded core, possibly representing additional growth during M3. Locally, intrafolial F2 folds are observed with axial planes parallel with the regional foliation, S3. Overall, F2 folds (Fig. 5) are quite rare in the study area.

D3 deformation

The study area is characterized by a penetrative regional foliation, S3, axial planar to the regional nappe structure (Figs. 2, 6), that exhibits an increase in dip from subhorizontal attitudes in the CMB to moderate southeast dips in the CSB (Fig. 3). This foliation, S3, is typically defined by aligned micas, primarily biotite with subordinate muscovite, and locally by alternating mafic and felsic banding that tends to cleave along the more micaceous layers. Most F3 folds are tight to isoclinal, with northwest vergence, northeast-trending axes, and subhorizontal axial surfaces that are subparallel with the regional folia-

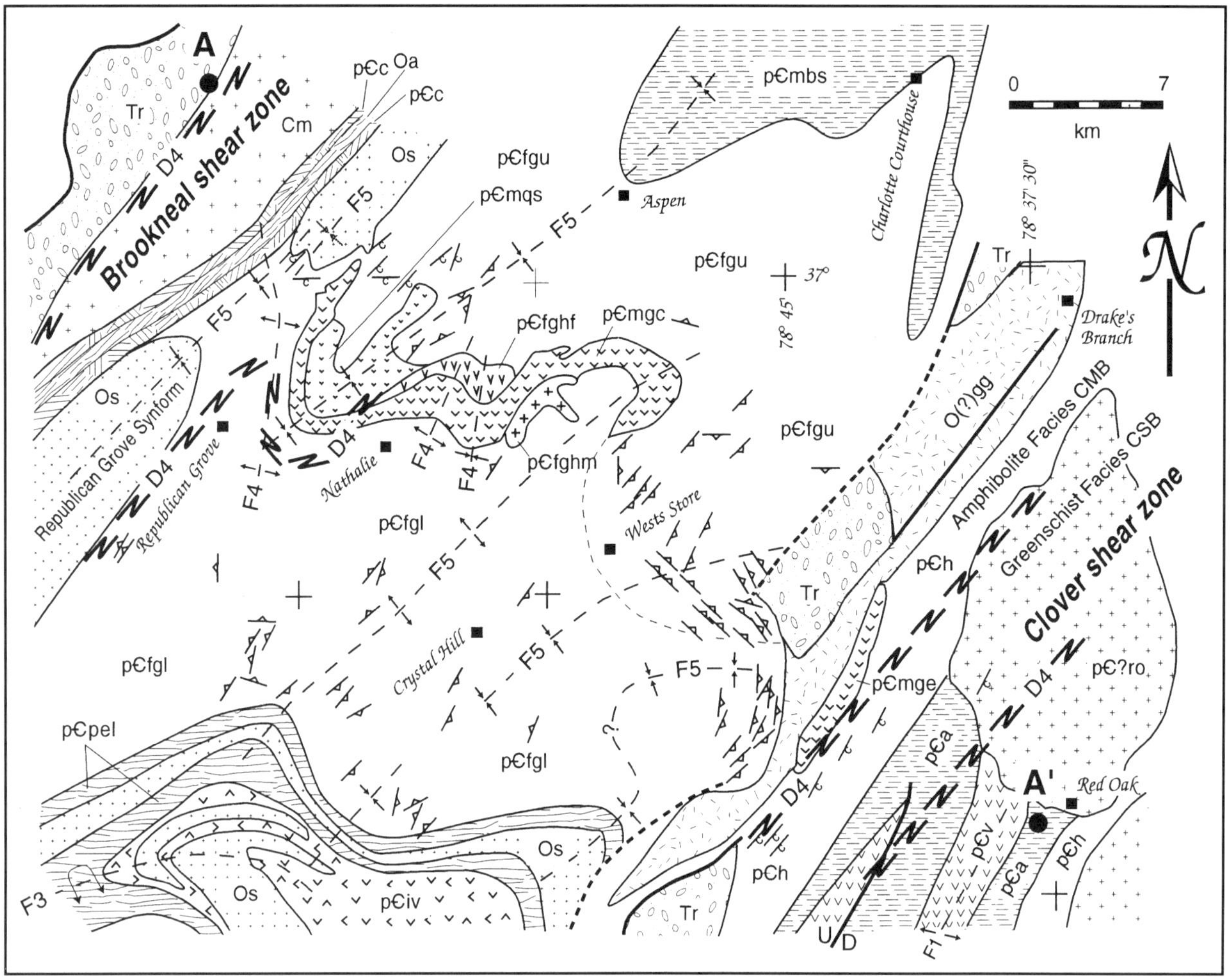

Figure 2. Geologic map of the south-central Virginia Piedmont. All boundaries are approximate. Area of map is shown in Figure 1(#5). Points A and A′ are end points of cross section shown in Figure 6. Geology from Laney (1917), Virginia Division of Mineral Resources (1963), Tobisch and Glover (1971), Gates (1981), and this study. Explanation on facing page.

tion. A regional mineral lineation, L3 (Fig. 7), consists primarily of aligned micas and is parallel to regional fold axes. Although the origin of most of the mineral lineations is not apparent, in some outcrops they are observed to be the intersection of S3 and the earlier cleavage, S2.

D4 deformation

Deformational event D4 is represented by the Brookneal (Gates et al., 1986) and Clover (Baird, 1988) shear zones that bound the CMB on the northwest and southeast, respectively (Fig. 2). D4 shearing was accompanied by local development of shear bands and sigmoidal fabrics defined by aligned micas and feldspar augen. A foliation, S4, is typically oblique (as much as 45°) to the regional foliation, S3, whereas the shear bands S4 (Fig. 2) are parallel with S3. The shear bands exhibit moderate to steep southeast dips in the northwesternmost part of the Brookneal shear zone (Fig. 8, Domain IVa), where the lithology is primarily an intrusive granite (Os, Fig. 2), but assume subhorizontal attitudes farther northeast (Fig. 8, Domain IVb) where the lithology is metavolcanic (pЄfgu, Fig. 2). The mineral extension lineation associated with shearing exhibits a strong northeast trend (Fig. 9).

Between the two shear zones, the rocks are *unsheared* except locally (Fig. 9, Domains IIb and IIIc), where shear bands are subhorizontal and movement sense is top to the southwest.

Within the Clover shear zone, the foliation related to the D4 shearing event is locally folded into kinks having amplitudes and wavelengths of as much as a few centimeters. On an equal-area net of F4 fold axes (Fig. 10), the kink axes (labeled "K" in Domain I) form a girdle defining a northwest-trending–southeast-plunging fold axis. The figure also defines two distinct trends for fold axes related to the shearing event (Domains IIIb and IIIc), one trending north to north-northeast and another northwesterly (the conjugate set of Tobisch and Glover [1971]?) (Table 1). Most F4 folds are broad warps or open folds with

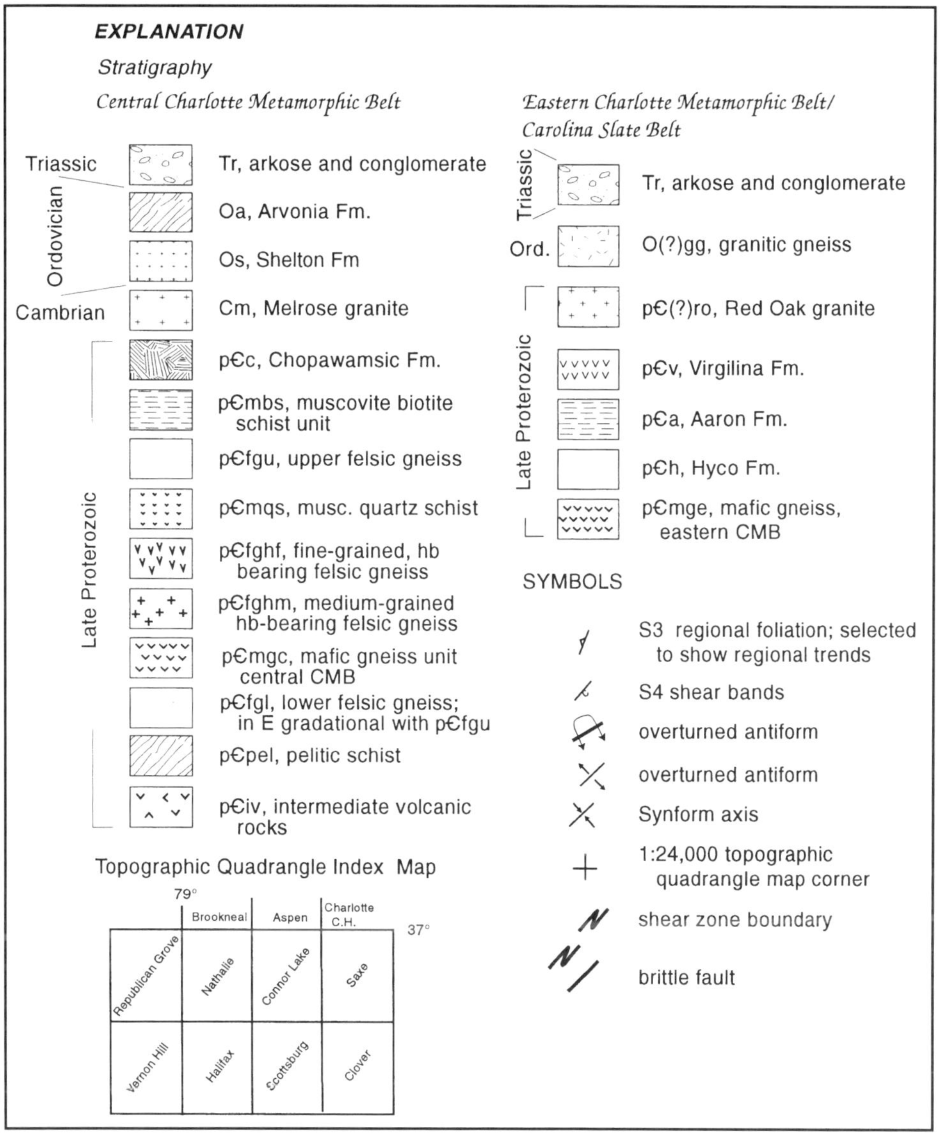

Figure 2 explanation

wavelengths on the order of a few to tens of meters. Axial surfaces for folds of this amplitude are difficult to determine but appear to be subvertical.

The Clover shear zone strikes across the Red Oak granite in the southeastern part of the area (Fig. 2). The granite is primarily massive southeast of the Clover zone, but there are local occurrences of 1- to 4-m-wide Sc shear zones. Some of these small zones trend northeast and are dextral like the Clover shear zone, whereas others trend northwest and exhibit sinistral movement. These are interpreted as small conjugate shear zones related to the Clover shear zone (cf. Tobisch and Glover, 1971).

Field relations shown on Figure 2 indicate the Red Oak is post-F1 (ca. 546- to 620-Ma Virgilinian deformation) and pre-D4 dextral shearing (Acadian or Alleghanian, Table 1). Other similar discordant granites in the area, such as the Roxboro (Glover and Sinha, 1973), are about 546 to 575 Ma (Table 1), or Latest Proterozoic. The lack of substantial offset in the Red Oak Granite suggests that the dextral displacement of the Clover shear zone is minimal, ca. 2 to 10 km.

D5 deformation

A late broad warping event is preserved in quadrangle-scale northeast-trending folds, such as the Republican Grove synform (Fig. 2). Figure 3 shows pronounced folding of S3 by the regional-scale F5 folds. This is especially prevalent in domains IVa, IVb, and IIa. The amplitude of F5 folding is too gentle to define in some of the S3 domains but is evident at the regional scale. F5 fold amplitude is greatest in the northwest CMB and progressively decreases to the southeast.

TABLE 1. CORRELATION OF DEFORMATION SEQUENCES IN CENTRAL AND SOUTH-CENTRAL VIRGINIA AND NORTHERN NORTH CAROLINA

Carolina	Carolina	Carolinia, Suture, and Laurentia	Carolina, Laurentia	Carolina, Laurentia	Probable Age
This chapter.	Tobisch and Glover (1971, modified* herein by Glover. Db: Obscure folds in oldest Charlotte metamorphic belt (CMB) unit (Basement under CMB?)	Gates (1981, 1987, this volume), Gates and Speer (1991), Gates et al. (1994).	Henika (1977)	Marr (1981)	
D1 Virgilina F1 fold in SE corner of Figure 2. S1 not found.	D1: *Carolina slate belt:* Virgilinian orogeny between 546 and 620 Ma; F1 open folds, faults, no S1 foliation or metamorphism (Glover and Sinha, 1973). *Charlotte belt;* premetamorphic folds without cleavage: Tobisch and Glover, 1971, p. 2212–2213, Figure 14, folds 1-5, p. 2217–2218; transected by pervasive foliation, S2, Figure 6 and p. 2212–2213; rare Virgilina F1 folds strongly overprinted by S2. Ca. 600-Ma Late Proterozoic Virgilinian orogeny between 575 and 620 Ma; Glover and Sinha (1973); between 546 and 620 Ma; Hibbard et al. (1995); Wortman and Sanson (1995).	Not in Laurentia. Not seen in Charlotte belt.	Not seen in Charlotte belt.	Not seen.	Late Proterozoic Virgilina Orogeny between 546 and 620 Ma.
D2: Local S2, oblique to and steeper than S3, in Charlotte belt from west to east; may be overturned to west, subhorizontal in central part and dips southeast at boundary with Carolina slate belt; local isoclines; rare intrafolial folds in Charlotte belt; relict S2 in garnets, possibly Cambrian.	D2: *Carolina slate belt;* D2 not seen. *Charlotte belt;* Local S2; synmetamorphic intrafolial folds (Tobisch and Glover, 1971, Fig. 5b); "early Virgilina-Halifax fold generation."	D1: Intrafolial folds in regional pervasive foliation, pre D2 or early D2. <534 ± 5-Ma orogeny (staurolite in Laurentian schist near Altavista; cut by 515-Ma Melrose Granite at Melrose, near Brookneal). Coeval with 570-535 Ma orogeny in the CSB near Columbia, SC. Dennis et al. (1993).	D1: NW-trending isoclinal folds, subhorizontal axial planes.	D1: Isoclinal, intrafolial folds in S2; post - ca.-550-Ma Chopawamsic Formation, pre-Arvonia (Late Ordovician–Early Silurian, ca. 440).	Ca. 535 Ma orogeny. Collision of CSB and CMB. Latest Proterozoic or Early Cambrian, Carolinian orogeny of Gates et al., 1994.
D3: Pervasive regional schistosity S3; isoclinal to tight folds; subhorizontal and gently dipping axial planes. Lineation, L3, parallel to regional F3 fold axes.	D3: "Virgilina-Halifax fold generation"; NE-trending axes; S3, regional schistosity and axial planes, subhorizontal to westward overturned in Charlotte belt, east dipping to vertical S3 in Carolina slate belt; Fold nappe-forming event. "Conformable syntectonic" Shelton Granite Gneiss; U/Pb zircon age = 463 Ma (Sinha et al., 1989); syn or pre S2.	D2: S2-pervasive foliation; F2 - locally tight asymmetric folds, moderately to steeply inclined, SE dipping, west vergent, SW plunging fold axes. Melrose Granite, emplaced in suture, is post D2(?) but yields 515 Ma zircon age (Sinha et al., 1989); see text.	D2: Tight to moderately open folds, gently to moderately inclined NE-trending axial surfaces.	D2: S2, pervasive regional schistosity; F2 tight and isoclinal to broad open folds; folds Late Ordovician to Silurian Arvonia Formation. Possibly S3—see comments on "D3 Taconian Orogeny" in text.	Pre-515 Ma if Melrose is correct or ca. syn-463 Ma if Shelton Granite Gn. age is correct; Taconian?

TABLE 1. CORRELATION OF DEFORMATION SEQUENCES IN CENTRAL-, SOUTH-CENTRAL VIRGINIA AND NORTHERN NORTH CAROLINA (continued)

Carolina	Carolina	Carolinia, Suture, and Laurentia	Carolina Laurentia	Carolina, Laurentia	Probable Age
D4 S4 foliation; N- and NW-trending open folds; N-trending set partially rotated toward NNE; vertical axial planes; Sc shear bands parallel with S3; mostly localized in Brookneal and Clover shear zones. Conjugate NE dextral and NW sinistral shear zones in Red Oak granite.	D4: "Milton–Hagers Mt. fold generation"; NW-trending axes, subvertical axial planes; S4 cleavage. Conjugate fold with D5? or pre D5?	D3: S3 NW-trending moderately SE-dipping axial planar crenulation cleavage increases in intensity toward Brookneal shear zone; L_{S2xS3} lineation in Falling River Granite. Dextral strike slip S-C fabric in Falling River Granite. Cuts Late Ordovician–Early Silurian Arvonia	D3: NE-trending open folds with near vertical axial planes; weak cleavage.	D3: Open, upright folds.	Acadian? Alleghanian Gates, 1981; Gates et al., 1986; Glover et al., 1983; Glover, 1989.
D5: NE-trending open folds; vertical axial planes; local week axial plane foliation F5 in parasitic folds on limbs of large folds.	D5: "Milton–Hagers Mt. fold generation"; NE-trending fold axes, subvertical axial planes. S5 cleavage. Conjugate folds with D4? or post D4?	D4: Crenulation cleavage, NE-trending, upright to SE dip; cuts S3.	D4: Open, NW-trending folds; NW-trending spaced cleavage.	D4: Gentle, open folds with NW-trending axial planes.	Alleghanian.

*Modification includes adding deformation phase numbers (D1, S2, etc.) to the relative time sequence of the type locality nomenclature used in the Tobisch and Glover (1971) paper. The Tobisch and Glover (1971, p. 2213) pre-metamorphic folding (D1 of this chapter) and "continuous" deformation of the Virgilina-Halifax fold sequence from pre- to synfoliation stages are herein suggested to be more episodic (D2 and D3 of this chapter).

No foliation accompanies D5 except for a weakly developed schistosity in the hinges of local parasitic folds on the limbs of the large F4 folds. The parasitic folds are close to tight folds with amplitudes and wavelengths of a few tens of centimeters. The axial surfaces of these folds form a girdle about the large-scale structures and define a northeast-trending axis. Sheared rocks near the southeast margin of the Brookneal shear zone have been folded about the regional-scale F4 fold axes (Fig. 8, Domains IVa, IVb). These quadrangle-scale folds commonly expose the conformable, synkinematic Shelton granite (463 ± 14 Ma by U-Pb, zircon, Hund, 1987) in their cores (Tobisch and Glover, 1971; Henika, 1980).

METAMORPHISM

No discernible metamorphism accompanied the D1 Virgilinian deformation in the CSB (Table 1), according to Glover and Sinha (1973). Coeval metamorphism, M1, probably also did not occur in the Charlotte belt because Tobisch and Glover reported no foliation accompanying the first fold generation in D1.

The first regional metamorphic event (here named M2 to conform with D2) accompanied D2 and is defined by assemblages in the regional foliations, S2 and S3. Mineral assemblages in S2 are the same as those in S3, suggesting that D2 and D3 were two events at the same grade or one protracted event (two stages) at the same grade.

The mineralogy in the D4 Brookneal shear zone is not significantly different from D2-D3 assemblages elsewhere in the CMB, but the occurrence of morphologically distinct garnets that overprint the sheared fabrics is evidence of their growth during and after D4, thus defining an M4 event.

D5 structures do not appear to correspond to an M5 metamorphic event.

Geothermobarometry

In monotonous quartzofeldspathic lithologies it is difficult to discern metamorphic gradients by petrographic examination. For instance, only four of 1,200 outcrops examined in the study area contained sillimanite. All four are in the central part of the area. The common occurrence of garnet, however, allowed a variety of thermobarometric techniques to be used to obtain pressure-temperature (P-T) data.

Polished sections for electron microprobe analysis were prepared from fresh samples of fine-grained biotite gneisses and mica schist. Mineral compositions were measured on the nine-spectrometer ARL-EMX electron microprobe in the Department of Geological Sciences at Virginia Polytechnic Institute and State University. Analytical procedures used are given in Solberg and Speer (1982). X-ray intensities were converted into weight percent oxides after Bence and Albee (1968). Mineral formulas were calculated using the computer program SUPERRECAL (SUPERRECAL prepared by J. C. Rucklidge of the Department of Geology at the University of Toronto).

Garnets were traversed from rim to rim. Care was taken to

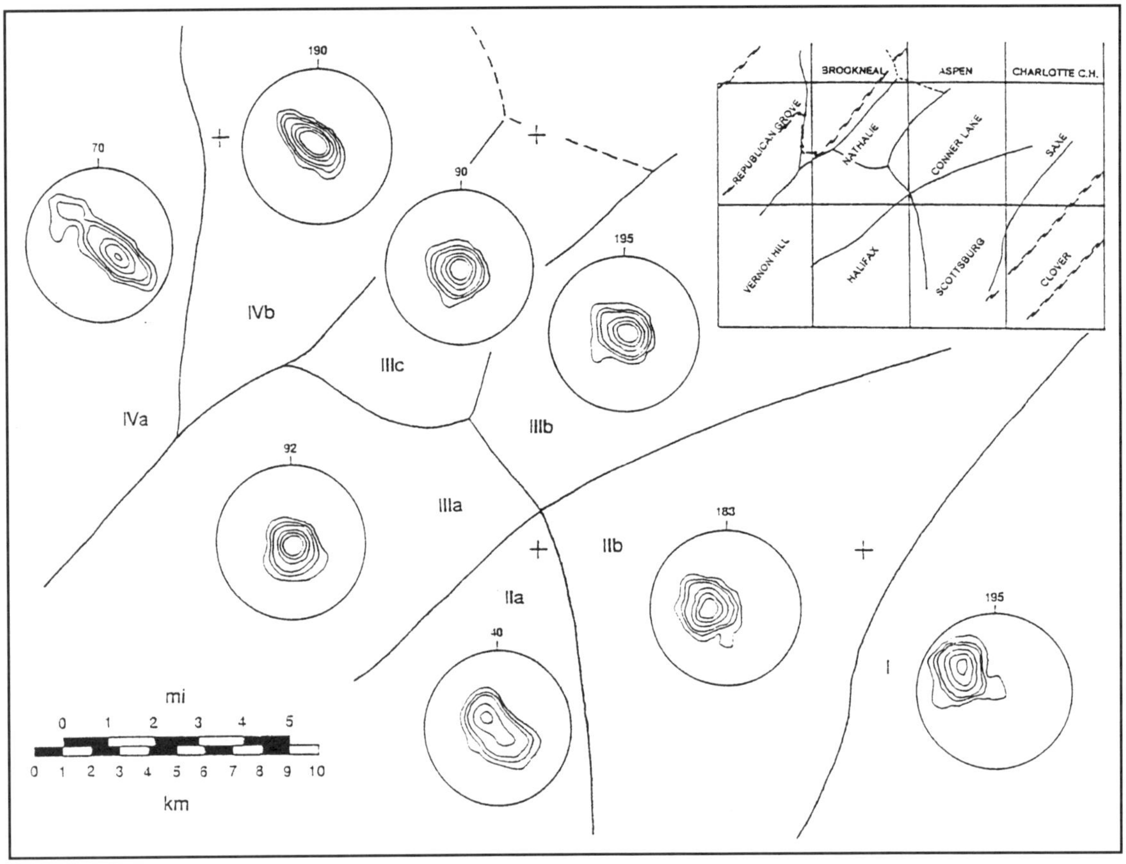

Figure 3. Poles to S3 foliation domains on equal-area net lower hemisphere plots. Region shown in this and similar diagrams is approximately the same area as Figure 2. Domain I in the CSB shows the moderate southeast dips of the foliation there. Throughout the CMB the subhorizontal S3 foliation is apparent. In domains IIa, IVa, and IVb the S3 foliation has been folded about a northeast-trending axis. Contours on the equal area nets are 1, 2, 3, 6, 9, 12 percent per 1 percent area. Number at top of the equal-area nets is the number of data points for that domain.

sample the most euhedral portions of the rim and to traverse as close to the core as possible. Tracy et al. (1976) noted that retrogression may occur on rims of garnets, as evidenced by peaks or troughs in the traverse results near the garnet rim. Loss of Mg from the garnet during retrogression would cause a corresponding increase of Mg in another coexisting ferromagnesian phase, such as biotite, giving an anomalously low temperature for the pair. This problem is partially ameliorated when the ratio of garnet to biotite is small, and the biotite effectively becomes an infinite reservoir (Tracy et al., 1976). In the samples used for geothermobarometry, garnet and biotite were the only ferromagnesian phases, retrograde rims on the garnets tended to be narrow, and the garnet-to-biotite ratio was typically low. Where there was no retrograde rim, the rim composition of the garnet was used; where there was a retrograde rim, a point was chosen just inside the rim as recommended by Tracy et al. (1976). The resulting temperature in the latter case is a minimum. Several measurements were made on the most homogeneous appearing portions of coexisting biotite, feldspars, and other minerals to ensure that zoned minerals were either excluded or accounted for. Biotite-garnet geothermometry temperatures were calculated after Thompson (1976) and Ferry and Spear (1978) using the garnet solution model of Hodges and Crowley (1985).

Pressures were determined using garnet-muscovite-plagioclase-biotite geobarometry (Ghent and Stout, 1981), garnet-rutile-almandine-ilmenite geobarometry (GRAIL, Bohlen et al., 1983), and quartz-garnet-almandine-plagioclase geobarometry (GASP, Ghent, 1976; Newton and Haselton, 1981; Ganguly and Saxena, 1984, 1985). Newton and Haselton (1981) stated that the four most complete studies involving use of the GASP geobarometer yielded an error of about ±1.1 kb for the reaction. Because of the monotonous mineralogy, most samples did not allow the use of multiple geobarometers.

Hodges and Crowley (1985) recommended that the above reactions for both temperature and pressure not be used for garnets with >25 mole percent spessartine. R. J. Tracy (personal communication, 1989), however, indicated that high Mn content has little effect, but that Ca content greater than 15 to 20 mole percent can affect the results. Only six garnets examined exceeded 15 mole percent Ca (Table 2).

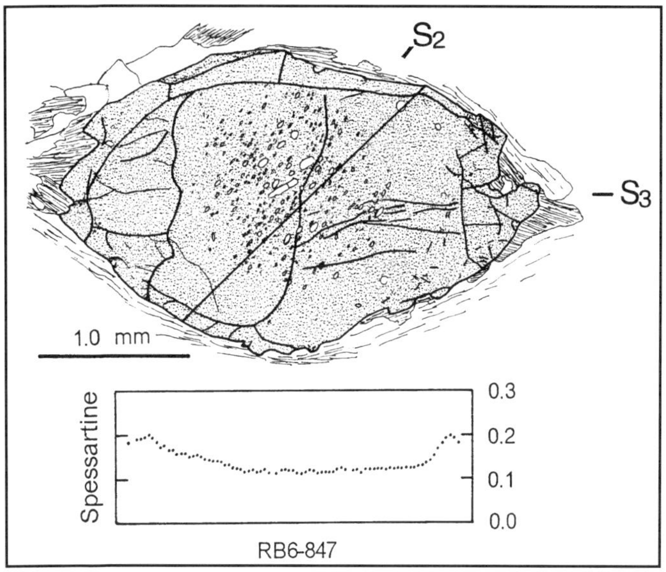

Figure 4. Drawing of a photomicrograph of a garnet nearly concordant with S3. Internal oblique alignment of quartz grains could represent the earlier S2 foliation. The garnet also appears to have a subrounded core (M2?) with a later overgrowth (M3?). Graph of spessartine content shown with scale in cations per formula unit. Microprobe traverse is shown by solid line from top to bottom.

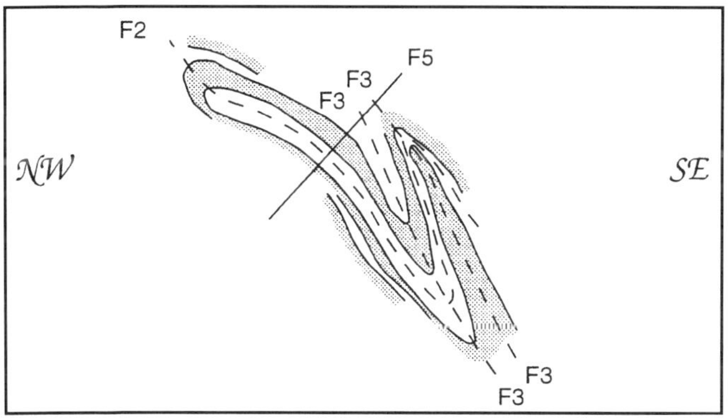

Figure 5. Drawing from a photograph of a cross section of an F2 fold refolded by F3, and warped by F5. Length of fold is approximately 1 m.

M2 metamorphism

As mentioned above, the rare M2 metamorphic mineralogy is indistinguishable from M3. It may be a separate event or the first stage in a protracted metamorphic deformation.

M3 metamorphism

Charlotte metamorphic belt. The mineral assemblages that define the S3 regional foliation in the CMB rocks are (1) metafelsites: quartz + plagioclase + K-spar ± muscovite ± biotite ± epidote ± garnet and plagioclase + quartz + biotite ± muscovite ± epidote ± hornblende ± garnet ± aluminum silicate; (2) metamafic rocks: hornblende + plagioclase + quartz + epidote ± biotite; and (3) mica schists: plagioclase + quartz ± biotite ± muscovite ± garnet.

Two morphologic types of garnet are present in the CMB. Morphologic type-one garnets (Fig. 11) are as much as 2 mm in diameter, subhedral to anhedral, and inclusion rich and have most fractures normal to the enclosing foliation. The garnets are semiconcordant with the S3 regional foliation. Near the D4 Brookneal shear zone, they exhibit varying degrees of rotation. Type-one garnets are thus assigned to M3. Chemically, morphologic type-one garnets are characterized by high almandine, moderate pyrope, and low spessartine and grossular contents (Table 2; Fig. 11). Morphologic type-two garnets (Fig. 12) are smaller and occur only in rocks overprinted by the Brookneal shear zone. They are discussed further below.

The results of geothermobarometry (Table 2) on type-one garnets indicate that M3 pressure-temperature conditions were about 560 to 580 °C at 4.6 to 4.8 kb.

Carolina slate belt. M3 metamorphism in the CSB is of greenschist facies. In the study area the original M3 amphibolite facies rocks at the CMB boundary were retrogressively overprinted by D4 Clover deformation zone greenschist shearing. Near the northwest margin of the Clover zone, relict garnet and hornblende now indicate M3 metamorphism of upper greenschist to lower amphibolite facies, probably near the original M3 CMB/CSB belt boundary. Thus Clover zone retrogression extended the present greenschist facies farther northwest than it was at the end of M3.

M4 metamorphism

Charlotte metamorphic belt. Within the Brookneal shear zone area (shown in the northwest in Fig. 2), the mineralogy of the quartzofeldspathic CMB assemblages appears to be little affected by the D4 shearing event and does not differ substantially from D3 regional event assemblages in these rocks. The mineral assemblages in the sheared CMB rocks are: (1) felsic gneisses: quartz + plagioclase + K-spar + muscovite ± biotite ± epidote ± garnet and plagioclase + quartz + biotite ± muscovite ± epidote ± hornblende ± garnet ± aluminum silicate; (2) metamafic rocks: hornblende + epidote + plagioclase + quartz ± sphene; and (3) mica schists: plagioclase + quartz + biotite + muscovite ± garnet.

Morphologic type-two garnets (Fig. 12) occur only in rocks overprinted by the Brookneal shear zone. These garnets are on the order of a few tenths of a millimeter long, euhedral to subhedral, inclusion free, and mostly fracture free. They displace or, more commonly, are superimposed on the sheared fabrics and overprint mica fish. Chemically, they are lower in almandine than morphologic type-one garnets. Pyrope is very low, whereas spessartine + grossular commonly equals almandine content (Table 2). Morphologic type-one and type-two garnets are clearly distinguished on the almandine-grossular-pyrope diagram (Fig. 13).

The results of geothermobarometry (Table 2) indicate that M4

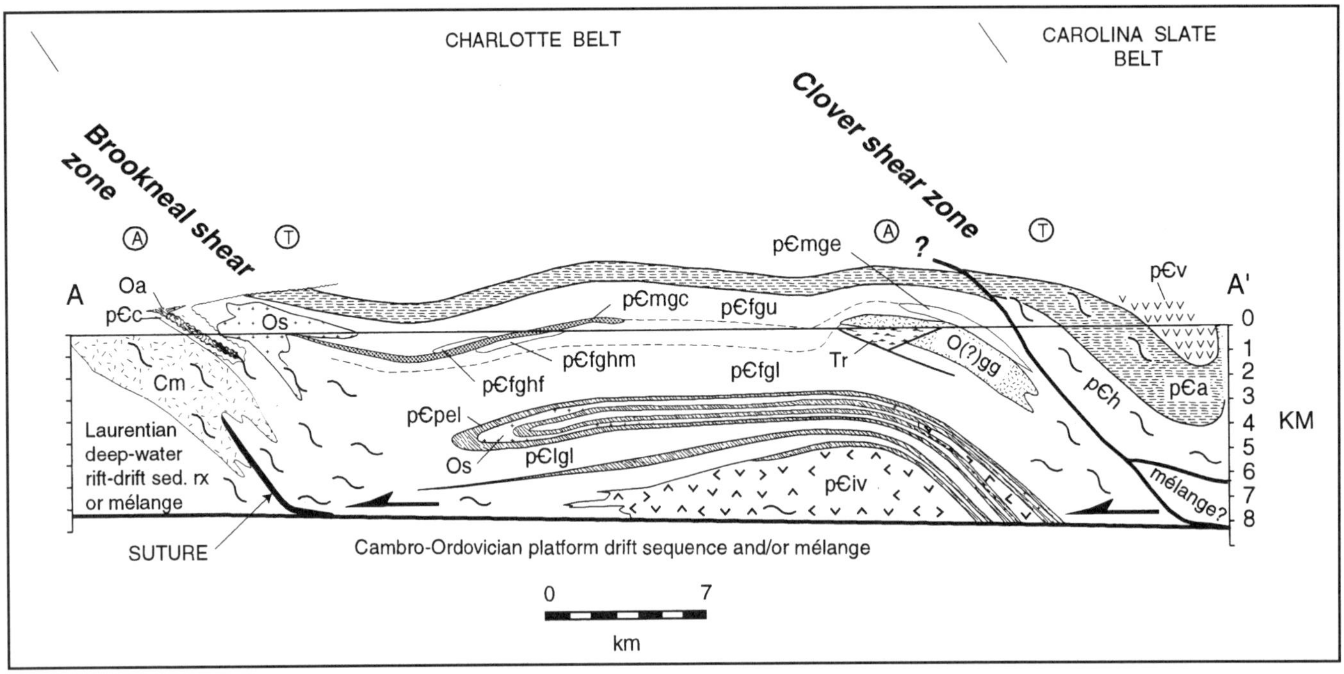

Figure 6. Cross section showing nappe structure of the area in Figure 2. Symbols as in Figure 2. Above-surface geology is projected from the northeast, and deep structure is projected from the southwest (Fig. 2). Presence of platform sequence below thrust is based on interpretation of Interstate 64 seismic line to the northeast (Pratt et al.,1988; Glover, 1989). Slice of mélange/basement (?) is included to take into account the Grenville Goochland nappe outcrop just northeast of the study area (Fig. 1). Vertical and horizontal scales are approximately the same. Zero depth is surface level. A = away; T = toward.

temperatures were comparable with those of M3 at 575 to 590 °C, whereas pressures were significantly higher at 6.6 to 7.1 kb.

Carolina slate belt. Lithologies of the CSB consist primarily of quartzofeldspathic rocks and schists, together with mafic rocks containing epidote and chlorite that have been overprinted by the Clover shear zone. The characteristic mineral assemblages are: (1) felsic metavolcanics: quartz + plagioclase + K-spar + muscovite ± biotite ± epidote ± garnet and plagioclase + quartz + muscovite + chlorite + epidote ± biotite; (2) greenstones: epidote + chlorite + quartz ± plagioclase; and (3) schists: plagioclase + quartz + muscovite + chlorite + epidote ± biotite.

DISCUSSION

Db possible basement folds

The "possible" folds (Table 1) seen by Tobisch and Glover (1971) in the oldest unit of their area have not been found by subsequent workers. This fold phase remains uncertain and may not exist.

D1, 546- to 620-Ma Virgilinian orogeny

This deformation (Table 1) involved buckle folding and faulting of *unmetamorphosed* layered volcanic rocks in the CSB of Glover and Sinha (1973). In the CMB, Tobisch and Glover (1971) also reported *premetamorphic* folds transected by S2-S3 schistosity (Table 1) developed before and during a protracted deformation in the same metamorphic interval. No other workers have reported pre-metamorphic folds in the CMB. The examples given by Tobisch and Glover (1971) in the CMB and by Glover and Sinha (1973) in the CSB are well documented and have recently been confirmed by Harris and Glover (1988) in central North Carolina. In this chapter we accept the Virgilinian orogeny as an event that occurred in both belts. It may have been an oblique-shear arc-rifting event (Harris and Glover, 1988) (Fig. 14).

D2 Early Cambrian Carolinian orogeny ca. 535 Ma

Timing and regional correlation. According to Marr (1981), his "D1" deformation in the Willis Mountain–Andersonville area (Fig. 1) affected the 550-Ma (U-Pb, zircon; Tilton et al., 1970; Higgins et al., 1971) Chopawamsic Formation but not the younger Arvonia (Latest Ordovician to Earliest Silurian). In the Tobisch and Glover sequence (Table 1) this deformation corresponds to the first foliation, which occurred during D2. Gates (Table 1) (Gates, 1981; 1987; this volume; Gates and Speer, 1991; Gates et al., 1994) also considers the first foliation to be D2 in the Brookneal area (Fig. 1). In our study area, D2 and S2 are the same as D2 and S2 of the Tobisch and Glover (1971) sequence as given

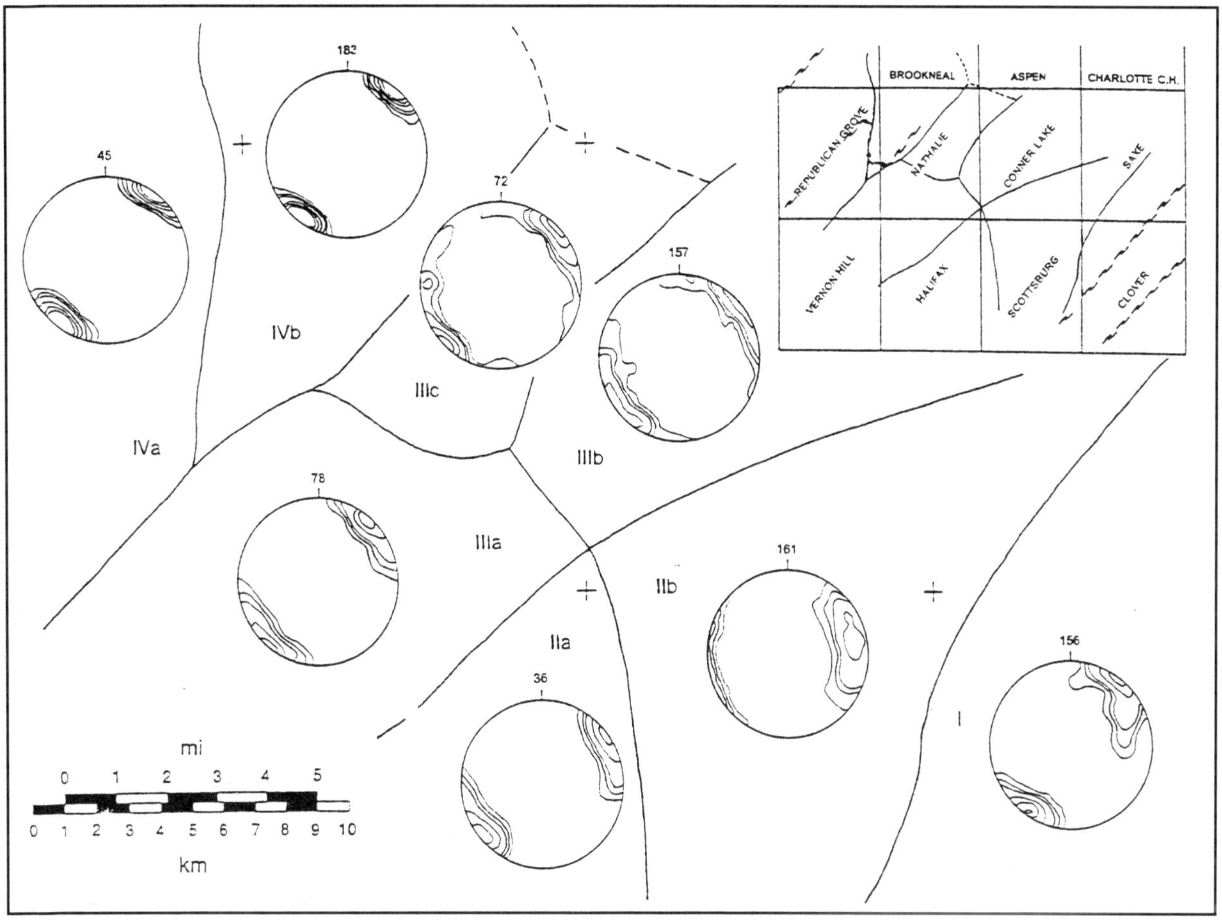

Figure 7. L3 mineral lineation domains on equal-area net lower hemisphere plots. Contours on the equal area nets are 1, 2, 3, 6, 9, 12 percent per 1 percent area. Number at top of the equal-area nets is the number of data points for that domain.

in Table 1. If these deformational events are correlative, as suggested in Table 1, then D2 is broadly constrained between ca. 550 Ma Chopawamsic volcanics and the approximate 440-Ma Late Ordovician to Early Silurian Arvonia Formation. Glover et al. (1992a, b), citing the Cambrian ages of plutons in the Pennsylvania, Maryland, and Virginia Piedmont that cut previously metamorphosed Carolinian and Laurentian rocks, proposed a Late Proterozoic or Early Cambrian collision of Carolinia and Laurentia. Gates et al. (1994) determined a 534 ± 5-Ma age of staurolite in Laurentian schists adjacent to the Carolina-Laurentian terrane suture at the northwestern edge of our study area that also dates the newly recognized Carolinian orogeny at Early Cambrian. A similar ca. 570- to 535-Ma age has been found for metamorphism and deformation in the Carolina terrane near Columbia, South Carolina (Dennis et al., 1993). 534 Ma apparently marks a cooling point in time on the P-T curve of thermal adjustment following collision at a somewhat earlier date.

M2 metamorphism. D2 fabrics and mineral assemblages are rare in the CMB because of the penetrative nature of D3. Mineral assemblages in S2 and S3 appear to be at the same metamorphic grade and are distinguished only by the structural fabric. In this chapter (Table 1) we propose that the age of M2 is about 535 Ma—Early Cambrian.

D3 Taconian orogeny

Timing and regional correlation. Tobisch and Glover (1971) and Henika (1980) mapped F2 folds (F3, this work) cored by conformable layers of syntectonic Shelton Formation granite. Hund (1987) reported a 463 ± 14-Ma (U-Pb, zircon) age for the Shelton, which falls in the 480- to 435-Ma range for the Taconian orogeny (Glover et al., 1983) and would seem to constrain D3 south of the study area to Taconian. However, Gates (1987) interpreted D2 structures (D3 of this study) as cut by the 515-Ma (Hund, 1987) Melrose Granite in the Brookneal area (Fig. 1, Table 1).

Is one of the ages wrong? Sinha et al. (1989) record at least five plutons in the Virginia-Maryland Piedmont in the range 520 to 500 Ma, so the Melrose age is not unique. Several studies using different isotopic systems have provided ages in this range for the central Appalachians. Inherited zircons might cause an

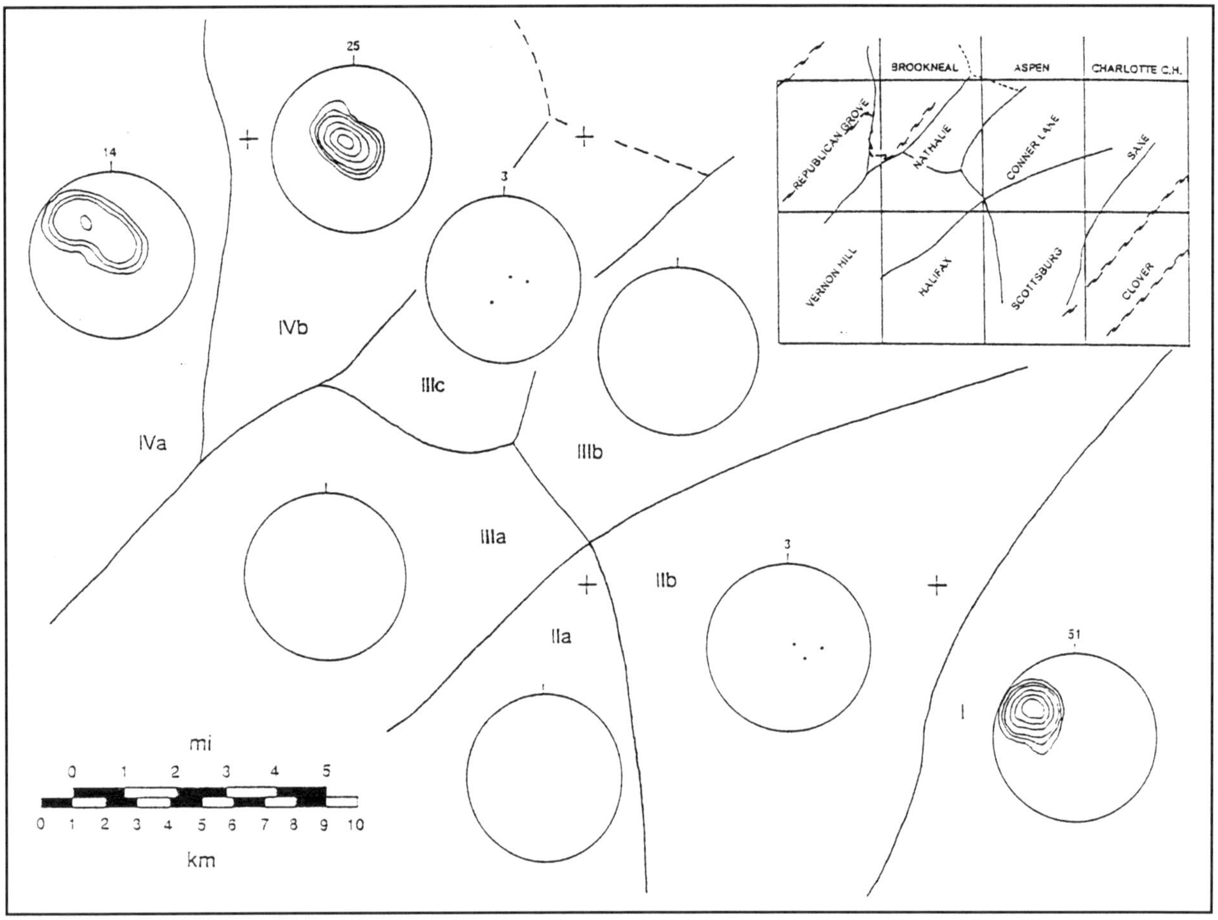

Figure 8. Poles to D4 slip surface domains on equal-area net lower hemisphere plots. Slip surfaces are generally parallel with the S3 foliation in a given domain (cf. Fig. 3). Contours on the equal area nets are 1, 2, 3, 6, 9, 12 percent per 1 percent area. Number at top of the equal-area nets is the number of data points for that domain.

older apparent age in some cases, but the Occoquan granite was dated at 494 by Rb-Sr whole-rock methods (Mose and Nagel, 1982), and the 500- to 490-Ma Baltimore gabbro was dated by both Nd-Sm (Shaw and Wasserburg, 1984) and Pb-U zircon methods (Sinha et al., 1989) with excellent agreement. Therefore 520 to 500 Ma seems to be a true age range of certain plutons in the region, and we are inclined to accept a 515-Ma age for the Melrose.

Is there more than one pervasive regional foliation? Most workers agree on the structures produced by D3 of this study (Table 1), except for Marr (1981), who believed that folding of the Late Ordovician–Early Silurian Arvonia Formation also took place during this event. Pervasive foliation in the Arvonia does not show the intrafolial folds and protracted history of the D2-D3 sequence in our area and therefore probably corresponds to our D4. Brown (1969) and Glover (1989) noted that the Arvonia Formation was deposited nonconformably upon the Chopawamsic volcanic rocks and Columbia granite following the Taconian orogeny. Accordingly, Arvonia folding must be post-D3, Silurian or younger. D2 and D3 appear to be phases in a protracted development of D3 regional schistosity (Tobisch and Glover, 1971), and the D3 phase may be less intense in the Melrose area near Brookneal. In this case D2 could be Cambrian (syn-534 ± 5-Ma metamorphism and pre-515-Ma Melrose granite). Earlier work also hinted at a Cambrian deformation: Fullagar and Dietrich (1976) found 520-, 531-, and 585-Ma whole-rock schist ages in the Evington and Ashe formations in Laurentian rocks west of the Melrose area. Gates and Mose (1987) got a 546 ± 21-Ma whole-rock Rb-Sr isochron on metapelites near Altavista, supporting the concept of a pre-Melrose—Latest Proterozoic to Cambrian orogeny, D2. Glover et al. (1992a, b) reviewed the evidence that several Virginia and Maryland 520- to 500-Ma plutons intrude schists and gneisses that were already deformed and metamorphosed to sillimanite grade before Cambrian intrusion.

The Taconian orogeny has usually been interpreted as the result of a collision between Carolinia and Laurentia in the early Paleozoic (ca. 480 to 435 Ma, Glover et al., 1983). The D2 event in the study area appears to be older than this time frame for the Taconian orogeny, yet younger than the ca. 546-to 620-Ma Virgilinian deformation. Therefore, D2 may represent a very early

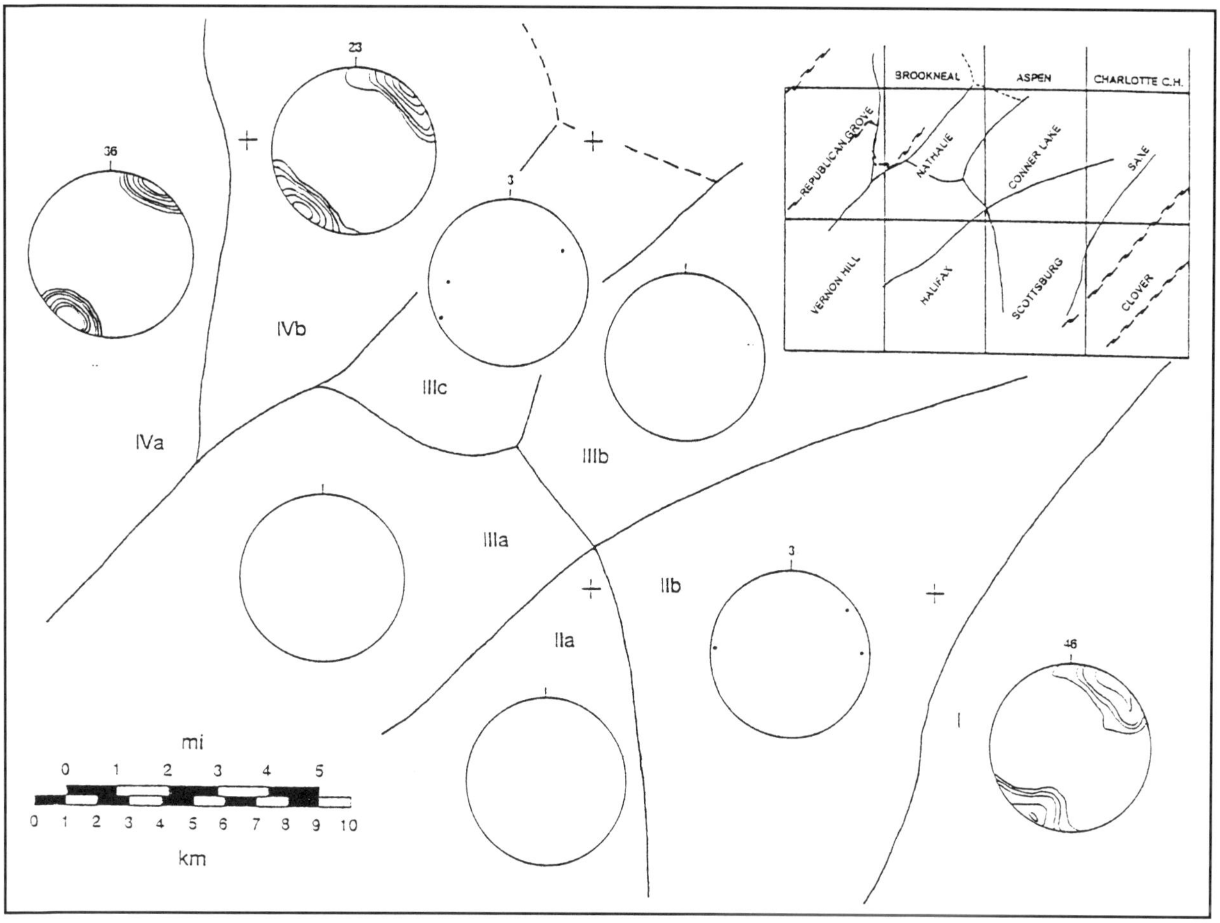

Figure 9. D3 mineral extension lineation domains on equal-area net lower hemisphere plots. Lineations are generally parallel with the northeast-trending Brookneal (domain IVa and IVb) and Clover (domain I) shear zones (cf. Fig. 2). Contours on the equal area nets are 1, 2, 3, 6, 9, 12 percent per 1 percent area. Number at top of the equal-area nets is the number of data points for that domain.

stage of the Taconian orogeny, but more probably it is a separate Late Proterozoic to Early Cambrian orogeny caused by the collision of the magmatic arc Carolinia with Laurentia (Fig. 14b).

F3 folds cored by 463 ± 14-Ma (U-Pb, zircon) synkinematic Shelton granite (Hund, 1987) that contains S3 (Table 1) are consistent with classic Taconian deformation and metamorphism and suggest that the D3 overprint to the south was stronger here than in the Brookneal area. The long time span of S2-S3 development may be about 70 m.y. This is nearly twice the duration of collision between India and the Asian continent. However, the Asian collision is still going on after at least 40 m.y. (Tapponier et al., 1982).

Nappe emplacement event. In the study area, the CMB is approximately 34 km across and characterized by S3 subhorizontal tectonic foliation (Fig. 3). Hobbs et al. (1976, p. 416–418) proposed two mechanisms that could produce subhorizontal foliations over a large area. The first involves regions of flat structure that served as soles of thrust sheets. Near the thrust plane, rocks would be highly strained, and a flat-lying mylonitic foliation would be present. The rocks of the central CMB, however, exhibit no significant shear strain, except along the margins where the belt is bounded by later shear zones. The S3 foliation is penetrative but not mylonitic as would be expected if a thrust were present in the shallow subsurface.

The second mechanism consists of the imposition of a moderate to steep foliation during a shortening event, followed by recumbent fold nappe emplacement. As the nappe develops, the foliation is eventually rotated to near horizontality. This mechanism is favored here as the best explanation of the subhorizontal foliation and is consistent with previous interpretations of the presence of nappes in the CMB (Tobisch and Glover, 1971). The scale of the structure (34 km) is comparable with that of large nappes in the Alps and elsewhere (Brown, 1968).

Cooper (1981) indicated that such fold nappes form by gravitational instability, rising and spreading laterally via ductile upwelling of the hot, buoyant infrastructure of an orogen. Intrusion of the syntectonic (Henika, 1980; Hund, 1987) Shelton Formation granite may have similarly assisted CMB nappe emplacement.

Supporting evidence for the antiformal fold-nappe forma-

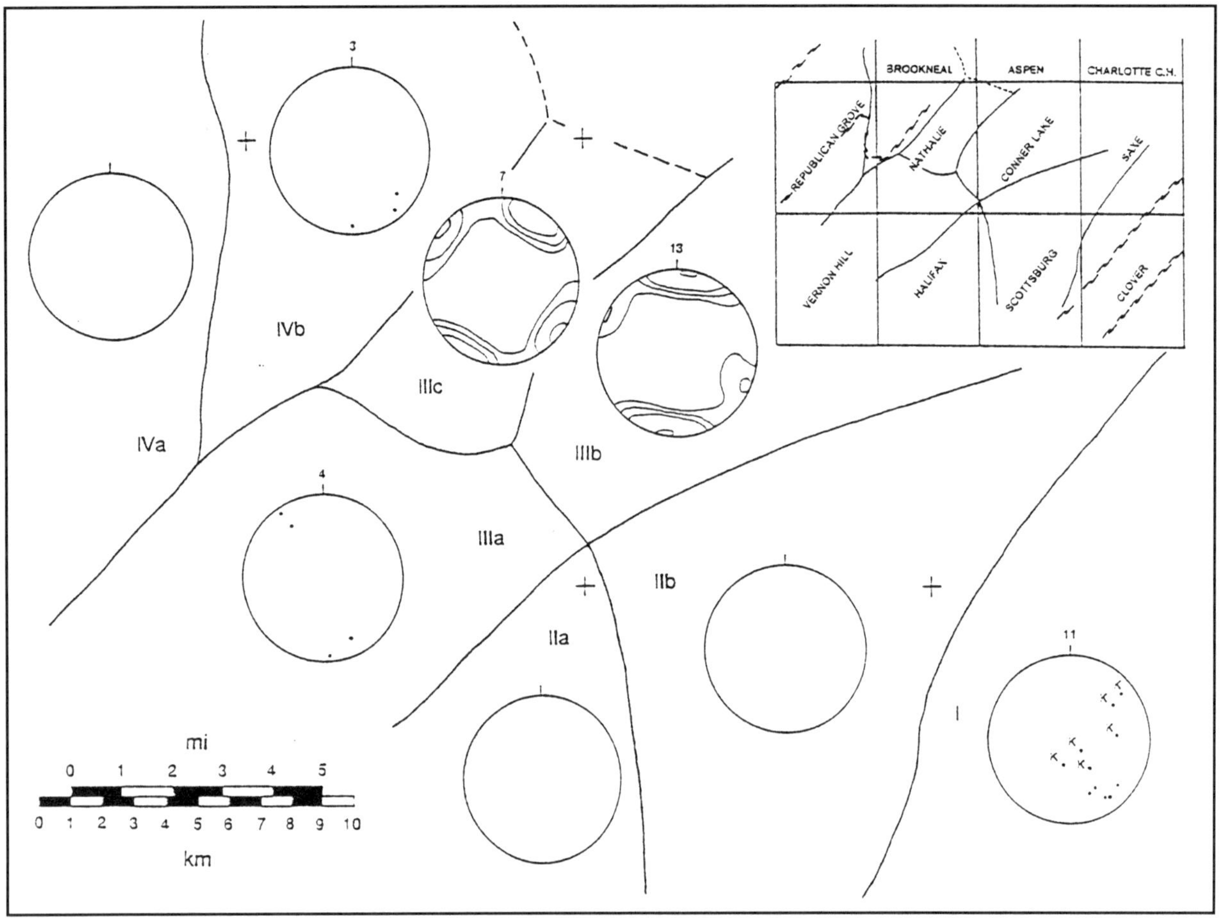

Figure 10. F4 fold axis domains on equal-area net lower hemisphere plots. Domains IIIb and IIIc show two distinct trends for F4 folds (see text). Domain I exhibits F4 kink axes (labeled K) rotated about a northwest trend. The data indicate different phases of D4 deformation. Contours on equal area nets are 1, 2, 3, 6, 9, 12 percent per 1 percent area. Number at top of equal-area nets is the number of data points for that domain.

tion includes the complementary synformal fold, the southeast-dipping Virgilina synclinorium in the southeastern part of the area shown in Figure 2 (synclinal structure cored by Virgilina Formation). Although the basic synclinal structure dates to the Virgilinian deformation (Glover and Sinha, 1973), it was tightened and enhanced during D2 and D3 when CSB and upper-level CMB rocks were transported northwestward over lower-level CMB rocks, folding the CMB into an F3 fold nappe (Fig. 6; see also "Virgilina-Halifax" fold generation of Tobisch and Glover, 1971; Table 1).

Gates (1987) mapped the Laurentian shelf rocks northwest of and adjacent to the present study area. The upright stratigraphy and inverted metamorphic gradient he found are consistent with hot nappe emplacement over that region.

Extent of the nappe structure. Although the CMB in south-central Virginia consists of a recumbent fold nappe, this situation apparently does not continue southwestward in the CMB. Griffin (1978) indicated that the structure near the South Carolina–Georgia border consists of a conformable CMB sequence underlying the CSB, both of which have been deformed into a series of accordion-like moderately to steeply dipping folds. In York County, South Carolina, Butler (1971) found that the average strike and dip of CMB rocks was N54E, 72SE. These findings contrast with the subhorizontal average foliation of the CMB in Virginia. It is possible that, to the south, the CMB either does not consist of a nappe or is a strongly refolded nappe.

Origin of the L3 mineral lineation. The regional mineral lineation, L3 (Fig. 7), is locally the result of the S2-S3 intersection. The origin of most D3 lineations, however, is not obvious.

M3 metamorphism. Garnet-biotite geothermometry yields conditions of 560 to 580 °C at 4.6 to 4.8 kb (Table 2) for CMB M3 assemblages. Assuming a pressure gradient of 0.3 kb per 1 km of overburden (Turner, 1981), these conditions correspond to a burial depth of about 15 to 16 km for the region during the Taconian orogeny. Using an average surface temperature of 20 °C, the temperature increase is 540 to 560 °C over 15 to 16 km, yielding a geothermal gradient range of 34 to 36 °C/km. This range is consistent with the low-pressure range for regional

TABLE 2. MICROPROBE DATA AND RESULTS FOR GEOTHERMOBAROMETRY

Sample	Event	Alm. Garnet	Pyro.	Spess.	Gross.	Biotite			Muscov.		Feldsp.		Ilmenite? Rutile?	AL_2SiO_5
		XFe	XMg	XMn	XCa	F/FM*	Xann†	Xphl§	Xmu**	XA1IV‡	Xan§§	Xilm***		
RB6- 6	M3										0.19			
RB6- 45	M3	0.661	0.079	0.132	0.128	0.545	0.451	0.379			0.36	Yes		
RB6- 52	M3	0.636	0.095	0.154	0.115	0.509	0.405	0.400			0.36			
RB6-105	M4(?)	0.439	0.040	0.356	0.165	0.632	0.517	0.311			0.19			
RB6-119	M4(?)	0.374	0.045	0.386	0.195	0.549	0.431	0.365	0.978	0.732				
RB6-127	M4(?)	0.426	0.021	0.318	0.236	0.743	0.599	0.212	0.980	0.761	0.19			
RB6-136	M4(?)	0.460	0.025	0.311	0.204	0.720	0.538	0.213	0.972	0.783	0.22			
RB6-148	M4(?)	0.432	0.035	0.351	0.182	0.668	0.539	0.274	0.974	0.748	0.22			
RB6-156	M4(?)	0.453	0.157	0.377	0.014	0.315	0.251	0.564	0.787	0.885	0.03			
RB6-175	M3	0.624	0.103	0.245	0.029	0.399	0.385	0.572	0.800	0.873	0.13			
RB6-223	M4(?)	0.528	0.058	0.200	0.214	0.583	0.485	0.352	0.905	0.798	0.24	Yes		
RB6-375	M4	0.519	0.123	0.336	0.022	0.384	0.316	0.506			0.13			
RB6-376b	M4	0.552	0.146	0.230	0.072	0.349	0.307	0.580			0.24	Yes		
RB6-465	M3										0.27			
RB6-524	M3	0.586	0.094	0.268	0.053	0.477	0.403	0.420	0.837	0.842	0.13			
RB6-550	M3										0.39			
RB6-553	M3	0.759	0.135	0.051	0.055	0.503	0.398	0.394			0.32	0.97	Yes	Sill
RB6-619	M3	0.785	0.116	0.057	0.041	0.552	0.436	0.356	0.906	0.902	0.24			Sill
RB6-621	M3										0.34			
RB6-755	M4(?)										0.34			
RB6-829	M3	0.752	0.081	0.118	0.050	0.593	0.471	0.317						
RB6-847	M3	0.749	0.144	0.061	0.045	0.464	0.372	0.433	0.895	0.903		0.97	Yes	Sill
RB7- 5	M3	0.416	0.037	0.469	0.078	0.609	0.491	0.328	0.966	0.727	0.16			
RB7-217	M3	0.784	0.116	0.046	0.055	0.530	0.411	0.366	0.905	0.909	0.27			Yes
RB7-222	M3	0.634	0.079	0.229	0.059	0.554	0.444	0.362			0.42			
RB7-242	M3	0.625	0.159	0.145	0.071	0.411	0.330	0.479			0.31		Yes	

*F/FM = Fe/(Fe+Mg) in biotite.
†Xann = mole fraction of annite component of biotite.
§Xphl = mole fraction of phlogopite component of biotite.
**Xmu = mole fraction of muscovite component in muscovite.
‡XA1IV = mole fraction of aluminum in the octahedral site in muscovite.
§§Xan = mole fraction of anorthite component of plagioclase.
***Xilm = mole fraction of ilmenite component in ilm.

metamorphism (Miyashiro, 1973) and is typical of a magmatic arc tectonic regime where heat is constantly added via plutonism.

The CSB in the study area is at greenschist facies metamorphic grade but retains evidence that the northwestern part experienced at least upper greenschist to lower amphibolite grade metamorphic conditions during M3. This conclusion is based on the presence of relict garnet partially altered to chlorite and relict hornblende partially altered to epidote within a kilometer or so of the boundary with the CMB. This gradient is a folded, originally vertical gradient related to upwelling and extrusion of the CMB during D2 and D3 growth and emplacement of the fold nappe. Farther southeast, the synclinal CSB was probably at greenschist facies during M3 also.

D4 Acadian-Alleghanian shearing/folding event

Timing and regional correlation. The youngest rock in the area affected by D4 shearing is the Late Ordovician to Early Silurian (Tillman, 1970; Glover, 1989) Arvonia Formation, which unconformably overlies the CMB sequence. Triassic basin rocks are only tilted and nonconformably overlie rocks near the Brookneal shear zone. Shearing in the Brookneal zone is thus broadly constrained to post–Late Ordovician/Early Silurian and pre-Triassic, or Acadian to Alleghanian. Glover et al. (1983) gave a 380- to 340-Ma range for the Acadian event and 330- to 230-Ma for the Alleghanian. Gates et al. (1986) reported that the 515-Ma (U-Pb, zircon) Melrose granite yielded an age of 300 Ma (Rb-Sr, biotite-whole rock) in portions of it affected by shearing. Assuming a 300 °C closure temperature for ^{87}Sr diffusion in biotite, they interpreted this date as a cooling age postdating most of the shearing. Harper et al. (1973) reported a 300-Ma conventional argon date for the Arvonia Formation about 70 km northeast, probably also reflecting the date of the shearing event. Thus the post-Melrose Brookneal shearing is probably Alleghanian.

By analogy, the Clover zone may also be Alleghanian. Gates (1988) reported, however, that Appalachian dextral faults range from 345 to 250 Ma (late Acadian through late Alleghan-

TABLE 2. MICROPROBE DATA AND RESULTS FOR GEOTHERMOBAROMETRY (continued - page 2)

Sample	Event	Temperature		G and S Pressure§		GRAIL**		QGAP‡
		ABT* (°C)	F and S† (°C)	Mg (kb)	Fe (kb)	Bohlen	H and C	N and H
RB6- 6	M3							
RB6- 45	M3	580	560					
RB6- 52	M3	595	580					
RB6-105	M4(?)	600	590					
RB6-119	M4(?)	590	570					
RB6-127	M3(?)	600	585	6.6	7.3			
RB6-136	M4(?)	580	560	6.3	6.8			
RB6-148	M4(?)	615	605	6.2	6.7			
RB6-156	M4(?)	585	570	6.8	7.0			
RB6-175	M3	495	460	3.7	3.7			
RB6-223	M4(?)	620	610	7.2	7.9			
RB6-375	M4	565	540					
RB6-376b	M4	565	540					
RB6-465	M3							
RB6-524	M3	565	530	5.7	5.9			
RB6-550	M3							
RB6-553	M3	615	605			6.7	6.4	5.4
RB6-619	M3	615	605	4.6	4.7			5.4
RB6-621	M3							
RB6-755	M4(?)							
RB6-829	M3	575	555					
RB6-847	M3	590	575			6.3	6.0	
RB7- 5	M3	550	525	4.2	4.5			
RB7-217	M3	595	580	5.0	5.2			5.5
RB7-222	M3	575	555					
RB7-242	M3	620	615					

*ABT = Garnet-biotite temperatures per the parameters of Thompson (1976).
†F and S = Garnet-biotite temperatures per the parameters of Ferry and Spear (1978).
§G and S = Pressures per the parameters of Ghent and Stout (1981); Mg and Fe calculated using the F and S temperatures.
**GRAIL = Pressures for garnet-rutile-aluminosilicate-ilmenite geobarometer per the parameters of Bohlen et al. (1983) and of Hodges and Crowley (1985); calculated using the F and S temperatures.
‡QGAP = Pressures for the quartz-garnet-aluminosilicate-plagioclase geobarometer per the parameters of Newton and Haselton (1981); calculated using the F and S temperatures.

ian), in agreement with the proposal by Glover et al. (1983, p. 225) that the Acadian and Alleghanian "could be two phases of one long protracted event." Based on this range of dates, the Clover zone, and thus D4, may have developed any time within this range.

Figure 15S shows the pole to the average S4 foliation in the least shear-strained (least affected by Sc shear bands) rocks as a triangle. Figure 15C shows the poles to slip planes in several 1- to 4-m left- and right-lateral shear zones measured in the Red Oak granite southeast of the main Clover zone. The bisector of the average planes of the sinistral and dextral zones, respectively, is shown as a triangle in Figure 15C. There is about a 4° angular difference between the pole to the average S4 and the bisector of the conjugate zones. Both approximate the principal shortening direction. Because of the similar orientations yielded by the two methods, the principal shortening direction in this region during D4 is therefore taken as approximately east-west. An east-west principal shortening direction is roughly consistent with that expected to have caused formation of the northeast-trending dextral shear zones (present directions) common along the eastern margin of North America. The data seem equally consistent with a north-northeast dextral transform shear couple, which seems the more likely solution from plate movements in this time period (Dalziel, 1992). The D4 deformational event is thus interpreted to be the result of Acadian or Alleghanian oblique shear (ca. 345 Ma). Ferrill and Thomas (1988) modeled oblique collision at this time, resulting in local shortening, transform movement, and extension simultaneously along different portions of the Devonian North American margin.

The Milton–Hager's Mountain fold generation (Table 1, D4-D5) was interpreted by Tobisch and Glover (1971) to have both northwest- and northeast-trending axes and cleavages that they believed to be limbs of conjugate folds, although both sets were never observed closer than within 6 m in the same outcrop. An alternate interpretation is that the northwest-trending set is equivalent to the northwest-trending F4 folds found in this study. The northeast-trending set would thus be equivalent to the later D5 event in the study area. Both Henika (1977) and Marr (1981)

TABLE 2. MICROPROBE DATA AND RESULTS FOR GEOTHERMOBAROMETRY; LOCATION AND LITHOLOGY OF SAMPLES* (continued - page 3)

Sample	Quadrangle	Location
6-6	Saxe	0.75 mi. SW of Watkins Bridge over Roanoke River, SW Saxe quadrangle: †Bt Ksp Qtz Plag Gneiss.
6-45	Aspen	N&W RR 475 ft. west of Highway 617: †Bt Qtz Plag Schist.
6-52	Aspen	N&W RR 500 ft. west of Highway 675 at Phenix: Bt Schist.
6-105	Brookneal	Childrey Creek 1,350 ft. east of creek intersection with Highway 632: Bt Gneiss.
6-119	Brookneal	Childrey Creek 4,850 ft east of creek intersection with Highway 501: Gneiss.
6-127	Brookneal	Snake Creek 1,300 ft. northwest of creek intersection with Highway 501: Gneiss.
6-136	Nathalie	Bentley Creek 3,050 ft west of intersection of Highways 636 and 501: †Bt Ep Ms Ksp Plag Qtz Gneiss.
6-148	Nathalie	Bentley Creek 2,700 ft. west of intersection of Highways 636 and 501: Gneiss.
6-156	Nathalie	Birch Creek 5,850 ft. west of intersection of Highways 645 and 501: Bt Ms Schist.
6-175	Nathalie	Little Childrey Creek 1,325 ft. north of creek intersection with Highway 634: †Bt Qtz Plag Gneiss.
6-223	Brookneal	Tributary to Snake Creek 2,450 ft. south of intersection of Highways 40 and 808: Ms Bt Schist.
6-375	Nathalie	Tributary to Birch Creek 3,975 ft. south of intersection of Highways 626 and 636: Bt Ms Gneiss.
6-376B	Nathalie	Birch Creek 4,550 ft. south of intersection of Highways 626 and 636: Bt Amphibole Gneiss.
6-465	Conner Lake	Unnamed tributary to Roanoke River about 3,000 ft. west of intersection of Highways 618 and 623; 950 ft. south of Roanoke River along tributary: Bt Gneiss.
6-524	Nathalie	Tributary to Armistead Branch to Catawba Creek 1,900 ft. southeast of intersection of Highways 627 and 644: Ms Gneiss.
6-550	Conner Lake	First northwest tributary of Hunting Creek west of Highway 619; 2,900 ft. northwest of tributary intersection with Highway 619: Bt Schist.
6-553	Conner Lake	First northwest tributary of Hunting Creek west of Highway 619; 1,700 ft. west of tributary intersection with Highway 619: Bt Schist.
6-619	Conner Lake	Unnamed tributary to Hunting Creek 2,525 ft. north of intersection of Highways 603 and 616 at Neal's Corner: Bt Gneiss.
6-621	Conner Lake	Unnamed tributary to Hunting Creek 3,775 ft. north of intersection of Highways 603 and 616 at Neal's Corner: Bt Gneiss.
6-755	Scottsburg	Tributary to Difficult Creek 1,975 ft. east of intersection of Difficult Creek and Southern RR: Gneiss.
6-829	Conner Lake	First tributary to East Prong of Difficult Creek 4,700 ft. west of intersection of Highways 603 and 616 at Neal's Corner: Bt Gneiss.
6-847	Conner Lake	First tributary to East Prong of Difficult Creek 2,500 ft. west of intersection of Highways 603 and 621 at Mount Laurel: Bt Gneiss.
7-5	Vernon Hill	3,125 ft. southwest of Millstone Church on Highway 501: Bt Schist.
7-217	Nathalie	E-W tributary to Difficult Creek 9,750 ft. southeast of intersection of Highways 603 and 626: Bt Gneiss.
7-222	Conner Lake	Tributary flowing from pond to Buckskin Creek 5,200 ft northeast of intersection of Highways 603 and 623: Bt Gneiss.
7-242	Halifax	Tributary to Winn Creek 3,800 ft. southeast of intersection of Highways 610 and 615: Bt Gneiss

*All of the rocks are quartzo-feldspathic with varying amounts of mica, usually biotite with subordinate muscovite, and all have a small amount of garnet. Rocks listed simply as gneiss have only a small amount of mica. Rocks with a biotite or muscovite prefix have a few to several percent mica. Gneiss and schist are used as relative terms to denote the relativve proportions of mica in the rocks. Gneiss contains only a few to several percent mica, whereas schist has tens of percent mica.
†A dagger by four rocks indicates point counting to determine mineral percentages.

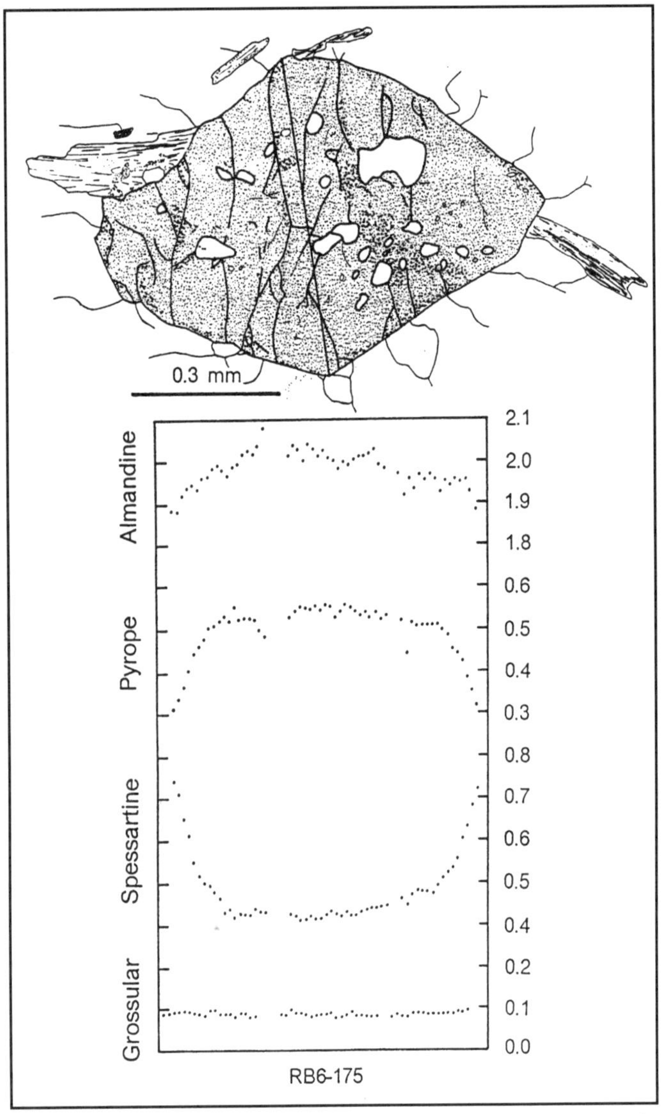

Figure 11. Morphologic type-one garnet. Typical garnet of this type is up to about 2 mm, subhedral to anhedral, inclusion rich, and fractured. Overall, almandine content is highest, pyrope is moderate, and spessartine and grossular contents are low. Microprobe traverse is shown by solid line from top to bottom. Scale in cations per formula unit. Note breaks in scale between molecular components.

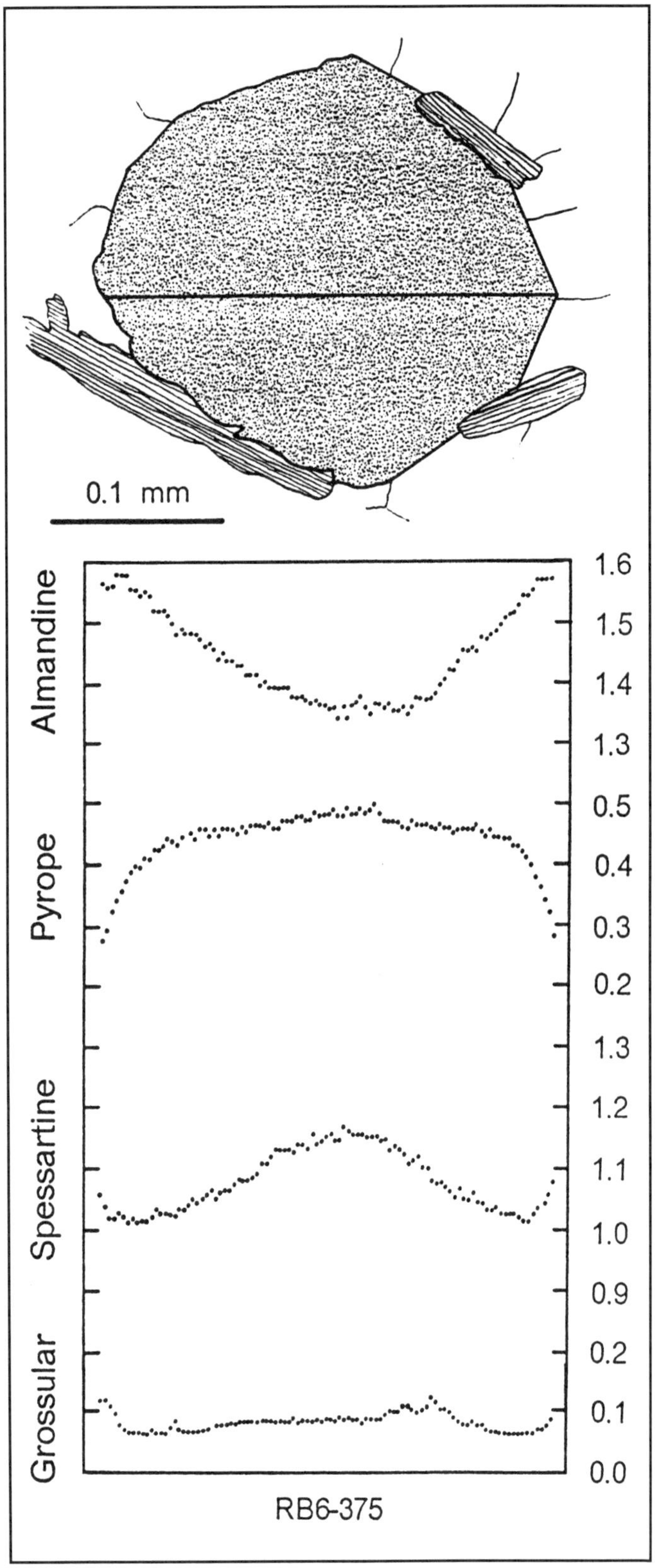

Figure 12. Morphologic type-two garnet. Typical garnet of this type is a few tenths of a millimeter across, euhedral to subhedral, inclusion free, and nonfractured. Overall, almandine content is lower than in type one, spessartine + grossular is subequal to almandine content, and pyrope is low. Microprobe traverse is shown by solid line from left to right. Vertical scale in cations per formula unit. Note breaks in scale between molecular components.

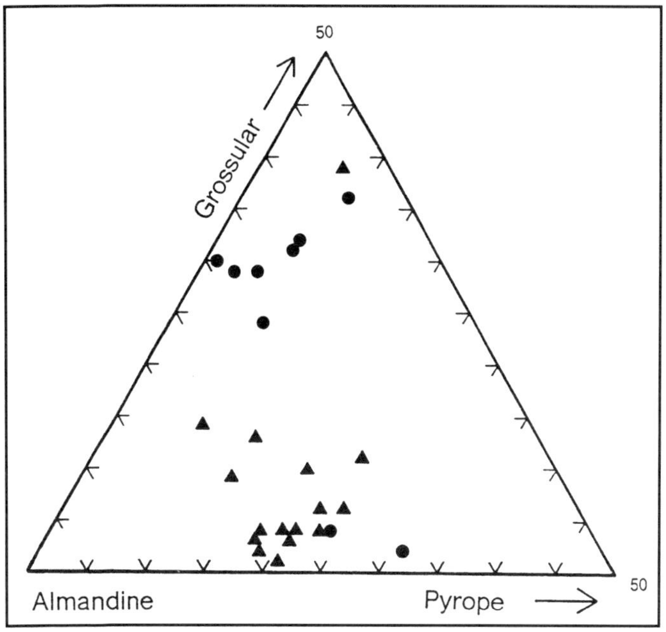

Figure 13. Almandine corner of an almandine-grossular-pyrope triangle. Filled triangles—morphology type-one garnets. Filled circles—Morphology type-two garnets. The two garnet types clearly plot in two distinct groups, excepting a couple of deviant samples in each field.

identify late northwest-trending fold sets (Table 1, their D4). Henika (1977) interpreted his F4 folds as related to post–Triassic-basin deformation. However, the small amount of Triassic basin investigation done in conjunction with the present study indicates that the strata are either horizontal or tilted to the southeast, which would produce northeast-trending, as opposed to northwest-trending, fold axes. It is therefore likely that the F4 folds of Henika (1977) and Marr (1981) correspond to D4 as defined in this study.

Dip of the Brookneal shear zone. Gates et al. (1986) determined that the Brookneal shear zone in the Melrose granite (Fig. 2) was oriented on average at N40E, 50SE. Within the Arvonia Formation, just 2 or 3 km southeast (Fig. 2), this study found that mylonitic shear planes dip about 30° to the southeast. In the sheared muscovite + biotite schist unit a few more kilometers southeast, shear planes are subhorizontal. In the latter two instances, mylonitic shear planes are parallel with S3. A possible explanation of the southeastward change in shear plane dip is that much of the Brookneal shear zone movement took place within the Melrose granite, but part of the strain was partitioned into slip that preferentially occurred along the S3 foliation in the adjacent CMB anisotropic rocks. Another possible explanation is that the flattening of the shear planes is a measure of the flattening of the Brookneal shear zone toward the south-

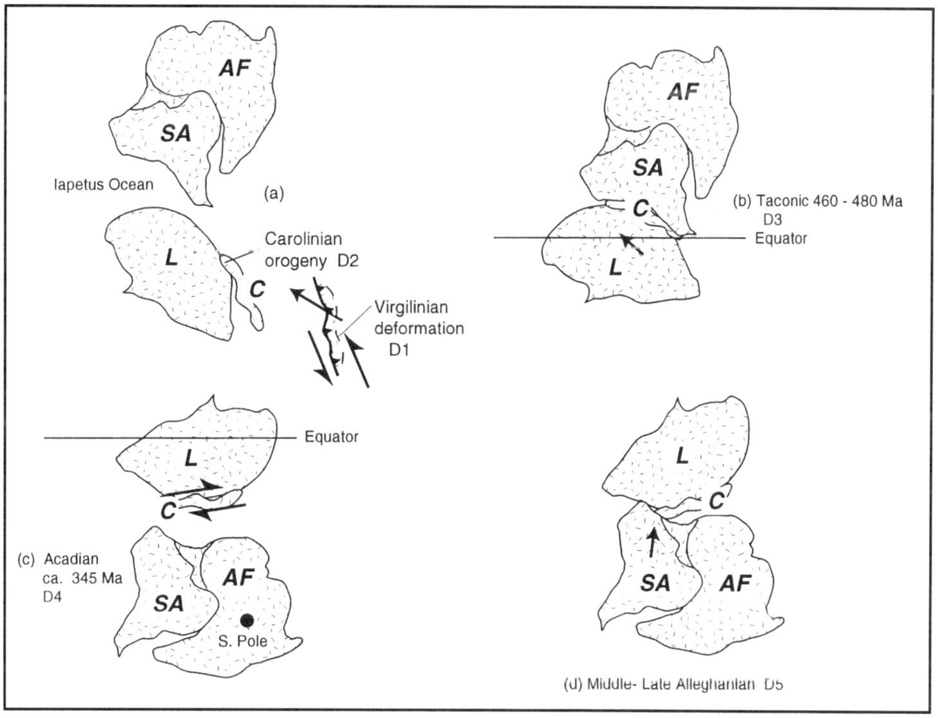

Figure 14. Schematic paleogeographic reconstruction illustrating movement in part consistent with the results of this chapter. Positions of major continental masses modified slightly from Dalziel (1992) and Salda and others (1992). a, Virgilinian deformation offshore (Table 1); Carolinia collision with Laurentia (>534 ±5 Ma, Table 1). b, Taconian (460 to 480 Ma) collision of South America and Carolinia/Laurentia. c, Possible Acadian (ca. 345 Ma) dextral shearing. d, Middle to late Alleghanian collision between Africa and Acadia. L—Laurentia; SA—South America; C—Carolinia; AF—Africa.

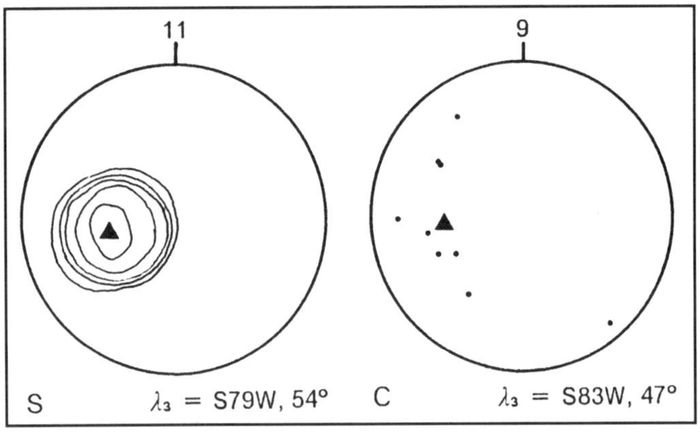

Figure 15. Equal-area net lower hemisphere plots used to determine the principal shortening direction during D4. S = foliation plane; C = shear plane. *Left*, Plot of S-plane (S4) foliation from the Clover shear zone from the least shear-strained rocks. Average pole to these planes (filled triangle) is assumed parallel to the principal shortening direction. *Right*, Poles to slip surfaces from small local left- (below the triangle) and right-lateral (above the triangle) shear zones in the Red Oak granite (Fig. 2) southeast of the main Clover shear zone. Filled triangle represents bisector of intersection of average planes to left-lateral and right-lateral zones. Interpreted to be a good estimate of principal shortening direction during D3. Angular difference between results of the methods is approximately 4°. Contours on equal-area nets are 1, 2, 3, 6, 9, 12 percent per 1 percent area. Number at top of equal-area nets is the number of data points for that domain.

east (Fig. 6). Along-foliation slip was particularly favored in the more micaceous, schistose lithologies.

Phases of D4 deformation. D4 consisted of several phases, some of which operated throughout the event. The earliest was formation of S-C (foliation and shear) fabrics and initiation of shearing in the Clover and Brookneal zones. Folding about a north-trending axis apparently occurred throughout the event (Fig. 10, Domain IIIb). Within the Brookneal shear zone these folds appear to have been rotated to a north-northeast trend (Fig. 10, Domain IIIc). Minor kinking within the Clover zone affects previously sheared rocks; as such, it must post-date shearing at least locally. A still later phase folded the kinks about a northwest-trending axis (Fig. 10, Domain I).

Origin of north- to northeast-trending F4 fold axes. The northerly F4 fold trend (Fig. 10, Domain IIIb) is interpreted to have formed normal to the principal shortening direction during the D4 event. To the northwest, in Domain IIIc (Fig. 10), the northerly trending folds of Domain IIIb (Fig. 10) have been rotated to a north-northeast trend toward parallelism with the Brookneal shear zone during shearing. The trends exhibited in domains IIIb and IIIc (Fig. 10) represent a continuum in space and time with more complete reorientation closer to the shear zone.

Origin of northwest-trending F4 fold axes. An east-west (present directions) principal shortening direction, however, is inadequate to account for the northwesterly set of F4 axes. The Brookneal shear zone, however, is a northeast-trending dextral zone where the southeastern block has moved southwest relative to the northwestern block. The northwest-trending set of F4 axes is therefore interpreted to have formed normal to the northeast-southwest component of movement along the zone, or normal to northeast-southwest transform shear zones (Fig. 14c). In Domain I of Figure 10, the cluster of five unlabled data points represents the northwest axis of folding about which the six kinks (labeled K) were folded.

M4 metamorphism. Based on results from the type-two garnets (Table 2), M4 metamorphic conditions in the CMB ranged up to about 575 to 590 °C at 6.6 to 7.1 kb. Assuming a ±50 °C error, the temperature was fairly uniform across the area now exposed in the CMB, apparently dropping off somewhat both to the northwest and southeast. These conditions correspond to a depth of burial of about 22 to 24 km as opposed to the calculated 15- to 16-km depth during M3. Assuming a 20 °C average surface temperature, this yields a geothermal gradient range of 23 to 26 °C/km compared with 34 to 36°C/km during M3. The greater depth of burial suggests that crustal thickening occurred between M3 and M4, whereas the lower geothermal gradient suggests that volcanic activity was minimal at the time. These data are consistent with the statement by Glover (1989) and data of Speer et al. (1994) that late Paleozoic plutons of the central and southern Appalachians are typical of crustal shortening and are not subduction related.

The occurrence of relict hornblende and garnet in the northwestern part of the CSB indicates that the M3 Taconian greenschist/amphibolite isograd lay somewhat southeast of the present CMB/CSB boundary. During D4, the greenschist Clover shear zone developed across this old isograd, overprinting both amphibolite and greenschist facies rocks and retrograding M3 amphibolite facies rocks in the zone to greenschist (Fig. 16). Amphibolite to greenschist facies retrogression during shearing is common and has also been noted, for example, by Bobyarchick and Glover (1979) in the Hylas shear zone and by Butler and Fullagar (1977) in the Gold Hill shear zone.

D5 regional warping event

Timing and regional correlation. The last compressional event to affect the area was D5 deformation. Because D4 is constrained to be late Acadian–Alleghanian, D5 is probably Alleghanian to late Alleghanian. Henika (1977) noted that his F3 structures were the dominant northeast-trending structures. They are equivalent to F5 in our study area (Table 1). The northeast-trending Milton–Hager's Mountain deformation (Tobisch and Glover, 1971; Table 1) is probably also equivalent to D5 here. Supporting this interpretation, Kreisa (1980) recognized an earlier north-trending set of folds in the CMB east of the Henika (1977) area, later folded into a major northeast-trending antiform. This finding accords well with F4 and F5 fold events recognized in this study and by Gates (1987) to the northwest.

No evidence was found to indicate that an M5 event distinct from M4 accompanied D5. Intuitively, however, there may have been some cooling and erosion of the area between D4 and D5.

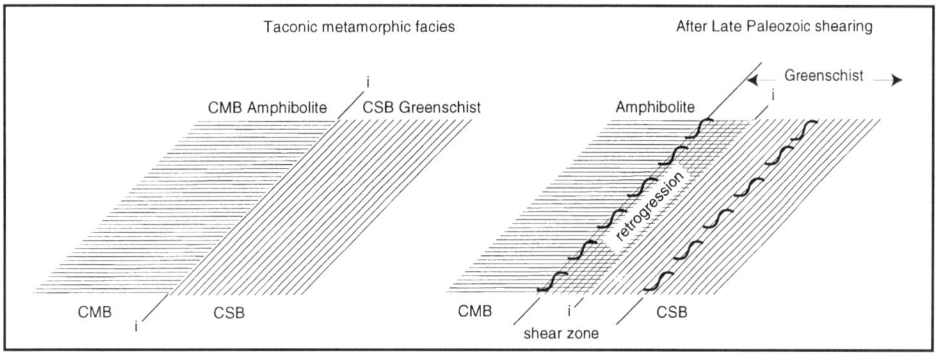

Figure 16. Map view diagram illustrating effect of late Paleozoic Clover shear zone shearing overprint on Taconian greenschist-amphibolite isograd. *Left*, Position of Taconian greenschist-amphibolite isograd (i) after Taconian orogeny. *Right*, Former placement of Taconian isograd (i) with respect to the Clover shear zone and location of postshearing isograd. Shearing took place under greenschist facies conditions, causing retrogression of part of the Taconian amphibolite facies rocks.

Relation between D4 shear zone orientation and F5 fold axes. In contrast to D4 (Fig. 14c) with an east-west compressional regime, D5 deformation suggests a northwest-southeast (present orientation) compressional regime (Fig. 14d). Therefore, the question is whether D4 and D5 resulted from the same tectonic event or whether some tectonic event such as continental repositioning prior to final docking occurred between the two folding episodes.

Considering the northeast trends of regional F5 folds, such as the Republican Grove synform (Fig. 2), the principal shortening direction for the D5 folding event appears to have been northwest-southeast. It therefore seems difficult to call upon a single strain field orientation to explain both D4 and D5 events. Either there were two discrete events, or F5 folds are reoriented F4 folds.

Northwest of the study area, Gates (1987) noted that preshearing F2 folds have been partially reoriented into parallelism with the Brookneal shear zone as a result of strain associated with that zone. This is also the case on the southeast side of the shear zone. That F4 fold axes have been partially reoriented into parallelism with the Brookneal shear zone can be seen by referring to the large F4 fold between the Republican Grove and the next synform to the southeast (Fig. 2) adjacent to the Brookneal shear zone and by noting the change in orientation of the north-trending F4 fold set in Domain IIIb to north-northeast in Domain IIIc (Fig. 10). Simple reorientation adjacent to the shear zone, however, does not appear an adequate explanation for the northeast trend of such large structures. Northeast-trending F5 folds are also recognized in the central CMB (Fig. 2) where no evidence of major shearing was found, and, as such, these cannot be reoriented F4 folds. Therefore, some shift in collisional orientation during docking of Africa and Laurentia may have occurred (Fig. 14d).

Edelman et al. (1987) noted a similar circumstance in the Brevard shear zone in South Carolina. The Brevard zone is oriented similarly to the Clover zone, as it is to most of the other shear zones of the eastern Piedmont fault system (Hatcher et al., 1977; Bobyarchick, 1981). Consistent with the findings in the present study area, Edelman et al. (1987) recognized two Alleghanian deformations, an earlier shearing event parallel with the Brevard zone, and a later northwest-directed translation of the Inner Piedmont rocks over the underlying platform sedimentary rocks normal to the Brevard zone. Again, the latter deformation seemingly requires a northwest-southeast–oriented compressional stress regime.

CONCLUSIONS

1. Five Paleozoic deformational events representing three or four orogenies are recorded in the Charlotte metamorphic and Carolina slate belts of south-central Virginia.

2. D1, the Virgilinian orogeny (between 620 and 546 Ma), occurs in both the Carolina slate belt and the Charlotte metamorphic belt. It formed folds and faults but as yet no ductile fabrics related to it have been found. The Virgilinian appears to be restricted to the Carolina terrane (Fig. 14a).

3. D2 was probably a separate orogeny (Carolinian orogeny of Gates et al., 1994) of Latest Proterozoic(?) or Early Cambrian age (ca. 534 Ma) resulting from collision between Carolinia and Laurentia (Fig. 14a). Possible age ranges of D1 and D2 nearly overlap, and they may be found to be the same or related in time and place.

4. D3 is the classic regional Taconian event representing continued (?) collision of Carolinia and Laurentia or more likely a separate collision outboard of Carolinia (Fig. 14b).

5. High geothermal gradients in the Charlotte metamorphic belt are consistent with a magmatic arc setting on Carolinia during D3.

6. D3 Carolinian fold nappes were emplaced on Laurentia at mid-crustal depths.

7. D4 dextral shear zones probably resulted from transpressional tectonics in Acadian(?) and Alleghanian time.

8. The Charlotte metamorphic belt is bounded on both the northwest and southeast by D4 shear zones. The northwest

(Brookneal) zone was localized by the Late Proterozoic to Cambrian suture, whereas the southeast (Clover) zone was localized by a structural flexure from subhorizontal to moderately east-dipping layering and foliations and by a reaction softening front related to greenschist/amphibolite retrogression.

9. D4 structures are consistent with an east-west principal shortening direction or northeast-southwest–trending transform shear.

10. Increasing pressures and declining temperatures between M3 and M4 suggest crustal thickening and a cessation of magmatic activity, respectively, between D3 and D4.

11. D5 is attributed to Alleghanian final docking of Gondwana and North America and produced broad, regional-scale folds throughout the area. Their northeast trend is consistent with a northwest-southeast principal shortening direction, indicating a change in orientation of this shortening direction between D4 and D5.

ACKNOWLEDGMENTS

Fieldwork for this study was funded by Nuclear Regulatory Commission contract NRC-04-85-106 to Glover, John Costain, Cahit Çoruh, and Gilbert Bollinger. Partial support was also provided by a Sigma Xi Grant-In-Aid of research and a Virginia Polytechnic Institute and State University Graduate Student Assembly grant to Baird. The manuscript was greatly improved through reviews by Ken Eriksson, Rick Law, Krishna Sinha, Robert Tracy, J. A. Speer, J. R. Butler, and James Hibbard. Thanks are also extended to Marge Dellers, program support technician, for logistical support during the fieldwork and to Kathy Hawkins, illustrator for the Orogenic Studies Lab, for advice on drafting the figures.

REFERENCES CITED

Baird, R. A., 1988, Shear zone geometry in previously foliated rocks—An example from the Clover shear zone, south central VA: Geological Society of America Abstracts with Programs, v. 20, p. 253.

Baird, R. A., 1991, Stratigraphy of the northern Charlotte metamorphic belt: Application to the Charlotte belt/Milton belt problem: Southeastern Geology, v. 32, p. 61–82.

Bence, A. E., and Albee, A. L., 1968, Empirical correction factors for the electron microanalysis of silicates and oxides: Journal of Geology, v. 76, p. 382–403.

Bobyarchick, A. R., 1981, The eastern Piedmont fault system and its relationship to Alleghanian tectonics in the central and southern Appalachian orogen: Journal of Geology, v. 89, p. 335–347.

Bobyarchick, A. R., and Glover, L., III, 1979, Deformation and metamorphism in the Hylas zone and adjacent parts of the eastern Piedmont in Virginia: Geological Society of America Bulletin, v. 90, p. 739–752.

Bohlen, S. R., Wall, V. J., and Boettcher, A. L., 1983, Experimental investigation and geological applications of equilibria in the system $FeO-TiO_2-Al_2O_3-SiO_2-H_2O$: American Mineralogist, v. 68, p. 1049–1058.

Brown, E. H., 1968, Metamorphic structures in part of the eastern Otago schists: New Zealand Journal of Geology and Geophysics, v. 11, p. 41–65.

Brown, W. R., 1969, Geology of the Dillwyn quadrangle, Virginia: Virginia Division of Mineral Resources Report of Investigations 10, 77 p.

Butler, J. R., 1971, Structure of the Charlotte belt and adjacent belts in York County, South Carolina: South Carolina Geology Notes, v. 16, p. 59–78.

Butler, J. R., and Fullagar, P. D., 1977, The Gold Hill fault zone in the Carolinas: Age of movement and southwestern extension: Geological Society of America Abstracts with Programs, v. 9, p. 125.

Cooper, M. A., 1981, The internal geometry of nappes: Criteria for models of emplacement, *in* McClay, K. R., and Price, N. J., eds., Thrust and nappe tectonics: Geological Society of London Special Publication 9, p. 225–234.

Dalziel, W. D., 1992, Antarctica; a tale of two supercontinents? Annual Reviews of Earth and Planetary Science, v. 20, p. 501–526.

Dennis, A. J., Wright, J. E., Barker, C. A., Pray, J. R., and Secor, D. T., Jr., 1993, Late Precambrian–Early Cambrian orgeny in the S. Carolina Piedmont: Geological Society of America Abstracts with Programs. v. 25, p. 484.

Edelman, S. H., Liu, A., and Hatcher, R. D., Jr., 1987, The Brevard Zone in South Carolina and adjacent areas: An Alleghanian orogen-scale dextral shear zone reactivated as a thrust fault: Journal of Geology, v. 95, p. 793–806.

Ferrill, B. A., and Thomas, W. A., 1988, Acadian dextral transpression and synorogenic successions in the Appalachian orogen: Geology, v. 16, p. 604–608.

Ferry, J. M., and Spear, F. S., 1978, Experimental calibration of the partitioning of Fe and Mg between biotite and garnet: Contributions to Mineralogy and Petrology, v. 66, p. 113–117.

Fullagar, P. D., and Dietrich, R. V., 1976, Rb-Sr isotopic study of the Lynchburg and probable correlative formations of the Blue Ridge and western Piedmont of Virginia and North Carolina: American Journal of Science, v. 276, p. 347–365.

Ganguly, J., and Saxena, S. K., 1984, Mixing properties of aluminosilicate garnets: Constraints from natural and experimental data, and applications to geothermo-barometry: American Mineralogist, v. 69, p. 88–97.

Ganguly, J., and Saxena, S. K., 1985, Mixing properties of aluminosilicate garnets: Constraints from natural and experimental data, and applications to geothermo-barometry: Clarifications: American Mineralogist, v. 70, p. 1320.

Gates, A. E., 1981, Geology of the western boundary of the Charlotte belt at Brookneal, Virginia [M.S. thesis]: Blacksburg, Virginia Polytechnic Institute and State University, 88 p.

Gates, A. E., 1987, Transpressional dome formation in the southwest Virginia Piedmont: American Journal of Science, v. 287, p. 927–949.

Gates, A. E., 1988, Alleghanian strain partitioning in the central and southern Appalachian orogen, Virginia: Geological Society of America Abstracts with Programs, v. 20, p. 266.

Gates, A. E., and Mose, D., 1987, Carboniferous dextral transpression and the resetting of Rb/Sr ages in the Bowen's Creek fault, northner extension of the Brevard zone: Eos (Transactions, American Geophysical Union), v. 68, p. 416.

Gates, A. E. and Speer, J. A., 1991, Allochemical retrograde metamorphism in shear zones: An example in metapelites, Virginia, U.S.A.: Journal of Metamorphic Geology, v. 9, p. 581–604.

Gates, A. E., Simpson, C., and Glover, L., III, 1986, Appalachian carboniferous dextral strike slip faults: An example from Virginia: Tectonics, v. 5, p.119–133.

Gates, A. E., Lanzirotti, A., and Glover, L., III, 1994, Isotopic evidence for Cambrian tectonism in the crystalline southern Appalachians, Carolina: Geological Society of America Abstracts with Programs, v. 26, p. 15.

Ghent, E. D., 1976, Plagioclase-garnet-Al_2SiO_5-quartz: A potential geobarometer-geothermometer: American Mineralogist, v. 61, p. 710–714.

Ghent, E. D., and Stout, M. Z., 1981, Geobarometry and geothermometry of plagioclase-biotite-garnet-muscovite assemblages: Contributions to Mineralogy and Petrology, v. 76, p. 92–97.

Glover, L., III, 1989, Tectonics of the Virginia Blue Ridge and Piedmont, *in* Glover, L., III, Evans, N. H., Patterson, J., and Brown, W. R., Tectonics of the Virginia Blue Ridge and Piedmont: International Geological Congress, 28th, Field Trip Guidebook T363, July 19–24, 1989: Washington, D.C., American Geophysical Union, p. 1–29.

Glover, L., III, and Sinha, A. K., 1973, The Virgilina deformation, a Late Precambrian to Early Cambrian orogenic event in the central Piedmont of Virginia and North Carolina: American Journal of Science, v. 273-A, p. 234–251.

Glover, L., III, Speer, J. A., Russell, G. S., and Farrar, S. S., 1983, Ages of regional metamorphism and ductile deformation in the central and southern Appalachian orogen: Lithos, v. 16, p. 223–245.

Glover, L., III, Gates, A. E., Wang Ping, and Valentino, D. W., 1992a, Anomalous heat source under the Laurentian margin during Cambrian collision with the Avalonian magmatic arc: Geological Society of America Abstracts with Programs, v. 24, p. 17.

Glover, L., III; Gates, A. E., Wang Ping, and Valentino, D. W., 1992b, Emplacement of bimodal magmas into the passive and active margins during collision in the central Appalachians about 500–520 Ma—The result of mantle delamination? Geological Society of America Abstracts with Programs, v. 21, p. 290.

Griffin, V. S., Jr., 1978, Detailed analysis of tectonic levels in the Appalachian Piedmont: Geologische Rundschau, v. 67, p. 180–201.

Harper, C. T., Russell, C. W., Sherrer, G., Stonebraker, J. D., and Kish, S., 1973, The Arvonia Slate—A useful standard for K-Ar dating: Geological Society of America, Abstracts with Programs, v. 5, p. 402.

Harris, C. W., and Glover, L., III, 1985, The Virgilina deformation: Implications of stratigraphic correlation in the CSB: Carolina Geological Society Field Trip Guide, November 16–17, 1985, p. 1–58.

Harris, C. W., and Glover, L., III, 1988, The regional extent of the ca. 600 Ma Virgilina deformation: Implications for stratigraphic correlation in the Carolina terrane: Geological Society of America Bulletin, v. 100, p. 200–217.

Hatcher, R. D., Jr., 1989, Tectonic synthesis of the U.S. Appalachians, *in* Hatcher, R. D., Jr., Thomas, W. A., and Viele, G. W., eds., The Appalachian-Ouachita Orogen in the United States: Boulder, Colorado, Geological Society of America, The Geology of North America, v. F-2, p. 511–535.

Hatcher, R. D., Jr., Howell, D. E., and Talwani, P., 1977, Eastern Piedmont fault system: Speculations on its extent: Geology, v. 5, p. 636–640.

Henika, W. S., 1977, Geology of the Blairs, Mount Hermon, Danville, and Ringgold quadrangles, Virginia, with sections on the Triassic system by Thayer, P. A.: Virginia Division of Mineral Resources Publication 2, 45 p.

Henika, W. S., 1980, Metamorphic and structural evidence of an intrusive origin for the Shelton Formation, *in* Price, V., Thayer, P. A., and Ranson, W. A., eds., Geological investigations of the Piedmont and Triassic rocks, Central North Carolina and Virginia: Carolina Geological Society Field Trip Guidebook, October 11–12, 1980, p. V1–V17.

Hess, H. H., 1962, History of ocean basins, *in* Engle, A. E. J., James, H. L., and Leonard, B. F., eds., Petrological Studies: A volume in honor of A. F. Buddington: New York, Geological Society of America, p. 599–620.

Hibbard, J., 1991, The Carolina slate belt—Milton belt contact in north central North Carolina: Geological Society of America Abstracts with Programs, v. 23, p. 44.

Hibbard, J. P., Shell, G. S., Wilkins, J. K., Samson, S., and Wortman, G., 1994, Multiple deformation at the western edge of the Carolina slate belt, north-central North Carolina: Geological Society of America Abstracts with Programs, v. 26, p. 19.

Hibbard, J., Wortman, G., and Samson, S. D., 1995, The Virgilina orogeny in the Carolina terrane: Geological Society of America Abstracts with Programs, v. 27, p. 62

Higgins, M. W., Sinha, A. K., Tilton, G. R., and Kirk, W. S., 1971, Correlation of metavolcanic rocks in the Maryland, Delaware, and Virginia Piedmont: Geological Society of America Abstracts with Programs, v. 3, p. 320–321.

Hobbs, B. E., Means, W. D., and Williams, P. F., 1976, An outline of structural geology: New York, John Wiley and Sons, 571 p.

Hodges, K. V., and Crowley, P. D., 1985, Geothermometry, geobarometry, and the Al_2SiO_5 triple point at Mt. Moosilauke, New Hampshire: American Mineralogist, v. 70, p. 702–709.

Horton, J. W., Jr., and Zullo, V. A., 1991, An introduction to the geology of the Carolinas, *in* Horton, J. W., Jr., and Zullo, V. A., eds., The geology of the Carolinas: Carolina Geological Society 50th anniversary volume: Knoxville, University of Tennessee Press.

Horton, J. W., Drake, A. A., and Rankin, D. W., 1989, Tectonostratigraphic terranes and their Paleozoic boundaries in the central and southern Appalachians: Geological Society of America Special Paper 230, p. 213–245. Reprint on file.

Horton, J. W., Drake, A. A., Rankin, D. W., and Dallmeyer, R. D., 1991, Preliminary tectonostratigraphic terrane map of the central and southern Appalachians: U.S. Geological Survey Miscellaneous Investigations Series Map I-2163, scale 1:2,000,000.

Hund, E. A., 1987, U-Pb dating of granites from the Charlotte belt of the central and southern Appalachian orogen [M.S. thesis]: Blacksburg, Virginia Polytechnic Institute and State University, 83 p.

Kreisa, R. D., 1980, Geology of the Omega, South Boston, Cluster Springs, and Virgilina quadrangles, Virginia: Virginia Division of Mineral Resources Publication 5, 22 p.

Laney, F. B., 1917, The geology and ore deposits of the Virgilina district of Virginia and North Carolina: Virginia Geological Survey Bulletin 14, 176 p.

Marr, J. D., Jr., 1981, Stratigraphy and structure (Triassic system by M. B. McCollum), *in* Geologic investigations of the Willis Mountain and Andersonville quadrangles, Virginia: Virginia Division of Mineral Resources Publication 29, p. 3–8.

Milton, D. J., 1984, Revision of the Albemarle Group, North Carolina: U.S. Geological Survey Bulletin 1537-A, p. A69–A72.

Miyashiro, A., 1973, Metamorphism and metamorphic belts: Surrey, England, Unwin Brothers, 492 p.

Morgan, W. J., 1968, Rises, trenches, great faults, and crustal blocks: Journal of Geophysical Research, v. 73, p. 1959–1982.

Mose, D. G., and Nagel, M. S., 1982, Plutonic events in the Piedmont of Virginia: Southeastern Geology, v. 23, p. 25–39.

Newton, R. C., and Haselton, H. T., 1981, Thermodynamics of the garnet-plagioclase-Al_2SiO_5-quartz geobarometer, *in* Newton, R. C., Navrotsky, A., and Wood, B. J., eds., Advances in physical chemistry: New York, Springer-Verlag, p. 131–147.

Pratt, T. L., Costain, J. K., Çoruh, C., and Glover, L., III, 1988, A geophysical study of the earth's crust in central Virginia: Implications for Appalachian crustal structure: Journal of Geophysical Research, v. 93, p. 6649–6667.

Ramsay, J. G., Casey, M., and Kligfield, R., 1983, Role of shear in development of the Helvetic fold-thrust belt of Switzerland: Geology, v. 11, p. 439–442.

Salda, L. D., Cingolani, C., and Varela, R., 1992, Early Paleozoic orogenic belt of the Andes in southwestern South America: Result of Laurentia-Gondwana collision?: Geology, v. 20, p. 617–620.

Samson, S. D., Wortman, G., and Hibbard, J., 1992, Nd isotopic composition of the northern Carolina slate belt and Milton belt, southern Appalachians, 1992: Geological Society of America Abstracts with Programs, v. 24, p. 62.

Secor, D. T., Jr., Samson, S. L., Snoke, A. W., and Palmer, A. R., 1983, Confirmation of the CSB as an exotic terrane: Science, v. 221, p. 649–651.

Shaw, H. F., and Wasserburg, G. J., 1984, Isotopic constraints on the origin of Appalachian mafic complexes: American Journal of Science, v. 284, p. 319–349.

Sinha, A. K., Hund, E. A., and Hogan, J. P., 1989, Paleozoic accretionary history of the North American plate margin (central and southern Appalachians): Constraints from the age, origin, and distribution of granitic rocks, *in* Hillhouse, J. W., ed., Deep structure and past kinematics of sccreted terranes: International Union of Geodesy and Geophysics, Geophysical Monograph 50, IUGG Volume 5, p. 219–238.

Solberg, T. N., and Speer, J. A., 1982, QALL, a 16-element analytical scheme for efficient petrologic work on an automated ARL-SEMQ: Application to mica reference samples: Microbeam Analysis—1982, p. 422–426.

Speer, J. A., McSween, H. Y., Jr., and Gates, A. E., 1994, Generation, segregation, ascent, and emplacement of Alleghanian plutons in the southern Appalachians: Journal of Geology, v. 102, p. 249–267.

Tapponnier, P., Peltzer, G., Le Dain, A. Y., and Armijo, R., 1982, Propagating extrusion tectonics in Asia: New insights from simple experiments with plasticine: Geology, v. 10, p. 611–616.

Thompson, A. B., 1976, Mineral reactions in pelitic rocks. II. Calculation of some P-T-X(Fe-Mg) phase relations: American Journal of Science, v. 276,

p. 425–454.

Tillman, C. G., 1970, Metamorphosed trilobites from Arvonia, Virginia: Geological Society of America Bulletin, v. 81, p. 1189–1200.

Tilton, G. R., Doe, B. R., and Hopson, C. A., 1970, Zircon age measurements in the Maryland Piedmont with special reference to Baltimore Gneiss problems, *in* Fisher, G. W., Pettijohn, F. J., Reed, J. C., Jr., and Weaver, K. N., eds., Studies of Appalachian geology—Central and southern: New York, Interscience Publishers, p. 429–434.

Tobisch, O. T., and Glover, L., III, 1971, Nappe formation in part of the central and southern Appalachian Piedmont: Geological Society of America Bulletin, v. 82, p. 2209–2230.

Tracy, R. J., Robinson, P., and Thompson, A. B., 1976, Garnet composition and zoning in the determination of temperature and pressure of metamorphism, central Massachusetts: American Mineralogist, v. 61, p. 762–775.

Turner, F. J., 1981, Metamorphic petrology: Mineralogical, field, and tectonic aspects: Washington, D.C., Hemisphere Publishing, 524 p.

Virginia Division of Mineral Resources, 1963, Geologic map of Virginia: Virginia Division of Mineral Resources, scale 1:500,000.

Wortman, G. L., And Samson, S. D., 1995, U-Pb zircon geochronology of the Milton and Carolina Slate belts, southern Appalachians: Geological Society of America Abstracts with Programs, v. 27, p. 98.

Manuscript Accepted by the Society April 30, 1996

Printed in U.S.A.

Geological Society of America
Special Paper 314
1997

Multiple reactivations of accreted terrane boundaries: An example from the Carolina terrane, Brookneal, Virginia

Alexander E. Gates
Department of Geology, Rutgers University, Newark, New Jersey 07102

ABSTRACT

The western boundary of the Carolina terrane at Brookneal, Virginia, is a fundamental zone of crustal weakness that underwent multiple periods of major faulting and controlled pluton emplacement and sedimentation. This boundary is the Taconian-Penobscottian(?) suture zone and an extension of the central Piedmont suture zone. It separates metasedimentary rocks of North America to the west from metavolcanic rocks of the Charlotte belt (Carolina terrane) to the east. In addition to marking an accreted terrane boundary, the zone was intruded by postkinematic plutons of bimodal composition (Melrose granite); the zone localized the restricted basin deposits of the Arvonia Formation during extensional or transtensional faulting, formed the 6-km-wide dextral strike-slip Brookneal zone with associated retrograde metamorphism during the Alleghanian orogeny, and localized deposits of the Danville Basin during Mesozoic extensional faulting. In addition, because the boundary marks the edge of Laurentia, Late Proterozoic rifting of the Grenville supercontinent must have been localized along the zone, which may in turn have reactivated a Grenville zone of weakness. Each activation of this terrane boundary mimics the large-scale plate motion of the activating event. Because such zones mimic the primary plate motions, they can be considered secondary plate margins, and their activity is a subset of the major tectonic cycles.

INTRODUCTION

Multiple reactivation of major fault zones is common in the Appalachians (Rodgers, 1970; Ratcliffe, 1971; Hatcher, 1978; Glover et al., 1983; Gates et al., 1986; Swanson, 1986). For example, the Ramapo fault in New York was active at least four to six times from Precambrian to Mesozoic (Ratcliffe, 1971, 1980; Costa and Gates, 1993). Because faults are zones of crustal weakness during orogenesis as well as extension, they can serve both as conduits for magma ascent and emplacement and as the cause of surface depressions and consequent depocenters. Therefore multiple plutons and localized sedimentary deposits can indicate the number, types, and ages of fault activity. The most convincing examples of fault constraints on sedimentation in the Appalachians are the bounding normal faults to the Mesozoic sedimentary basins that were in turn reactivated from middle and late Paleozoic reverse faults (Ratcliffe, 1971, 1980; Swanson, 1986) and strike-slip faults (Bobyarchick and Glover, 1979; Swanson, 1986). These multiply active faults not only formed structural discontinuities but also controlled the distribution of Mesozoic sedimentation. Multiply active faults have also been shown to control some igneous activity in the Appalachians. Gates et al. (1988) and Speer et al. (1994) have shown that Alleghanian dextral strike-slip faults controlled the ascent, emplacement, and possibly the generation of late Paleozoic granitoid and gabbroid plutons in the southern Appalachians.

Goldstein (1989) showed that one of the most frequently reactivated fault zones in the northern Appalachians corresponds to the Avalon terrane boundary. His study proposes that multiply active zones of faulting can be established along the boundaries of accreted terranes. This chapter supports Goldstein's (1989) proposal that the accreted terrane boundaries can control the locus of deformation during orogenic events but also proposes that deposition and igneous intrusion may be similarly controlled. The western boundary of the Carolina terrane is herein proposed to be a fundamental zone of crustal

Gates, A. E., 1997, Multiple reactivations of accreted terrane boundaries: An example from the Carolina terrane, Brookneal, Virginia, *in* Glover, L., III, and Gates, A. E., eds., Central and Southern Appalachian Sutures: Results of the EDGE Project and Related Studies: Boulder, Colorado, Geological Society of America Special Paper 314.

weakness in the southern Appalachians that was multiply activated during all major tectonic events. The central- to southern-Virginia segment of this boundary will be shown to be the Taconian-Penobscottian(?) suture zone, a conduit for magma, and a control on deposition in the area. In conjunction with these rock-forming events, the faults along the zone were active at least four times from Ordovician to Triassic.

STRATIGRAPHY

The Charlotte belt is generally considered to be part of the Carolina terrane (Williams and Hatcher, 1982; Horton et al., 1989). Rocks of the Charlotte belt in Virginia (Henika and Thayer, 1977; Conley, 1978) (Fig. 1), however, are considered to be part of the Milton terrane (Horton et al., 1989) and not part of the Carolina terrane. Hibbard (1991) considered the Milton terrane to be an extension of the Inner Piedmont terrane. Baird (1991) proposed that the Charlotte and Milton belts are part of the Charlotte metamorphic belt that stretches from northern Virginia to Georgia and is a higher grade equivalent of the Carolina slate belt. Based on mapping in southern Virginia (Henika and Thayer, 1977; Conley, 1985) and supporting lithologic and tectonic studies (Gates, 1981; Glover et al., 1989), the Charlotte belt rocks at Brookneal are herein considered part of the Carolina terrane.

A more vigorous controversy, however, concerns the rocks to the west of the Carolina terrane in southern Virginia. The Smith River (allochthon) terrane (Fig. 1) (Conley and Henika, 1973) is considered to be an exotic terrane that was accreted to North America during the Taconian orogeny (Conley, 1985; Horton et al., 1989; Drake et al., 1989). Espenshade (1954), Redden (1963), Kaldy (1977) and Gates (1986), however, mapped across the proposed northwest Smith River–North America (Blue Ridge terrane) boundary of Conley (1978, 1985) and found a continuous and typical Late Proterozoic to early Paleozoic Laurentian margin stratigraphy. Gates and Speer (1991) found support for Taconian nappe emplacement in the area, based on the petrology of metapelites. Therefore, at least part of the Smith River terrane could be far-traveled, distal passive margin deposits and facies equivalents of the Blue Ridge sequence. In this description, Blue Ridge terminology is used but with the understanding that the stratigraphy for the Smith River terrane is as yet unresolved.

Charlotte belt

The western boundary of the Charlotte belt is exposed in Brookneal, Virginia (Figs. 1 and 2). Here, the Charlotte belt is dominantly composed of layered metavolcanic and metavolcaniclastic rocks intruded by metamorphosed mafic, ultramafic, and granitoid plutons. The rocks grade from relatively weakly deformed greenschist facies rocks in the western part of the field area to relatively strongly deformed amphibolite facies rocks to the east. Inferences on protoliths and paleoenvironments are made exclusively using results from the western part of the Brookneal area.

The metavolcanic-metavolcaniclastic sequence consists of interbedded mafic, felsic, and metagraywacke layers. The mafic layers are medium-grained greenstones to amphibolites and range in thickness from 2 cm to 3 m. The greenstones are composed of chlorite, actinolite, epidote, biotite, titanite, and opaques, and amphibolites are andesine-hornblende schists. The contacts of the greenstones with felsic units are commonly gradational on top and sharp on the bottom.

The felsic layers are white fine-grained quartz-feldspar porphyries in the western map area and quartzofeldspathic gneiss in the eastern map area. Individual layers range in thickness from 1 cm to 1 m and are commonly discontinuous. The quartz-feldspar porphyry is composed of 1- to 1.5-mm euhedral plagioclase and rare quartz and alkali feldspar phenocrysts in a matrix of recrystallized quartz and plagioclase with accessory biotite, muscovite, epidote, chlorite, actinolite, and apatite (Fig. 3a). Other rocks with similar character to the felsic rocks contain up to 25% amphibole, chlorite, and biotite.

The metagraywackes are medium-grained biotite schists with variable compositions. Beds range in thickness from 2 cm to 1 m and locally exhibit scoured bases. The metagraywackes are composed of quartz and plagioclase with variable amounts of biotite, amphibole, chlorite, and muscovite and accessory K-spar, garnet, epidote, apatite, and zircon. Quartzite also occurs within the metavolcanic sequence as thinly layered 1 to 2-m sequences. Quartz constitutes 95 to 98% of the rock, with minor to moderate amounts of magnetite and hematite at the layer contacts.

Discontinuous, 1 to 50-m concordant and discordant metaultramafic layers and pods occur within the metavolcanic sequence. These rocks are composed of 80 to 90% coarse-grained, randomly oriented to weakly aligned hornblende with minor amounts of chlorite, epidote, and opaques. The major intrusive body in the sequence, however, is the concordant Falling River granite. It is a medium-grained, lineated leucocratic monzogranite that, on average, contains 34% quartz, 33% microcline, and 25% albite. Varietal phases include muscovite, biotite, garnet, allanite, and opaques. The biotite and garnet are commonly altered to chlorite, and the allanite is rimmed by epidote. The lineations are composed of rods of quartz and microcline aggregates wrapped in mica. The Falling River granite has a preliminary U/Pb zircon age of 470 Ma (A. Sinha and L. Glover, III, personal communication, 1982).

Evington Group

The rocks of the Evington Group (Brown, 1958), in sharp contrast to those of the Charlotte belt, are exclusively metasedimentary (Figs. 1, 2, and 4). The unit in the Brookneal area is likely the Candler Formation of Espenshade (1954) and is an aluminous schist interpreted to be a metapelite. Major minerals include sillimanite, staurolite, kyanite, garnet, and biotite in addition to quartz, oligoclase, and accessory tourmaline and opaques. These rocks are progressively retrograded to chlorite-biotite schist toward the southeast. The metapelite is locally intruded by concordant, 3- to 5-m thick garnet-tourmaline-beryl, two-mica, two-feldspar granite and pegmatite dikes that are observed to be locally derived where they originate from migmatite segregations.

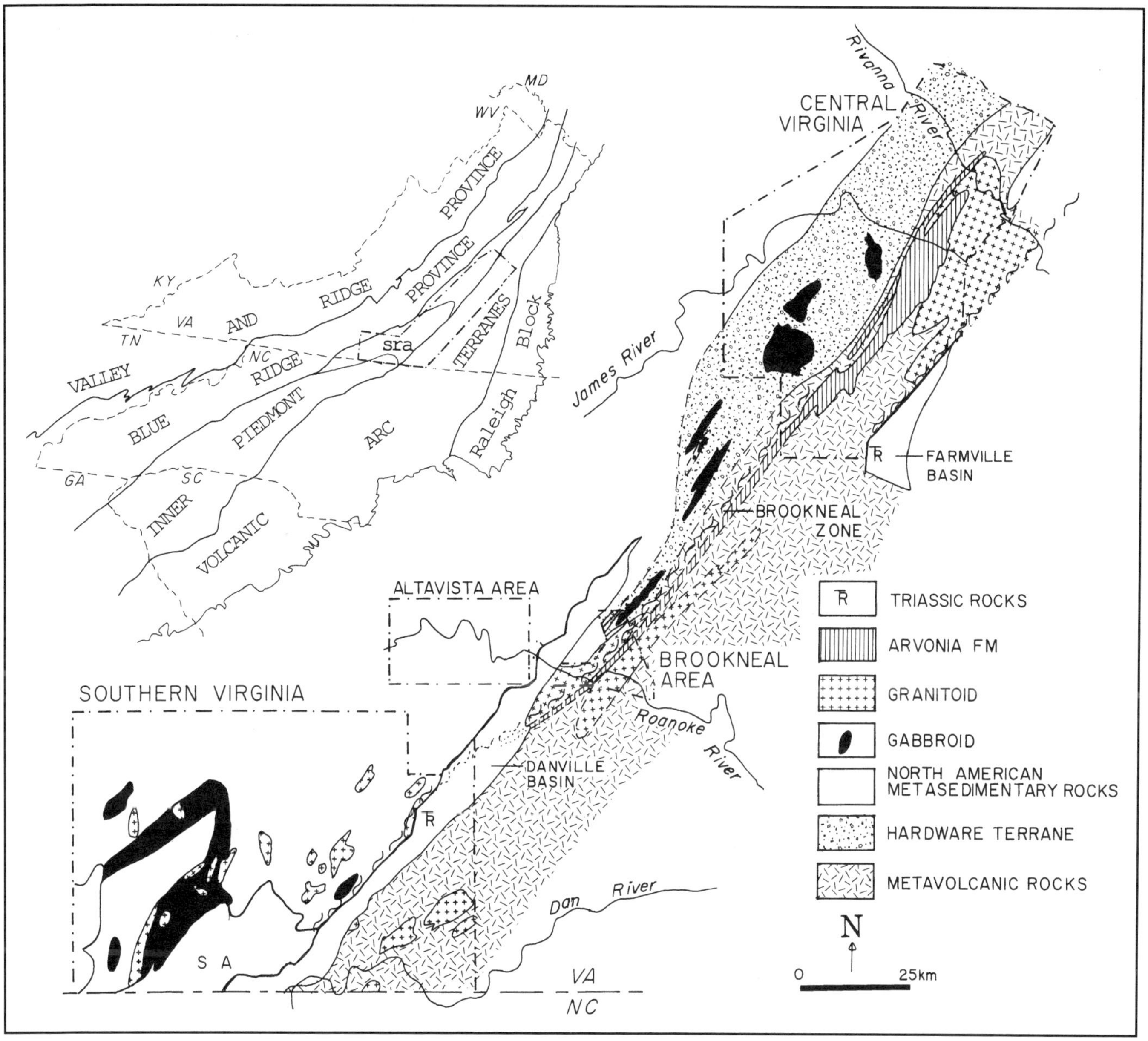

Figure 1. Geologic map of the central and southern Virginia Piedmont and location tectonic map of the southern Appalachians. Metavolcanic rocks include Charlotte belt (Milton terrane) in southern Virginia and Chopawamsic Formation (Chopawamsic terrane) in central Virginia. North American metasedimentary rocks include Lynchburg Formation, Catoctin Formation, and Evington Group in central Virginia and Brookneal area and Bassett Formation and Fork Mountain Formation in southern Virginia. Gabbroids include Diana Mills and unnamed plutons in central Virginia, hornblende metagabbro in Brookneal, and Rich Acres Formation in southern Virginia. Granitoids include Columbia Granite (Hatcher Complex) in central Virginia, Melrose Granite (west) and Falling River Granite (east) in Brookneal, and Leatherwood Granite (west) and Shelton Granite (east) in southern Virginia. Boxed areas (dash-dot) have been mapped in detail. SA = Sauratown Mountains Anticlinorium; sra = Smith River allochthon (modified from Gates, 1981, after Smith et al., 1964; Ern, 1968; Brown, 1969; Marr, 1980a, b; Conley, 1985; Gates, 1986; Glover et al., 1989).

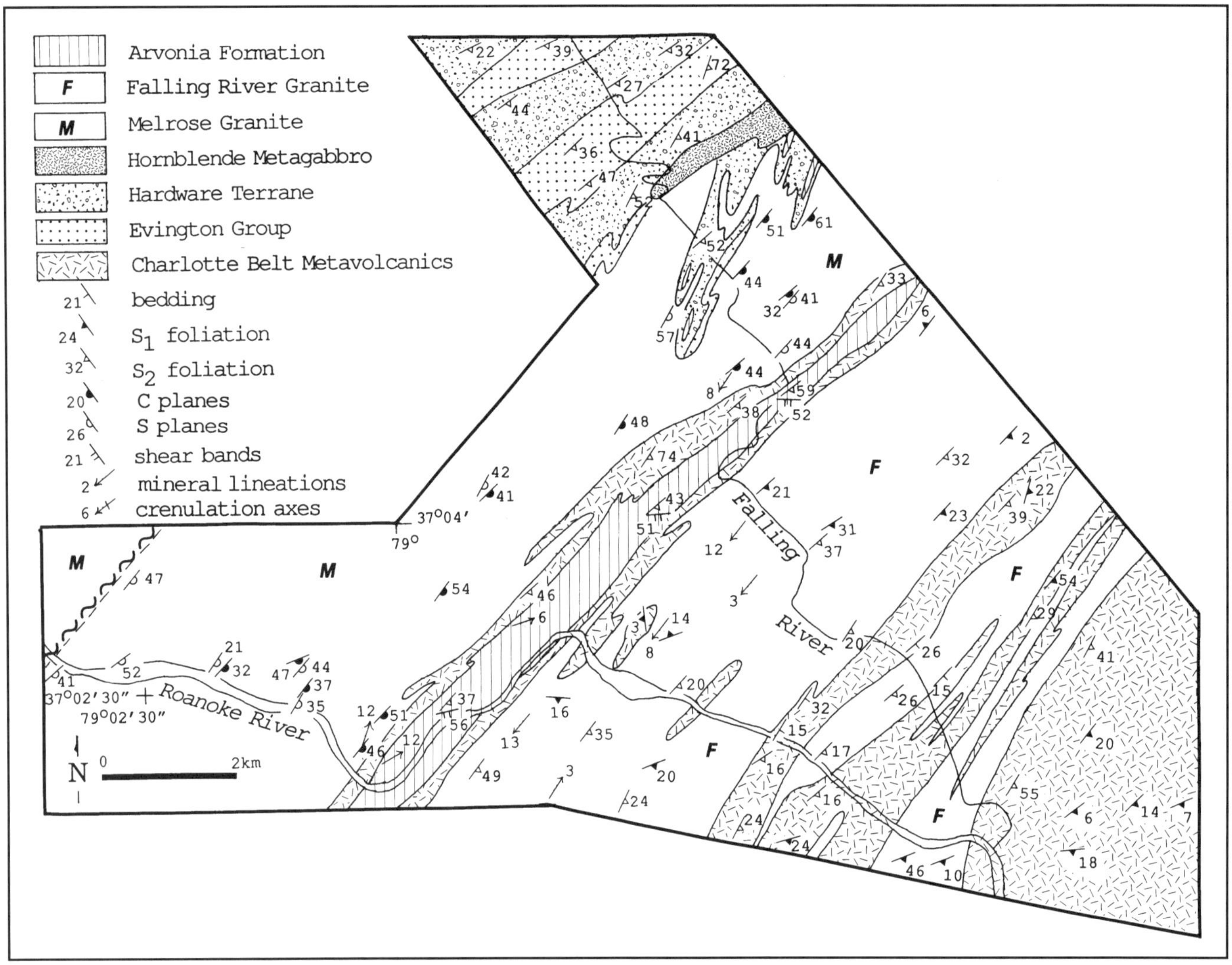

Figure 2. Geologic map of the Brookneal area/Roanoke River traverse (see text for description of units).

Toward the west, the metapelites are interlayered with two discrete calcsilicate- quartzite sections of tens of meters in thickness that thicken to the west (Gates, 1986) (Fig. 4). The metapelite sequence is underlain by a sequence of quartzite and metagraywacke that contains graded bedding and cross-bedding as well as a discontinuous layer of amphibolite. The Evington Group rests on metabasalts that are amygdaloidal in part and equivalents of the Late Proterozoic Catoctin Formation. The metabasalts in turn overlie very coarse- to medium-grained metaclastic rocks of the Lynchburg Group.

Hardware terrane(?)

A massive coarse sand to pebble biotite schist structurally overlies the metapelite where it comprises the cores of synclines in the northwestern Brookneal area (Figs. 2 and 4). This unit is likely an extension and the southwesternmost limit of the Hardware terrane of central Virginia (Evans, 1984) (Fig. 1). Scarcity of outcrop precludes the documentation of unit thickness. Where contacts are observable, bedding ranges from 10 cm to 3 m; some of the thinner layers exhibit normal grading from granule metaconglomerate to metasiltstone. The relict clasts in the metasandstones and conglomerates are 2 to 6 mm and are composed primarily of quartz (Fig. 3b), some of which is highly rutilated (Fig. 3c), normally zoned oligoclase, microcline, and rock fragments with granitic composition. Plagioclase dominates over microcline in most samples by a ratio of 2:1 to 3:1, but in the coarsest granule conglomerate, microcline is the exclusive feldspar. The matrix of the metasandstone is fine-grained quartz, plagioclase, biotite, amphibole, epidote, muscovite, apatite, opaques, and tourmaline in varying quantities.

Melrose Granite

The Melrose Granite (Jonas, 1927) is a late- to post-tectonic pluton with respect to deformation and metamorphism of the Evington Group. In the western part of the pluton, the rock is a coarse-grained, hypidiomorphic, biotite-titanite quartz mon-

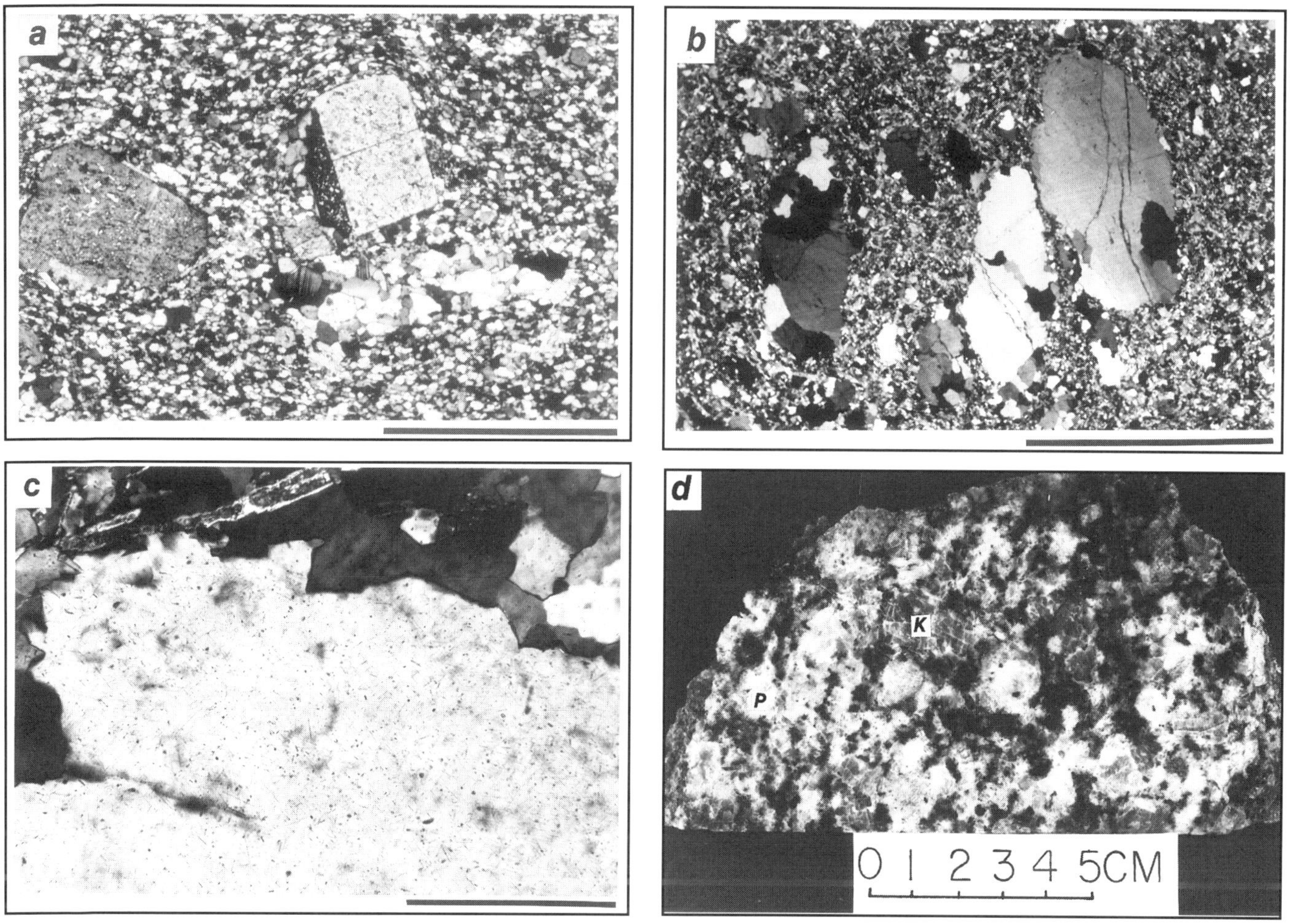

Figure 3. Photographs and photomicrographs of rocks from Brookneal. a, Preserved igneous texture of a dacitic crystal tuff from the low-grade portion (western side) of the Charlotte belt (Milton terrane). Scale bar = 5 mm. b, Photomicrograph of granule metaconglomerate-biotite schist (Hardware terrane). Matrix is biotite, quartz, plagioclase, epidote, apatite, and minor muscovite. Scale bar = 1 cm. c, Photomicrograph of rutilated quartz granule from Hardware granule metaconglomerate. Scale bar = 0.5 mm. d, Photograph of undeformed Melrose granite. K = alkali feldspar; P = plagioclase; mafics are biotite and titanite.

zonite to quartz diorite (Streckheisen, 1973 classification) (Figs. 2 and 3d). The main components are 1- to 3-cm oligoclase, perthitic microcline, and lesser amounts of quartz. Mafic phases include 0.5- to 1-cm sagenitic biotite and titanite, and accessory phases include zircon, apatite, and magnetite. Xenoliths of Evington Group metapelite are common along the western part of the pluton, and inclusions of Charlotte belt lithologies occur in the eastern parts of the pluton. Eastward across the pluton, the composition changes gradually to a monzogranite, and in conjunction, deformation increases into the Brookneal shear zone (Gates et al., 1986). The Melrose Granite is mylonitic and ultimately ultramylonitic within the zone. The Melrose Granite has a U/Pb zircon age of 516 Ma (Sinha et al., 1989), but Rb/Sr biotite–whole-rock ages are 300 ± 5 Ma (Gates et al., 1986).

Apparently cointrusive with the Melrose Granite is a dike of coarse-grained hornblende metagabbro with a rim of fine-grained amphibolite. The dike occurs in the northwestern map area, is 10 to 25 m thick, thickens to the north, and extends about 15 km along strike (Jonas, 1927). Randomly oriented, 1.0- to 1.5-cm hornblende grains comprise 60 to 75% of the coarse-grained facies with interstitial andesine and accessory titanite, zircon, magnetite, and apatite. The fine-grained facies exhibits strong foliation and retrogression. It is composed of 85% pale amphibole with plagioclase, epidote, biotite, chlorite, titanite, and apatite. Blocks of the hornblende metagabbro occur as xenoliths in the Melrose Granite, but dikes with similar composition and character intrude the Melrose Granite. At the contact between the main dike and the Melrose Granite, the two rock types intermingle, forming swirl patterns, and fragments of each rock type can be found enclosed within the other.

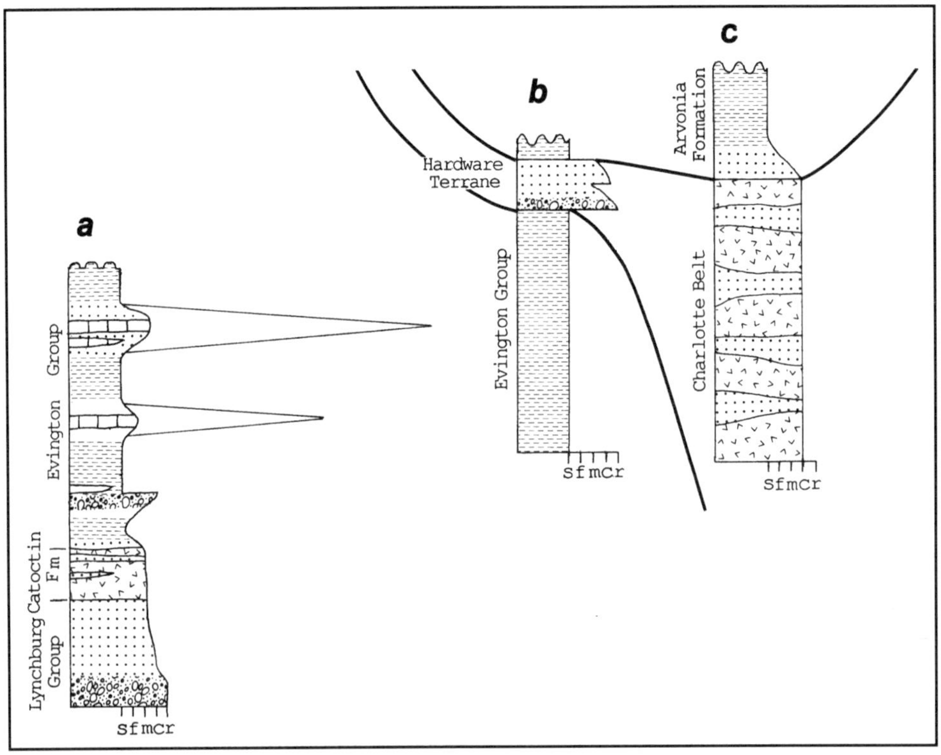

Figure 4. Stratigraphic columns for sequences of interpreted protoliths along the Roanoke River: a, From the Altavista area (from Gates, 1986); b, from Brookneal to the west of the suture zone; and c, from Brookneal to the east of the suture zone. Thicknesses are relative and approximate because of structural complexity. No sequence is recognized in the Charlotte belt rocks. Columns equated approximately on time lines, but unidentified faulting and unaccounted fault throw may introduce significant errors. Heavy lines are faults or unconformities. Abbreviations: s = shale; f = fine-grained sandstone; m = medium-grained sandstone; c = coarse-grained sandstone; r = conglomerate. Patterns: dashes = shale; dots = sandstone; circles and dots = conglomerate; random Vs = volcanics; brick pattern = carbonates.

Arvonia Formation

The Arvonia Formation forms a northeast-trending belt across the central map area and occurs as outliers overlying Hardware terrane and Charlotte belt rocks (Figs. 2 and 4). The main belt of rock continues northeast and connects with the Arvonia syncline in central Virginia (Fig. 1). Trilobites from the Arvonia Formation in its type locality indicate a latest Ordovician age (Tillman, 1970), but regional correlations indicate that deposition may have continued to earliest Silurian (Glover et al., 1989).

The Arvonia Formation in Brookneal consists of greenschist facies graphitic metapelites, metasandstones, and metacarbonates. Along the western edge of the main belt, quartzite to metagraywacke and chlorite-muscovite schist alternate in 3-cm- to 1-m-thick cycles. The aluminous schist is interpreted as metapelite. Locally, the cycles are graded, indicating that the metasandstones comprise the base of the section, but deformation is intense and sedimentary structures are not generally observable. Toward the east, the chlorite-muscovite schist is gradually replaced by layered sequences of calc-silicate and marble. The calc-silicate contains chlorite, biotite, tremolite, calcite, dolomite, quartz, plagioclase, muscovite, and apatite in varying quantities. The marble is calcitic with minor quartz, zircon, apatite, and tourmaline.

A massive graphitic "button" schist replaces the layered sequence in the middle of the belt and continues to the eastern edge. The silver-black schist contains 60 to 75% muscovite with disseminated graphite and lesser amounts of quartz, biotite, chlorite, titanite, and hematite. The schist contains abundant pods of vein quartz and rare boudins and rootless folds of calcsilicate. The graphitic schist is highly contorted and sheared and lies within the Brookneal zone.

Mesozoic rocks

Sedimentary rocks of the Danville Basin occur along the western edge of the map area (Fig. 1). Coarse-grained sandstone, conglomerate, and fanglomerate are red on the eastern side of the basin but variably gray toward the west. The rocks range from massive and poorly sorted to relatively well sorted and bedded. The clasts are composed of all units in the area but are dominated by Melrose Granite. Paleontologic studies indicate that these rocks are of Late Triassic age (Henika and Thayer, 1977).

Pristine, mafic dikes with north-south trend cut the Brookneal area. The dikes are diabasic and undeformed and are associated in the sequence of events with the formation of the Danville Basin.

STRUCTURAL GEOLOGY

The structural geology of the Brookneal area is complex and varies considerably across the area (Fig. 2). The oldest structures found in the Charlotte belt are isoclinal, recumbent to shallowly southeast-dipping intrafolial folds that occur in amphibolite facies rocks in the southeasternmost map area (Fig. 5a) and formed during a D1E deformational event. These intrafolial folds have shallow to horizontal southwest-trending fold axes and are synchronous with the pervasive foliation in the metavolcanic and metavolcaniclastic rocks (Fig. 6a). These older folds are refolded into type-3 refolds (Ramsay, 1967) by tight, moderately to steeply inclined southeast-dipping, westward vergent asymmetric folds (Fig. 5b) of the D3 deformational event. These folds have shallowly southwest-plunging fold axes. Moderately southeast-dipping axial planar crenulation cleavage (Fig. 6b) is associated with these later folds in some lithologies, which increases in intensity toward the Brookneal zone. The intersection of the second cleavage and the older pervasive foliation produces the strong mineral lineations in the Falling River granite (Fig. 5c). The recumbent folds are less common toward the northwest, whereas the upright folds increase in tightness and intensity toward the northwest and the Brookneal zone. As shown by Baird (1991), the area of recumbent folds to the southeast has been uplifted relative to the area of inclined folds toward the northwest.

The Evington Group metapelites exhibit a strong metamorphic foliation and contain recumbent isoclinal and rootless isoclinal folds formed during the D2W deformational event. The lower limbs of these folds are consistently sheared out and indicate a westward transport direction (Fig. 5d). The pervasive foliation in these rocks is defined by the middle and upper amphibolite facies mineral assemblages. Gates and Speer (1991) have shown that the concomitant deformation and metamorphism result from westward-directed nappe emplacement in a two-stage event. D1W deformational features include aligned inclusions in the cores of porphyroblasts but otherwise were obliterated during D2W. The isoclinal folds in the metapelites are refolded with the overlying metasandstone unit into shallowly to moderately southwest-plunging folds that are of the same generation of those described earlier in the Charlotte belt rocks (D3). These folds are overturned to the northwest and have wavelengths of approximately 2 km. Local and late northwest-striking, steeply dipping crenulation cleavages (D4) cross these folds in the northwestern map area.

All rock types are deformed in the Brookneal shear zone, a D3 structure, across the central map area. The Melrose Granite forms a type I S-C mylonite (Lister and Snoke, 1984) with a consistent dextral strike-slip shear sense (Gates et al., 1986). The Arvonia Formation displays a pervasive foliation and disharmonic folding in the calcsilicate units. The graphitic schist contains reverse shear cleavages (RSC; Dennis and Secor, 1987) and shear bands, all of which indicate a consistent dextral transcurrent sense of shear (Gates et al., 1986). The buttons in the graphitic schist are composed of bundles of RSC's bounded by shear bands (Fig. 5e). The Evington Group rocks are progressively retrograded to greenschist facies assemblages and are mylonitic as are the metasandstones (Gates, 1981). The Falling River granite also developed mylonitic fabric toward its western edge, which displays S-C fabric, and rotated σ-type porphyroclasts (Passchier and Simpson, 1986), all indicating dextral strike-slip displacement.

Late stage brittle faults occur within the western, undeformed part of the Melrose Granite in the vicinity of the Danville Basin (D5). The faults are steeply southeast dipping, show normal displacement, and are filled with hematite cemented cataclasite (Fig. 5f). These faults appear to be associated with the Danville Basin; however, no basin-bounding normal fault was observed in the study area.

PERIODS OF ACTIVATION

Suturing of the Carolina Terrane

The Charlotte belt rocks at Brookneal are exclusively of volcanic or volcaniclastic origin (Fig. 4). Although bimodal in appearance, the metavolcanic rocks clearly contain andesites-dacites as the felsic component instead of rhyolites. The rocks are therefore consistent with generation in a volcanic arc (Fig. 7a). The abundance of amphibole, epidote, magnetite, and plagioclase as the exclusive feldspar in the metasedimentary rocks is evidence for a volcaniclastic origin. These rocks are dominantly metagraywackes and form conformable layers or fill channels in the metavolcanic rocks. The metagraywackes are interpreted as parts of a turbidite fan complex that developed within the volcanic pile. The metamorphosed predominantly mafic and ultramafic dikes are interpreted as feeders for the volcanic rocks.

The Charlotte belt rocks have been traced southward to the Danville area by reconnaissance methods and support the correlations of Jonas (1927). The metavolcanic rocks in Danville show the same decreasing metamorphic grade toward the northwest (Henika, 1975). In the areas of lower metamorphic grade, volcanic protoliths are revealed by relict igneous textures similar to those in Brookneal (Henika and Thayer, 1977). These rocks are intruded by the Shelton granite, which yielded a 429-Ma Rb/Sr whole-rock age (Kish et al., 1979) and a 463 ± 14 Ma U/Pb age on zircon fractions (Hund, 1987) and resembles the 470 Ma Falling River granite (A. Sinha and L. Glover, III, personal communication, 1983). Conley and Henika (1973), and Conley (1978) determined that the Danville metavolcanic sequence is a direct northern extension of the Charlotte belt portion of the Carolina terrane (Horton et al., 1989), and Baird (1991) showed that the sequence extends to the Brookneal area.

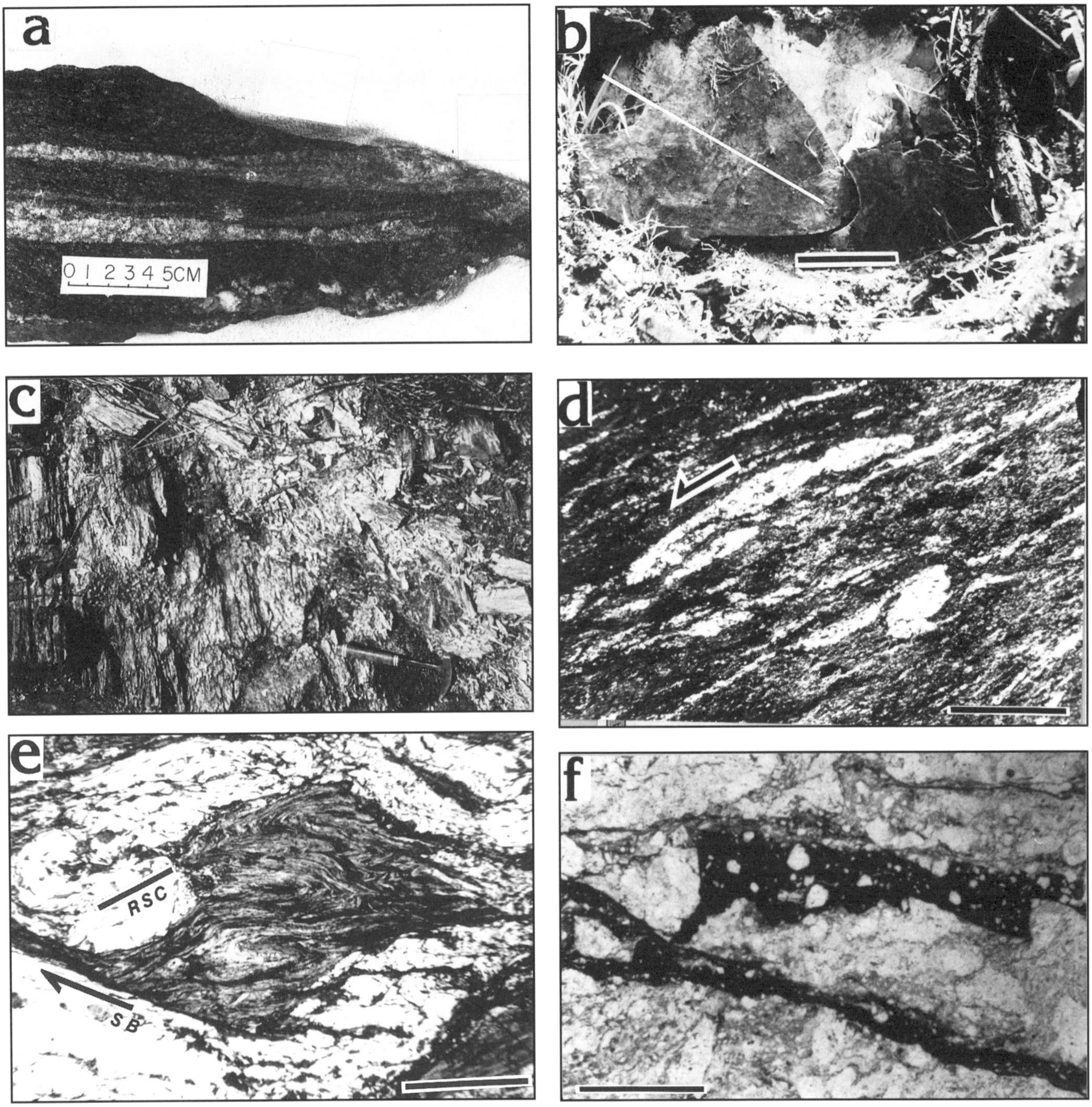

Figure 5. Photographs and photomicrographs of structural features from the Brookneal area, Virginia. a, Photograph of an early isoclinal fold from the amphibolite facies Charlotte belt rocks in the southeastern map area. b, Photograph of a southeast-dipping overturned fold in metadacite. These folds overprint the isoclinal folds toward the Brookneal zone. Bar marks the axial plane; bar scale = 10 cm. c, Weathered quartz and feldspar aggregate rods that form the subhorizontal lineations in the Falling River granite. d, Rootless isoclines in metapelitic gneiss of the Evington Group show consistent westward transport (arrow) with the lower limb sheared out. Scale bar = 10 cm. e, Photomicrograph of a button from sheared graphitic Arvonia Formation shows dextral movement sense. Arrows show offset direction; RSC = reverse shear cleavages (Dennis and Secor, 1987); SB = shear bands; scale bar = 5 mm. f, Photomicrograph of hematite-rich cataclasite-filled fracture that cross cuts the mylonitic foliation within the Brookneal zone. Scale bar = 5 mm.

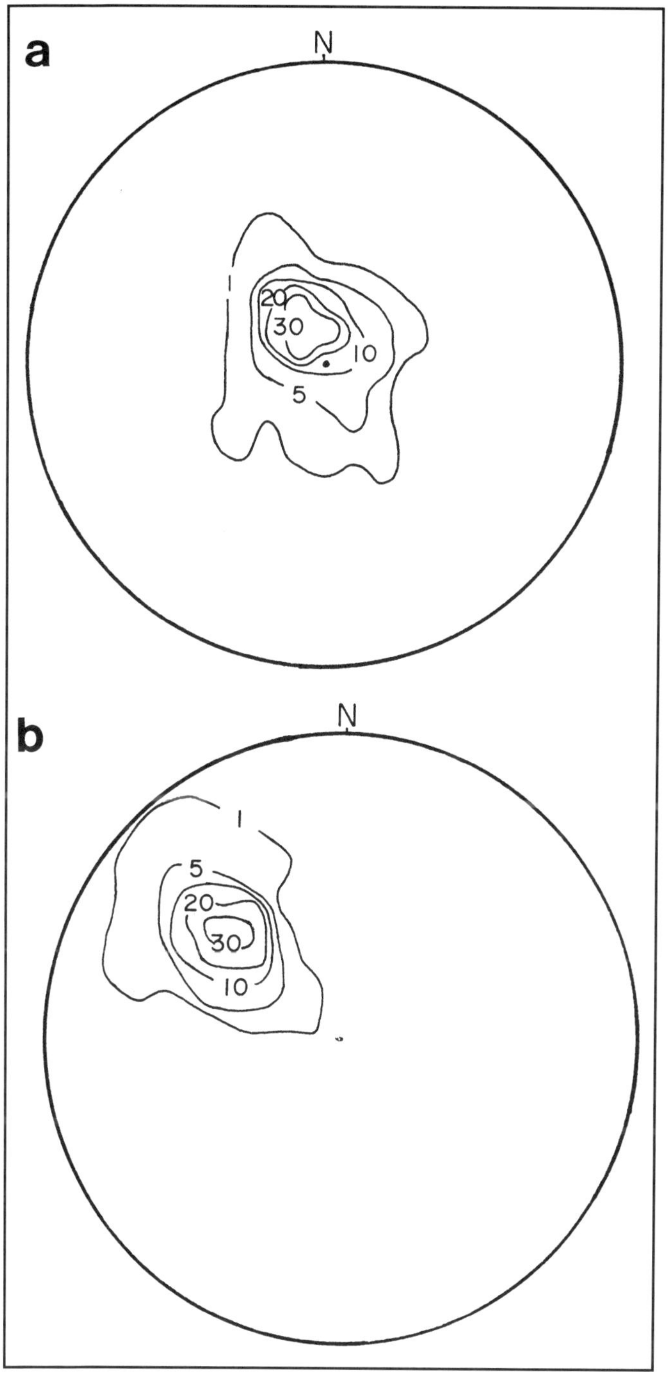

Figure 6. Equal area plots of poles to foliations in the Charlotte belt rocks. a, Pervasive axial planar schistosity to the isoclinal folds (Fig. 5a) from the non-overprinted southeasternmost map area. N = 47. b, Axial planar crenulation cleavage to the southeast-dipping folds (Fig. 5b) that overprint the early isoclinal folds toward the Brookneal zone.

The Charlotte belt rocks have also been traced into the Chopawamsic Formation (Southwick et al., 1971) of central Virginia by detailed (Marr, 1980a, b) and reconnaissance mapping. This correlation supports the findings of Jonas (1927) but requires that the Chopawamsic terrane of Williams and Hatcher (1982) and Horton et al. (1989) be an extension of the Carolina terrane rather than a distinct sequence (Glover, 1989). The Chopawamsic Formation is composed of greenschist to amphibolite facies intermediate-mafic metavolcanics with metagraywacke and intruded by small ultramafic bodies. It has been dated at 570 to 525 Ma using U/Pb zircon analyses (Seiders et al., 1975; Higgins et al., 1971, 1977; Pavlides, 1981).

Redden (1963), Kaldy (1977), and Gates (1986) showed that the rocks of the Evington Group (Brown, 1958) could be traced directly from their type section to the metapelites in Brookneal (Fig. 4). These rocks are interlayered with and overlie metabasalts of the 570-Ma Catoctin Formation (Badger and Sinha, 1988) that in turn overlie Late Proterozoic Lynchburg Group metasediments (Rankin, 1975; Wehr and Glover, 1985). The Evington Group rocks are therefore Cambrian to Cambro-Ordovician in age at latest and of apparent North American affinity (Brown, 1970). The Evington Group rocks are proposed to be offshore equivalents of the Valley and Ridge section of that age (Brown, 1970; Gates, 1986; Patterson, 1988) (Fig. 7a), though they are herein interpreted as quite distal.

The metasandstones that overlie the Evington Group in Brookneal are enigmatic. They could be a previously unidentified part of the Evington Group but more likely are lateral equivalents of the Hardware terrane (Evans, 1984) of central Virginia (Figs. 1, 4, and 7a). Brown (1976) and Bland and Brown (1977) recognized that the rocks to the west of the Arvonia syncline represent a tectonic mélange. Evans (1984) found that the graywacke to the west of the mélange is part of the same terrane and proposed that it represents forearc prism deposits in front of the Chopawamsic arc (Carolina terrane) similar to the Potomac terrane (Drake, 1985) of northern Virginia and Maryland.

The metasandstones of Brookneal are most likely part of the Hardware terrane. They clearly indicate a mixed source. Some of the pebble conglomerates contain microcline as the exclusive feldspar, whereas in most rocks, plagioclase dominates over alkali feldspar by a 2:1 or 3:1 ratio. Most quartz grains are inclusion free, but grains concentrated with rutile needles (Fig. 3c) occur sparsely within the sandstones. In addition, lithic clasts are common as is epidote, some of which appears detrital. The presence of amphibole clearly distinguishes these metaclastics from those in the Evington Group. Laurentia likely provided the microcline, rutilated quartz and some of the clear quartz because there is no source of these minerals in Charlotte belt rocks, whereas the Carolina terrane likely provided most of the plagioclase, fragments, clear quartz, and components of epidote and amphibole.

The accretion of the Carolina terrane to Laurentia constitutes the Taconian or Penobscottian(?) orogeny and took place

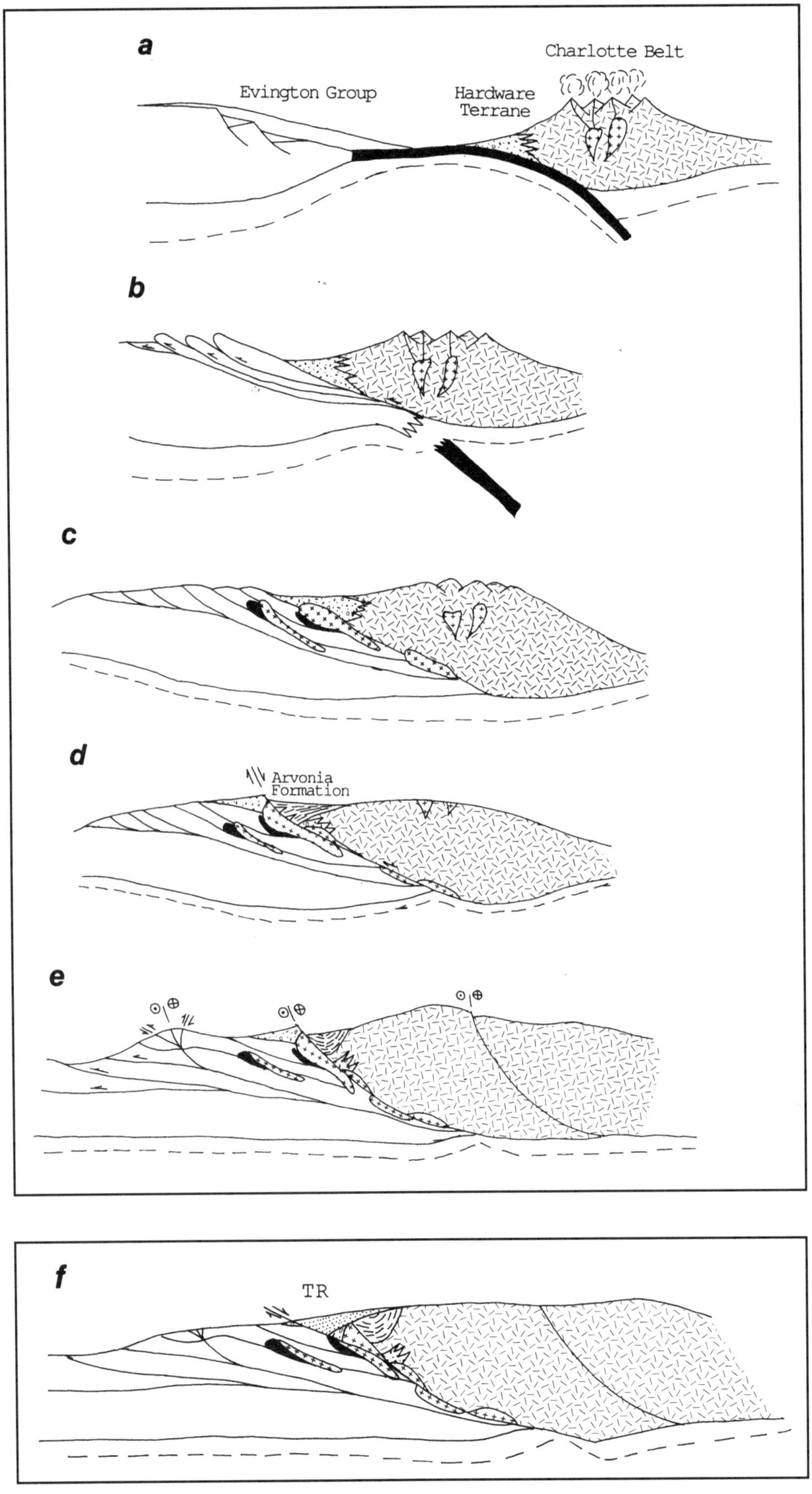
a
Evington Group
Hardware Terrane
Charlotte Belt
b
c
d
Arvonia Formation
e
f
TR

Figure 7. Model for periods of activity in the Brookneal area. Left (West) Continent (no pattern = North America; right (east) continent (random dashes) = Carolina (Carolina terrane island arc); Brookneal zone separates the two continents. a, Subduction of North America beneath the Carolina terrane island arc. Deposition of Evington Group, Charlotte belt, and Hardware terrane rocks. b, Taconian, Penobscottian, or Carolinian suturing of Carolina terrane arc with North America and westward thrust faulting and nappe emplacement. c, Intrusion of granitoids (Melrose granite) and gabbroids (Hornblende metagabbro) into or near the suture zone. d, Reactivation of suture zone by extension (possible core complex) or transtension forming a restricted basin into which the Arvonia Formation was deposited. e, Alleghanian orogeny causing dextral strike-slip reactivation of suture zone and forming 6-km-wide Brookneal mylonite zone, folding of Arvonia Formation, formation of positive flower structures to the west in Altavista, and westward directed thrust faulting. f, Mesozoic extension, normal reactivation of the suture zone, and deposition of the Danville Basin siliciclastic rocks (TR).

along the suture between the Charlotte belt and the Evington Group (Fig. 7b). Petrology and trace element geochemistry of the metavolcanic rocks indicate that they originated in an island arc setting (Glover and Sinha, 1973; Bland, 1978; Bland and Blackburn, 1980). These data and the location of metallogenic deposits on the west side of the belt indicate an east-dipping subduction zone (Bland and Blackburn, 1980; Glover et al., 1989). The location of the Hardware terrane forearc deposits on the west side of the volcanic deposits and total lack of arc-related volcanics in the Evington Group support this polarity. The southeast dip of the faults and documented westward-directed nappes (Conley and Henika, 1973; Conley, 1978; Gates and Speer, 1991) are also consistent with subduction of the Laurentian plate beneath the Carolina terrane. This tectonism produced the recumbent isoclinal folds in Charlotte belt (D1E) and the metamorphism and early west direct thrust fault–related deformation (D1W and D2W) in the Evington Group. Rb/Sr biotite–whole-rock dating in Evington Group rocks yields 440 Ma (Odom and Russell, 1975) and is interpreted to record uplift and exhumation of the metamorphic pile. A six-point whole-rock Rb/Sr age for metapelites from the Altavista area is 546 ± 21 Ma (Gates and Mose, 1987), which could record the age of deposition and diagenesis or the age of metamorphism. Recent U/Pb dating of staurolites from Evington Group metapelites yields two ages of metamorphism, an older 535 ± 10 Ma age and a younger 454 ± 6 Ma age (Lanzirotti and Hansen, 1997). These data support earlier findings that there was a Cambrian tectonic event followed by later shearing, retrograde metamorphism, and the resetting of isotopic systems (Gates et al., 1994). These metamorphic ages indicate that the Evington Group is Early Cambrian at youngest and cannot be the offshore equivalent of most of the early Paleozoic Valley and Ridge section as proposed. A 535-Ma accretion of the Carolina terrane to Laurentia is inconsistent with both the Penobscottian and Taconian events. Therefore, this event might be better designated the "Carolinian" orogeny because it involves the accretion of Carolinia. The later event is more likely the result of the Taconian orogeny and is recorded in the rocks throughout the Piedmont, Blue Ridge, and early Paleozoic Valley and Ridge. Little evidence of this tectonism exists in Brookneal, but it is well represented to the west in Altavista (Gates and Speer, 1991; Gates et al., 1994) (Fig. 1).

Magma conduit

Subsequent to the accretion of the Carolina terrane to Laurentia, the suture zone served as a plane of weakness for magmatic ascent (Fig. 7c). The Melrose granite and associated hornblende metagabbro intruded the suture zone after the cessation of activity. A U/Pb age on zircon of 516 Ma was determined for the Melrose granite (Sinha et al., 1989). Because the Melrose Granite stitches the suture zone, this age further indicates that suturing occurred during the Cambrian as part of the Carolinian orogeny. On the other hand, there are cases of inherited lead in zircon in the central Appalachian Piedmont (Higgins et al., 1977). The Rb/Sr whole-rock ages and staurolite geochronology from the Evington Group support the early age of accretion.

Other bimodal plutonism occurs along the Carolina terrane boundary in Virginia (see Sinha et al., 1989) (Fig. 1). In central Virginia, the Columbia granite and Diana Mills mafic complex (Ern, 1968) appear cointrusive. The Columbia granite intrudes Chopawamsic metavolcanic rocks to the east of the suture zone and yields an Rb/Sr whole-rock age of 454 ± 9 Ma (Mose and Nagel, 1982) and a U/Pb zircon age of 445 ± 25 Ma (Higgins et al., 1977). The Columbia granite is petrologically similar to the Falling River granite rather than the Melrose granite. The Diana Mills pluton lies to the west of the suture zone unattached to the granite but intruded at the same point in the tectonic history of the area (i.e., post-Taconian[?], pre-Alleghanian). In southern Virginia, the Martinsville igneous complex includes the Leatherwood granite and the Rich Acres Formation, which is composed of peridotite and norite (Conley and Henika, 1973). These bodies are clearly cointrusive but occur far to the west of the Carolina terrane suture zone. An Rb/Sr whole-rock age on the Leatherwood granite is 464 ± 20 Ma (Odom and Russell, 1975), but U/Pb zircon ages are 516 Ma (Sinha et al., 1989). Other apparently cointrusive mafic and granitoid plutons of Cambrian to Ordovician age, such as the 459-Ma Garrison complex (Pavlides, 1976), occur along the suture zone to the north into Maryland (see Drake et al., 1989; Glover, 1992).

The origin of these plutons is debatable. Pavlides (1981) used these plutons in conjunction with other evidence to propose a westward-dipping subduction zone. The mafic plutons are proposed to have formed in a back-arc environment or the base of the arc (Sinha et al., 1989). Alternative models are delamination (Glover, 1992), postcollisional rifting associated with the development of metamorphic core complexes, transtension resulting from oblique subduction, or readjustment of the plates subsequent to collision. Constraints on these models are imposed by overlying metasedimentary units, as discussed below.

Control on sedimentation

The basal metasandstone in the Arvonia Formation in Brookneal is minor in comparison with the volume of graphitic schist and metacarbonates that dominate the unit (Fig. 4). The graphitic schist is interpreted to have been a highly carbonaceous mud that was deposited under euxinic conditions (Fig. 7d). The calcsilicate and marble were calcareous mud and limestone respectively and are consistent with deposition in a low-energy environment. The Arvonia Formation likely began as a high-energy deposit associated with the opening of a basin, but the abundant muds and carbonaceous and calcareous components indicate that the basin was restricted and likely protected from normal marine conditions.

The Brookneal section of the Arvonia Formation is the southernmost exposure because the belt disappears beneath the Mesozoic Danville Basin to the south (Fig. 1). To the north, in central Virginia, the Arvonia Formation exhibits evidence for normal marine conditions. The slate, which dominates the sequence, contains trilobite, brachiopod, crinoid, and bryozoan fossils (Darton, 1892; Stose and Stose, 1948; Brown, 1969; Tillman, 1970). These fossils and regional relations indicate that deposition lasted from Late Ordovician (Tillman, 1970) through Early Silurian (Glover et al., 1989) and in shallow-water conditions (Brown, 1969). The sequence also contains several important coarse-grained siliciclastic units. The basal unit of the Arvonia Formation, the Bremo Member, is a granule conglomerate with clasts of vein quartz, phyllite, and schist (Brown, 1969). The Bremo Member occurs along the west side of the belt and is a facies equivalent of slate to the east. It contains cross-bedded quartzite with flow direction from the north and northwest and quartz pebble conglomerate with pebbles up to 2 cm (Brown, 1969). The Buffards Formation forms the cap to the sequence in possible unconformable contact with the underlying Arvonia Formation. It is approximately 450 to 500 m thick and composed of conglomeratic and pyroclastic phyllite. The grain size coarsens to the northwest and is interpreted to indicate source direction (Brown, 1969). The Buffards Formation has also been proposed to form the base of the Arvonia Formation (Stose and Stose, 1948).

Previously, the Arvonia Formation was proposed to be an intermontane successor basin subsequent to the Taconian orogeny (Brown, 1969, 1970; Gates, 1981). However, the rocks in Brookneal can be interpreted to have been deposited in a restricted, apparently linear basin along a previously active fault. Evidence for restricted conditions is present in Arvonia Formation rocks for about 60 km along the Carolina terrane suture zone. In the type locality, normal marine conditions prevailed; however, several prominent conglomeratic units were deposited. These deposits represent periods of high-energy deposition and appear spatially restricted and episodic to catastrophic in the otherwise low-energy environment. These features are more consistent with a fault-controlled actively subsiding basin than with an intermontane basin. In addition, the source direction both by distribution and sedimentary structures is to the northwest. Considering that the structures in this part of the Piedmont dip to the southeast, the coarse material and flow direction would have been from the shallow dip-slope rather than the steep strike-slope. The flow direction and distribution of coarse material are also consistent with normal or transtensional reactivation of the old southeast-dipping thrust faults because the steeper slope would lie on the northwest side of the basin (Fig. 7d).

The cause of the extension or transtension is speculative. It may have occurred as a result of the Taconian orogeny. Subsequent to the orogeny, gravitational collapse may have occurred similar to the development of metamorphic core complexes in the western United States. Alternatively, the Taconian orogeny appears to have occurred later in the northern Appalachians than in the southern Appalachians (Drake et al., 1989). The Arvonia Formation may have been deposited in a transtensional basin that resulted from tectonic escape of the southern Appalachians in response to the collision to the north. A final possibility is that the basin is a product of Acadian tectonism and is transtensional as the result of oblique subduction (see Fitch, 1972) prior to the continental collision. Assuming the younger ages, the bimodal plutonism closely predates the development of the sedimentary basin. This magmatism may have served as a precursor to extension or transtension.

Alleghanian dextral transcurrent faulting

Alleghanian dextral transcurrent faulting is well documented in the Brookneal area (Gates et al., 1986) and southern Appalachians in general (Gates et al., 1986, 1988; Vauchez, 1987; Hatcher et al., 1989). The Carolina terrane boundary was reactivated during this event and formed the Brookneal zone (Gates et al., 1986), a 6-km-wide dextral transcurrent mylonite zone (Fig. 7e). The eastern 4 km of the Melrose granite formed a type I S-C mylonite (Lister and Snoke, 1984) with a minimum offset of 17 km (Gates et al., 1986) and more likely >50 km. The Arvonia Formation was sheared into a button schist with buttons composed of bundled dextral reverse shear cleavages (Dennis and Secor, 1987) separated by shear bands. The Falling River granite also exhibits a type I S-C mylonite along its western boundary. In addition, the upright folds and retrogression in the Charlotte belt, Evington Group, and metasandstone units are concentrated to the area around the Brookneal zone and are attributed to the Alleghanian orogeny. Gates et al. (1986) showed that shearing on the Brookneal zone lasted from ca. 324 to 300 Ma. Rb/Sr biotite–whole-rock ages within the Evington Group rocks to the west range from 288 to 310 Ma and Rb/Sr whole-rock analyses of rocks within a strike-slip shear zone yield 275 Ma (Gates and Mose, 1987).

Alleghanian deformation continued along the Carolina terrane boundary into central Virginia. Metamorphism and accompanying deformation of the higher-grade eastern Arvonia Formation occurred between 324 and 287 Ma (Smith et al.,

1964). The western low-grade slate yielded a whole-rock K-Ar age of 300 Ma (Harper et al., 1973). Deformation in the area has previously been considered only in terms of compressional tectonics (Smith et al., 1964; Ern, 1968; Brown, 1969; Conley, 1978; Evans, 1984). Reconnaissance mapping in the Andersonville area to the south of Arvonia, however, reveals dextral shear banding in the Arvonia Formation similar to that in Brookneal. In addition, Brown and Griswold (1970) found superposed folding in the main Arvonia syncline similar to that found in the Altavista area to the west of Brookneal. Gates (1987) showed that these multiple deformations resulted from Alleghanian dextral transpression.

Triassic-Jurassic normal faulting

The Danville Triassic Basin lies to the west of the Carolina terrane boundary in the Brookneal area (Fig. 7f). At the eastern edge of the basin, the Melrose Granite exhibits thin brittle fractures filled with cataclasite. This brittle deformation must post date the purely ductile Alleghanian shearing in granite.

Although the Carolina terrane boundary was not reactivated during the Mesozoic at Brookneal, the Brookneal zone and suture zone have been mapped by reconnaissance methods south to the Danville Basin (Fig. 1). The zone disappears beneath the Triassic sedimentary rocks and merges with the Chatham fault (Meyertons, 1963), which bounds the west side of the Danville basin. The Chatham fault has been shown to be a reactivated mylonite zone (Robinson, 1979). The mylonitization is proposed to be approximately coeval with the intrusion of 321 Ma pegmatites (Conley, 1985), which is within the age range of Alleghanian mylonitization on the Brookneal zone (Gates et al., 1986). Boulders and cobbles in the border conglomerate have been observed in reconnaissance to be composed of type I S-C mylonite that is very similar to mylonitic Melrose Granite.

DISCUSSION

The establishment of fundamental zones of crustal weakness may be controlled by plate tectonic processes. The western boundary of the Carolina terrane in southern Virginia is not only the site of Early Paleozoic terrane accretion but was reactivated during every subsequent major tectonic event in the southern Appalachians. If deformation had been more intense during any of these events, splitting of North America into separate continents might have occurred along this zone either as the result of rifting or transtension. Applying this principle to the pre-Carolinian history of the area, late Precambrian rifting of the Grenville supercontinent also may have occurred along this boundary because it was the site of the Iapetus ocean basin. The late Precambrian rifting could have similarly reactivated a zone of crustal weakness that was formed during the Grenville orogeny. The zone may therefore have been established long before the formation of any of the rocks currently exposed along it. In this case, the Carolina terrane boundary may have had at least six periods of tectonic activity.

The multiple reactivations that occur along the boundaries of accreted terranes may be regarded as subsets of major orogenic cycles (Wilson, 1966). Tectonic activity on zones of crustal weakness has more comprehensive implications than simply movement on old faults. The activity can be extensional or transtensional and cause subsidence, thereby localizing deposition. The zone can serve as a conduit for magma ascent and emplacement, resulting in extensive plutonism or even volcanism. Fluid movement along the zone during deformation can cause the localization of metamorphism and alteration. Because the activity can be so extensive in terms of deformation and the mixture of rock types produced, care must be exercised not to interpret one of these secondary boundaries as a primary plate margin.

Major rifting-accretion cycles occur at widely spaced intervals, but the zones of weakness undergo episodic activation at more closely spaced intervals within the major cycles. Because the smaller events are not necessarily cyclical, an analogy with transgressive and regressive cycles and subcycles in sea-level curves is inappropriate. Furthermore, because the opening of the Atlantic Ocean did not occur along the former Iapetan margin in southern Virginia, there is a complication in the major cycle. However, the subsets in this study are shown to mimic the major cycles not only in terms of deformation but also with the development of sedimentary basins and magmatic activity. As would be expected with a subset, these features form at a smaller scale than the full (Wilson, 1966) cycles.

ACKNOWLEDGMENTS

Funding for this project was provided through grants DE-AC05-78ET27001 from the U.S. Department of Energy and NRC-04-75-237 from the U.S. Nuclear Regulatory Commission to L. Glover, III, and J. Costain, and work was performed at the Orogenic Studies Laboratory at Virginia Polytechnic Institute and State University. The manuscript benefited greatly from discussions with A. Drake, Jr. Thanks to L. Glover, III, for reviewing an early version of the manuscript and to A. Goldstein and P. Muller for thorough reviews that improved the manuscript.

REFERENCES CITED

Badger, R. L., and Sinha, A. K., 1988, Age and Sr isotopic signature of the Catoctin volcanic province: Implications for subcrustal mantle evolution: Geology, v. 16, p. 692–695.

Baird, R. A., 1991, Stratigraphy of the north Charlotte metamorphic belt, south-central Virginia: Application to the Charlotte belt/Milton belt problem: Southeastern Geology, v. 32, p. 61–82.

Bland, A. E., 1978, Trace element geochemistry of volcanic sequences of Maryland, Virginia and North Carolina and its bearing on the tectonic evolution of the central Appalachians [Ph.D. thesis]: Lexington, University of Kentucky, 328 p.

Bland, A. E., and Blackburn, W. H., 1980, Geochemical studies on the greenstones of the Atlantic seaboard volcanic province, *in* Wones, D. R., ed., The Caledonides in the U.S.A., International Geologic Correlation Program, Project 27: Caledonide Orogen: Virginia Polytechnic Institute and State University Memoir 2, p. 263–270.

Bland, A. E., and Brown, W. R., 1977, Evington Group (?) of the Arvonia district: Rear arc sedimentation and volcanism: Geological Society of America Abstracts with Programs, v. 9, p. 120.

Bobyarchick, A. R., and Glover, L., III, 1979, Deformation and metamorphism in the Hylas zone and adjacent parts of the eastern Piedmont in Virginia: Geological Society of America Bulletin, v. 90, p. 739–752.

Brown, W. R., 1958, Geology and mineral resources of the Lynchburg quadrangle, Virginia: Virginia Division of Mineral Resources Bulletin 74, 99 p.

Brown, W. R., 1969, Geology of the Dillwyn quadrangle, Virginia: Virginia Division of Mineral Resources, Report of Investigations 10, 77 p.

Brown, W. R., 1970, Investigation of the sedimentary record in the Piedmont and Blue Ridge of Virginia, *in* Fisher, G. W., Pettijohn, F. J., Reed, J. C., and Weaver, K. N., eds., Studies in Appalachian geology—Central and southern: New York, Interscience Publishers, p. 335–349.

Brown, W. R., 1976, Tectonic mélange(?) in the Arvonia slate district of Virginia: Geological Society of America Abstracts with Programs, v. 2, p. 142.

Brown, W. R., and Griswold, T. B., 1970, Superposed folding in the Arvonia slate district, Virginia: Geological Society of America Abstracts with Programs, v. 2, p. 198.

Conley, J. F., 1978, Geology of the Piedmont of Virginia, interpretations and problems, *in* Conley, J. F., ed., Contributions to Virginia geology: Virginia Division of Mineral Resources Publication 7, p. 115–149.

Conley, J. F., 1985, Geology of the southwestern Virginia Piedmont: Virginia Division of Mineral Resources Publication 59, 33 p.

Conley, J. F., and Henika, W. S., 1973, Geology of the Snow Creek, Martinsville East, Price and Spray Quadrangles, Virginia: Virginia Division of Mineral Resources Report of Investigations 33, 71 p.

Costa, R., and Gates, A. E., 1993, Multiple activations of the Ramapo fault system, New Jersey, *in* Puffer, J. H., ed., Geologic traverse across the Precambrian rocks of the New Jersey Highlands: Geological Association of New Jersey Field Guide and Proceedings, v. 10, p. 168–195.

Darton, N. H., 1892, Fossils in the "Archean" rocks in central Piedmont, Virginia: American Journal of Science, Ser. 3, v. 44, p. 50–52.

Dennis, A. J., and Secor, D. T., 1987, A model for the development of crenulations in shear zones with applications from the southern Appalachian Piedmont: Journal of Structural Geology, v. 9, p. 809–817.

Drake, A. A., Jr., 1985, Metamorphism in the Potomac composite terrane, Virginia-Maryland: Geological Society of America Abstracts with Programs, v. 17, p. 566.

Drake, A. A., Jr., Sinha, A. K., Laird, J., and Guy, R. E., 1989, The Taconic orogen, *in* Hatcher, R. D., Jr., Thomas, W. A., Viele, G. W., eds., The Appalachian-Ouachita Orogen in the United States: Boulder, Colorado, Geological Society of America, The Geology of North America, v. F-2, p. 101–177.

Ern, E. H., 1968, Geology of the Buckingham Quadrangle, Virginia: Virginia Division of Mineral Resources Report of Investigations 15, 45 p.

Espenshade, G. H., 1954, Geology and mineral resources of the James River–Roanoke River manganese district, Virginia: U.S. Geological Survey Bulletin 1008, 155 p.

Evans, N. H., 1984, Latest Precambrian to Ordovician metamorphism and orogenesis in the Blue Ridge and western Piedmont, Virginia Appalachians [Ph.D. thesis]: Blacksburg, Virginia Polytechnic Institute and State University, 313 p.

Fitch, T. J., 1972, Plate convergence, transcurrent faults and internal deformation adjacent to southeast Asia and the western Pacific: Journal of Geophysical Research, v. 77, p. 4432–4461.

Gates, A. E., 1981, Geology of the western boundary of the Charlotte belt at Brookneal, Virginia [M.S. thesis]: Blacksburg, Virginia Polytechnic Institute and State University, 88 p.

Gates, A. E., 1986, The tectonic evolution of the Altavista area, Virginia [Ph.D. thesis]: Blacksburg, Virginia Polytechnic Institute and State University, 256 p.

Gates, A. E., 1987, Transpressional dome formation in the southwest Virginia Piedmont: American Journal of Science, v. 287, p. 927–949.

Gates, A. E., and Mose, D. G., 1987, Carboniferous dextral transpression and the resetting of Rb/Sr ages in the Bowens Creek fault, northern extension of the Brevard zone; Eos (Transactions, American Geophysical Union), v. 68, p. 416.

Gates, A.E., and Speer, J.A., 1991, Allochemical retrograde metamorphism in shear zones: An example in metapelites, Virginia, USA: Journal of Metamorphic Geology, v. 9, p. 581-604.

Gates, A. E., Simpson, C., and Glover, L., III, 1986, Appalachian Carboniferous dextral strike-slip faults; An example from Brookneal, Virginia: Tectonics, v. 5, p. 119–133.

Gates, A. E., Speer, J. A., and Pratt, T. L., 1988, The Alleghanian southern Appalachian Piedmont: A transpressional model: Tectonics, v. 7, p. 1307–1324.

Gates, A. E., Lanzirotti, A., Glover, L., III, and Mose, D., 1994, Isotopic evidence for Cambrian tectonism in the crystalline southern Appalachians: The Carolinian, not the Taconian: Geological Society of America Abstracts with Programs, v. 26, p. A15.

Glover, L., III, 1992, A new tectonic model for the central and southern Appalachians, in Glover, L., III, Çoruh, C., Costain, J. K., and Bollinger, G. A., eds., Piedmont seismic reflection study: A program integrated with tectonics to probe the cause of eastern seismicity: U.S. Nuclear Regulatory Commission Report NUREG/CR-5731, p. 1–53.

Glover, L., III, and Sinha, A. K., 1973, The Virgilina deformation, a Late Precambrian to Early Cambrian(?) orogenic event in the central Piedmont of Virginia and North Carolina: American Journal of Science, v. 273-A, p. 234–251.

Glover, L., III, Speer, J. A., Russell, G. S., and Farrar, S. S., 1983, Ages of regional metamorphism and ductile deformation in the central and southern Appalachians: Lithos, v. 16, p. 223–245.

Glover, L., III, Evans, N. H., Patterson, J. G., and Brown, W. R., 1989, Tectonics of the Virginia Blue Ridge and Piedmont: International Geological Congress, 28th, Field Trip Guidebook T363, 59 p.

Goldstein, A. G., 1989, Tectonic significance of multiple motions on terrane-bounding faults in the northern Appalachians: Geological Society of America Bulletin, v. 101, p. 927–938.

Harper, C. T., Russell, C. W., Sherrer, G., Stonebraker, J. D., and Kish, S., 1973, The Arvonia slate—A useful standard for K-Ar dating: Geological Society of America Abstracts with Programs, v. 5, p. 402.

Hatcher, R. D., Jr., 1978, Tectonics of the western Piedmont and Blue Ridge, southern Appalachians: Review and speculation: American Journal of Science, v. 278, p. 276-304.

Hatcher, R. D., Jr., Thomas, W. A., Geiser, P. A., Snoke, A. W., Mosher, S., and Wiltschko, D. V., 1989, Alleghanian orogen, *in* Hatcher, R. D., Jr., Thomas, W. A., and Viele, G. W., eds., The Appalachian-Ouachita Orogen in the United States: Boulder, Colorado, Geological Society of America, The Geology of North America, v. F-2, p. 233–318.

Henika, W. S., 1975, Progressive regional metamorphic gradient in rocks southeast of the Danville Triassic basin, Virginia: Geological Society of America Abstracts with Programs, v. 7, p. 499.

Henika, W. S., and Thayer, P. A., 1977, Geologic maps of the Blairs, Mount Hermon, Danville and Ringgold Quadrangles, Virginia: Virginia Division of Mineral Resources Publication 2, 45 p.

Hibbard, J., 1991, The Carolina slate belt–Milton belt contact in north central North Carolina: Geological Society of America Abstracts with Programs, v. 23, p. 44.

Higgins, M. W., Sinha, A. K., Tilton, G. R., and Kirk, W. S., 1971, Correlation of the metavolcanic rocks in the Maryland, Delaware and Virginia Piedmont: Geological Society of America Abstracts with Programs, v. 3, p. 320.

Higgins, M. W., Sinha, A. K., Zartman, R. E., and Kirk, W. S., 1977, U-Pb zircon dates form the central Appalachian Piedmont: A possible case of inherited radiogenic lead: Geological Society of America Bulletin,

v. 88, p. 125–132.

Horton, J. W., Jr., Drake, A. A., Jr., and Rankin, D. W., 1989, Tectonostratigraphic terranes and their Paleozoic boundaries in the central and southern Appalachians, *in* Dallmeyer, R. D., ed., Terranes in the Circum-Atlantic Paleozoic orogens: Geological Society of America Special Paper 230, p. 213–245.

Hund, E. A., 1987, U-Pb dating of granites from the Charlotte belt of the southern Appalachians [M.S. thesis]: Blacksburg, Virginia Polytechnic Institute and State University, 83 p.

Jonas, A. I., 1927, Geologic reconnaissance in the Piedmont of Virginia: Geological Society of America Bulletin, v. 37, p. 837–846.

Kaldy, W. J., 1977, Geology and geophysics of the Sherwill dome of the southwestern Piedmont of Virginia [M.S. thesis]: Lexington, University of Kentucky, 73 p.

Kish, S. A., Butler, J. R., and Fullagar, P. D., 1979, The timing of metamorphism and deformation in the central and eastern Piedmont of North Carolina: Geological Society of America Abstracts with Programs, v. 11, p. 184.

Lanzirotti, A., and Hanson, G. N., 1997, U-Pb dating of major and accessory minerals formed during metamorphism and deformation of metapelites: Contributions to Mineralogy and Petrology, (in press).

Lister, G. S., and Snoke, A. W., 1984, S-C mylonites: Journal of Structural Geology, v. 6, p. 617–638.

Marr, J. D., Jr., and Sweet, P. C., 1980a, Geology of the Willis Mountain Quadrangle, Virginia: Virginia Division of Mineral Resources Publication 5, scale 1:24,000.

Marr, J. D., Jr., and Sweet P. C., 1980b, Geology of the Andersonville Quadrangle, Virginia: Virginia Division of Mineral Resources Publication 26, scale 1:24,000.

Meyertons, C. T., 1963, Triassic formations of the Danville basin: Virginia Division of Mineral Resources Report of Investigations 6, 65 p.

Mose, D. G., and Nagel, M. S., 1982, Plutonic events in the Piedmont of Virginia: Southeastern Geology, v. 23, p. 25–39.

Odom, A. L., and Russell, G. S., 1975, The time of regional metamorphism of the Inner Piedmont, North Carolina, and Smith River allochthon: Inferences from whole-rock ages: Geological Society of America Abstracts with Programs, v. 7, p. 522.

Passchier, C. W., and Simpson, C., 1986, Porphyroclast systems as kinematic indicators: Journal of Structural Geology, v. 8, p. 831–843.

Patterson, J. G., 1988, Deformation in portions of the distal continental margin to ancestral North America: An example from the westernmost internal zone, central and southern Appalachian orogen, Virginia: Tectonics, v. 8, p. 535–554.

Pavlides, L., 1976, Piedmont geology of the Fredericksburg, Virginia area and vicinity: Geological Society of America, Northeast-Southeast Sections Joint Meeting, Guidebook for Field Trips 1 and 4, 44 p.

Pavlides, L., 1981, The central Virginia volcanic-plutonic belt: An island arc of Cambrian (?) age: U.S. Geological Survey Professional Paper 1231-A, 34 p.

Ramsay, J. G., 1967, Folding and fracturing of rocks: New York, McGraw-Hill, 568 p.

Rankin, D. W., 1975, The continental margin of eastern North America in the southern Appalachians: The opening and closing of the proto-Atlantic Ocean, *in* Ostrom, J. H., and Orville, P. M., eds., Tectonics and mountain ranges: American Journal of Science, v. 275-A, p. 298–336.

Ratcliffe, N. M., 1971, The Ramapo fault system in New York and adjacent northern New Jersey: A case of tectonic heredity: Geological Society of America Bulletin, v. 82, p. 125–142.

Ratcliffe, N. M., 1980, Brittle faults (Ramapo fault) and phyllonitic ductile shear zones in the Ramapo seismic zones, New York and New Jersey, and their relationship to current seismicity, *in* Manspeizer, W., ed., Field studies of New Jersey geology and guide to field trips: New York State Geological Association, v. 52, p. 278–311.

Redden, J. A., 1963, Stratigraphy and metamorphism of the Altavista area, *in* Geological excursions in southwestern Virginia: Virginia Polytechnic Institute Engineering Extension Series Guidebook II, p. 77–99.

Robinson, G. R., Jr., 1979, Pegmatite cutting mylonite—Evidence supporting pre-Triassic faulting along the western border of the Danville Triassic basin, southern Virginia: Geological Society of America Abstracts with Programs, v. 11, p. 210.

Rodgers, J., 1970, The tectonics of the Appalachians: New York, Wiley-Interscience, 271 p.

Seiders, V. M., Mixon, R. B., Stern, T. W., Newell, M. F., and Thomas, C. B., Jr., 1975, Age of plutonism and tectonism and a new minimum age limit on the Glenarm Series in the northeast Virginia Piedmont near Occoquan: American Journal of Science, v. 275, p. 481–511.

Sinha, A. K., Hund, E., and Hogan, J. P., 1989, Paleozoic accretionary history of the North American plate margin (central and southern Appalachians): Constraints from age, origin and distribution of granitic rocks, *in* Hillhouse, J. W., ed., Deep structure and past kinematics of accreted terranes: American Geophysical Union Geophysical Monograph 50, p. 219–238.

Smith, J. W., Millici, R. C., and Greenberg, S. S., 1964, Geology and mineral resources of Fluvanna County, Virginia: Virginia Division of Mineral Resources Bulletin 79, 62 p.

Southwick, D. L., Reed, J. C., Jr., and Mixon, R. B., 1971, The Chopawamsic Formation—A new stratigraphic unit in the Piedmont of northeastern Virginia: Contributions to stratigraphy, U.S. Geological Survey Bulletin 1324-D, p. D1–D11.

Speer, J. A., McSween, H. Y., and Gates, A. E., 1994, Generation, segregation, ascent, and emplacement of Alleghanian plutons in the southern Appalachians: Journal of Geology, v. 102, p. 249–267.

Stose, G. W., and Stose, A. J., 1948, Stratigraphy of the Arvonia slate, Virginia: American Journal of Science, v. 246, p. 393–412.

Streckheisen, A. L., 1973, Plutonic rocks: Classification and nomenclature recommended by the IUGS subcommision on the systematics of igneous rocks: Geotimes, v. 18, p. 26–31.

Swanson, M. T., 1986, Preexisting fault control for Mesozoic basin formation in eastern North America: Geology, v. 14, p. 419–422.

Tillman, C. G., 1970, Metamorphosed trilobites from Arvonia, Virginia: Geological Society of America Bulletin, v. 81, p. 1189–1200.

Vauchez, A., 1987, Brevard fault zone, southern Appalachians: A medium-angle dextral, Alleghanian shear zone: Geology, v. 15, p. 669–672.

Wehr, F., and Glover, L., III, 1985, Stratigraphy and tectonics of the Virginia–North Carolina Blue Ridge: Evolution of a late Proterozoic–early Paleozoic hinge zone: Geological Society of America Bulletin, v. 96, p. 285–295.

Williams, H., and Hatcher, R. D., Jr., 1982, Suspect terranes and accretionary history of the Appalachian orogen: Geology, v. 10, p. 530–536.

Wilson, J. T., 1966, Did the Atlantic close and then reopen?: Nature, v. 211, p. 676–681.

MANUSCRIPT ACCEPTED BY THE SOCIETY APRIL 30, 1996

Printed in U.S.A.

Geological Society of America
Special Paper 314
1997

The mafic-ultramafic association in the Virginia Blue Ridge cover rocks: Rifting sequence or ophiolitic mélange?

Ping Wang and Lynn Glover, III
Department of Geological Sciences, Virginia Polytechnic Institute and State University, Blacksburg, Virginia 24061-0420

ABSTRACT

Current tectonic models of the Blue Ridge propose that at least part of the post-Grenville sequence is ophiolitic mélange that was obducted onto Laurentia, thereby implying a terrane boundary or suture of Late Proterozoic to Ordovician age within the Blue Ridge. However, the tectonic lithofacies assemblage of the post-Grenville Moneta/Lynchburg/Catoctin sequence suggests a rifted margin model of Late Proterozoic age.

The Late Proterozoic Lynchburg Group in central and northern Virginia has an intact stratigraphy in which a single stratigraphic unit (i.e., graphitic schist) can be traced throughout a large region. A newly recognized rock unit in the Lynchburg, the Moneta Formation, occurs at the base of the Group. The formation is dominated by mafic and felsic volcanic rocks, with intercalated lenses of conglomerate and sandstone, interpreted as alluvial fan deposits. The igneous rocks of the Moneta are lithologically and geochemically bimodal, somewhat alkaline, and similar to rift-related magma suites.

The clastic rocks within the Lynchburg all share a common suite of quartzo feldspathic detrital grains that are similar to the Blue Ridge basement rock from which they were derived. Intrusive contacts between the mafic-ultramafic rocks and their sedimentary host rocks were observed at a number of places. Where mafic and ultramafic rocks occur together, they locally alternate, whereas in other places ultramafic rocks occur in the center and mafic rocks near the margin of the mafic-ultramafic body. The field relations do not support a disrupted or intact ophiolite stratigraphy. The microstructures of the Lynchburg are relatively simple; S_0 (bedding), S_1 (pervasive cleavage), and S_2 (spaced cleavage). We did not find mélange in the Lynchburg sequence.

A number of tectonomagmatic discriminant diagrams have been used to support the Blue Ridge ophiolite/suture tectonic model, and these diagrams are considered one of the most important arguments for the presence of ophiolites. We find that these diagrams fail to distinguish rift-related basaltic rocks from those formed in other tectonic settings. It is suggested, therefore, that geochemical data alone may give an incorrect tectonic classification and that a knowledge of field relations is of paramount importance.

Wang, P., and Glover, L., III, 1997, The mafic-ultramafic association in the Virginia Blue Ridge cover rocks: Rifting sequence or ophiolitic mélange? *in* Glover, L., III, and Gates, A. E., eds., Central and Southern Appalachian Sutures: Results of the EDGE Project and Related Studies: Boulder, Colorado, Geological Society of America Special Paper 314.

We conclude that the mafic-ultramafic suite and stratigraphically associated terrigenous clastic rocks in the Virginia Blue Ridge cover are part of an intrusive-extrusive sequence of mafic and ultramafic rocks related to a Late Proterozoic Iapetan rifting event rather an ophiolite-bearing sequence associated with a suture formed during the Taconian (mid-Ordovician) or earlier orogeny.

INTRODUCTION

Since Hess (1939, 1955) first recognized two mafic-ultramafic belts in the Appalachians, the origin and tectonic significance of the belts have been topics of controversy. Before the birth of plate tectonics, explanations for these mafic-ultramafic rocks ranged from fractional crystallization of basaltic magmas to diapiric intrusion (Hess, 1933, 1939, 1955; Bloomer and Werner, 1955; Brown, 1958; Espenshade, 1954). Since 1950, geophysical exploration of the ocean floor and the Deep Sea Drilling Project have led to a better understanding of the lithology, structure, and composition of oceanic lithosphere. With this new knowledge, it did not take long to recognize similarities between some of the ultramafic-mafic associations in the Blue Ridge Appalachians and the newly constructed geophysical models of oceanic lithosphere and their possible obducted equivalents called "ophiolite."

The Blue Ridge suture model, based mostly on studies in the northern and southern Appalachians, considers at least part of the post-Grenville basement sequence to be collisional ophiolitic mélange, thus creating a terrane boundary, usually in this case considered a suture (Fig. 1), of Late Proterozoic to Ordovician age (Hatcher et al., 1984; Abbott and Raymond, 1984; Conley, 1985; Hatcher, 1989; Rankin, 1989; Horton et al., 1989; Stanley and Ratcliffe, 1985; Keppie and Dallmeyer, 1989). The major observations that have been used to justify these models

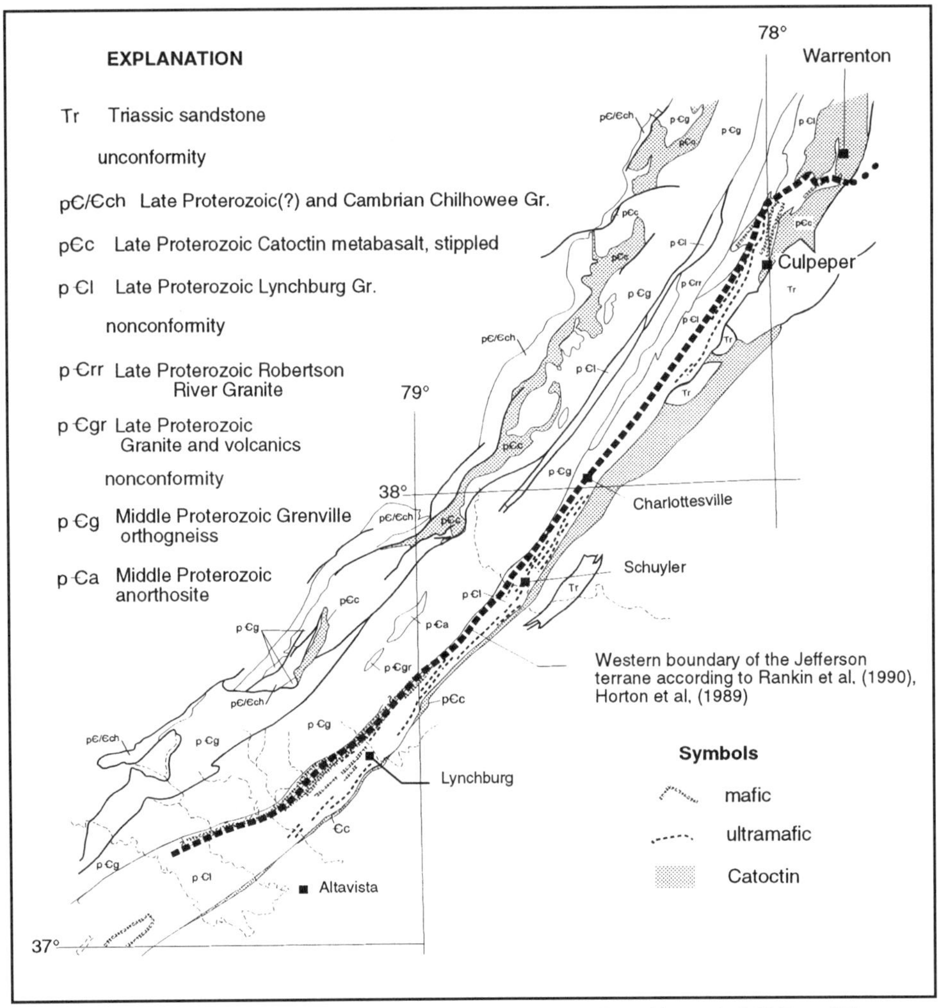

Figure 1. Virginia Blue Ridge, showing terrane boundary of Rankin et al. (1990), and Horton et al. (1989).

include the following: (1) Coeval mafic and ultramafic rocks exist in a linear distribution along the Blue Ridge and its analogs in the northern Appalachians and are generally interpreted as ophiolites. (2) Some of the mafic rocks plot in midocean ridge basalt (MORB) and/or ocean island basalt (OIB) and island arc basalt (IAB) fields on several tectonic setting discrimination diagrams using both major and trace elements. (3) Intense deformation in the rocks where most studies were carried out has commonly been interpreted as tectonic mélange structure.

However, recent studies of the Virginia post-Grenville cover rocks, the Lynchburg Group, show that they are a typical rifted-margin sequence. In northern Virginia, sedimentological studies of the Lynchburg (Wehr, 1983; Wehr and Glover, 1985; Kline, 1991) have shown that the clastic component is made up of braided river–delta–submarine fan deposits that would be typical of a rifted-continental-margin sequence derived from erosion of the Virginia Blue Ridge Middle Proterozoic Grenville basement rocks. "Melange structure" has not been demonstrated anywhere in the Lynchburg Group.

These two views—collisional mélange versus rifted-margin sequence—could not be further apart, and the choice between them is a first-order problem of Appalachian plate modeling. Therefore, it is the purpose of this study to determine which hypothesis is correct by examining the mafic-ultramafic rocks in detail, including their field relationships, geochemical characteristics, and contacts with country rocks as well as the paleoenvironments of the country rocks and the deformation characteristics of the sequence.

GENERAL GEOLOGY OF THE STUDY AREA

The study area is located on the southeastern limb of the Blue Ridge anticlinorium in central Virginia. The cover rocks unconformably overlying the basement are metamorphosed volcanic and subvolcanic rocks, conglomerates and sandstones of the basal Lynchburg Moneta Formation, and the metasedimentary rocks and mafic-ultramafic rocks of the higher Lynchburg Group formations I, II, and III (Fig. 2). A general summary of the stratigraphy, petrology, metamorphism, and structure of the post-Grenville, Lynchburg/Catoctin/Evington sequence is given in Table 1. The Lynchburg Group, in its type locality at Lynchburg City and to the southwest in the Otter River area, contains abundant mafic and ultramafic rocks. Well-exposed contact relations between mafic and ultramafic rocks, as well as with the clastic host rocks, can be observed in a number of places. This area is an ideal location in which to conduct research on paleotectonic reconstructions because the post Grenville sequence in central and northern Virginia is much less deformed and metamorphosed than equivalents to the north and south, where the ophiolitic mélange tectonic models were generated (Hatcher et al., 1984; Rankin et al., 1990; Horton et al., 1989; and numerous others). Detailed field mapping at 7.5-min scale was done in the Lynchburg City and Otter River areas and connected by reconnaissance (Fig. 2).

STRUCTURE, STRATIGRAPHY, AND PETROLOGY OF THE HOST ROCKS

The lower Lynchburg Moneta Formation nonconformably overlies basement rocks. A syndepositional faultline scarp occurs at the base of the Moneta about 6.5 km southwest of the James River (Fig. 2). Basement boulders up to 1.5 m in diameter occur adjacent to the scarp, and basal conglomerates with boulder-sized basement gneiss clasts are found in several places. The formation is dominated by mafic (50 to 70%) and felsic (10 to 15%) metaigneous rocks and is intercalated with lense-shaped conglomeratic (some boulder-sized) rocks (10 to 15%) and fine-grained sandstones (about 5 to 15%). Some of the sandstones are compositionally transitional to the closely associated mafic rocks and appear to be mafic volcaniclastic and volcaniclastic-epiclastic sedimented layers. The Moneta mafics (amphibolites) are fine- to coarse-grained, massive, shallow intrusive sills/dikes or schistose volcaniclastic rocks. Moneta felsites are fine to very fine grained, and some show relict porphyritic texture. The clasts in the conglomeratic rocks are composed of basement gneiss fragments, quartz, and feldspar grains. The clasts are subangular to subrounded and are poorly sorted. The clast size varies from 2 mm to as large as 150 cm. The matrix is mainly fine- to coarse-grained arkose or feldspathic quartz arenite with some clay (now mica). Based on the evidence of syndepositional faulting, deposition on a nonconformity, intercolated bimodal volcanics, arkosic clast compositions, and lens-shaped field geometry (Fig. 2) of these conglomeratic rocks, an alluvial fan depositional environment in a nonmarine rift basin is postulated.

As the host for most of the post-Grenville mafic-ultramafic bodies, the upper Lynchburg Group (formations I, II, and III) has a distinctive, undisrupted stratigraphy in which rock units (such as the Lynchburg II graphitic schist) can be traced 50 km or more throughout the region and beyond. The sedimentary host rock sequence shows a record of continuous basin evolution (Table 1 and Fig. 3). The cover sequence in central Virginia is similar to that demonstrated in northern Virginia by Wehr (1985). Regular bedding can be seen at each outcrop of metasedimentary rocks. Other primary sedimentary structures, such as graded bedding and scoured surfaces, have also been observed in many places. Lynchburg clastic rocks share a common suite of distinctive detrital grains (feldspathic quartz arenites) that are fully compatible with derivation from the Blue Ridge basement granitoid gneisses. Transitional (mafic/quartz arenite mixed) compositions and conformable contacts exist among effusive (mafic lava), pyroclastic, and epiclastic rocks, and terrigeneous sandstones occur in many places. The sedimentary environments represented by the cover sequence are interpreted to be alluvial fan-slope apron–channelized submarine fan–distal submarine fan (the Evington Group). There is no intense shearing, and observations at scales from outcrop to regional map (Fig. 2) show no chaotic bedding or mixture of different rock types that would be expected in a mélange.

On a regional scale, the Lynchburg Group has the simple

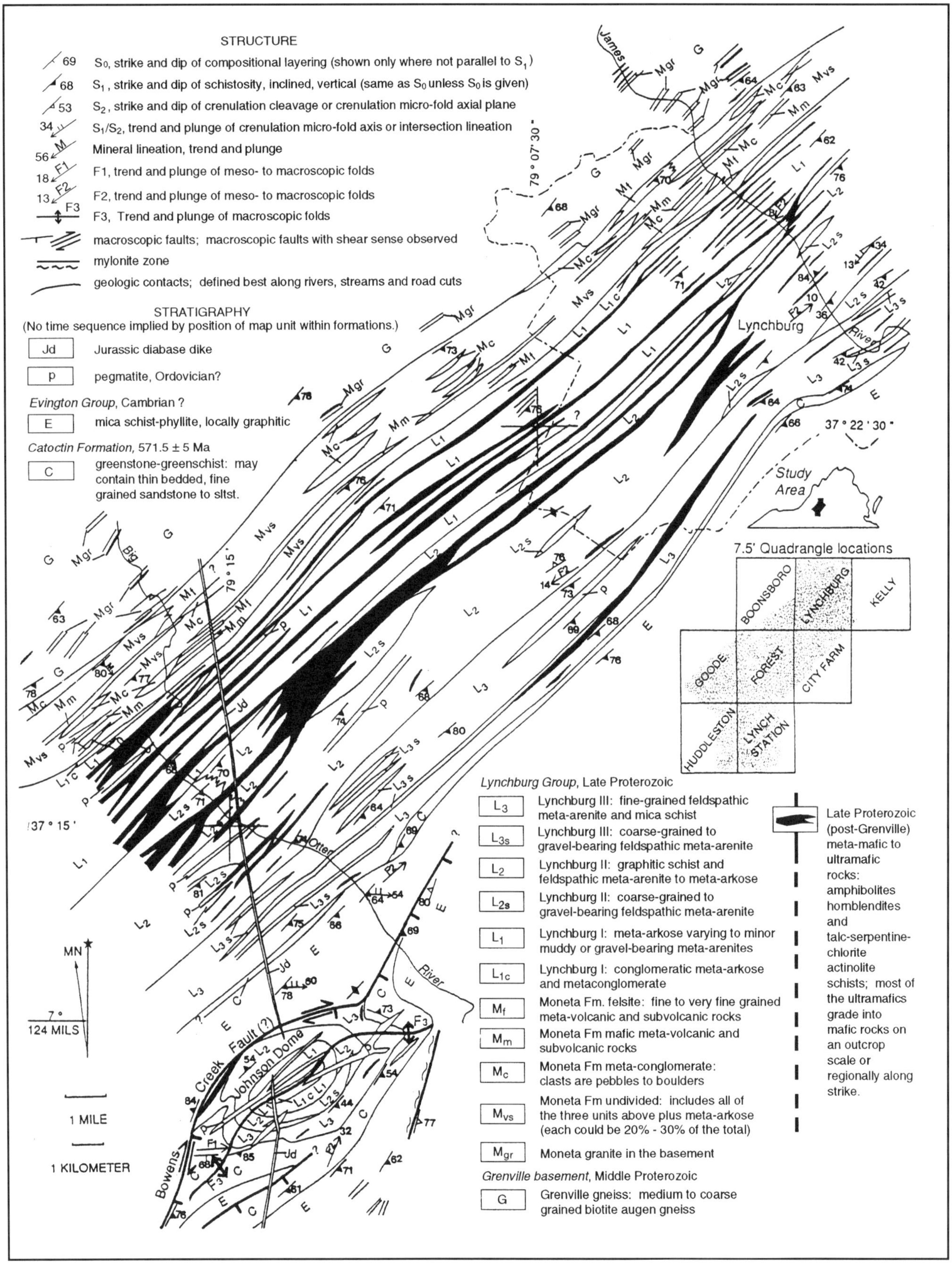

Figure 2. Geologic map of Lynchburg–Otter River area, central Virginia.

structural style of a southeast-dipping monoclinal sequence, although small-scale folds do exist locally. The folds are small-scale F1 isoclines and F2 open folds with well-preserved bedding and other primary sedimentary structures, such as lamination and graded beds. The microstructures of most of the sedimentary rocks are relatively simple, that is, S_0 (bedding), S_1 (pervasive cleavage), and S_2 (spaced crenulation cleavage), which was observed only in the pelitic rocks of Lynchburg II and higher.

Several authors have proposed that a suture or terrane boundary occurs within the study area (Rankin et al., 1990; Horton et al., 1989; Hatcher, 1989; Keppie and Dallmeyer, 1989). To test this model, a corridor including outcrop along both sides of the James River was mapped in detail. Our study shows that the rock record is unbroken, contacts between different rock units are conformable, major contacts can be traced from this corridor for tens of kilometers, and the whole sequence is essentially monoclinally southeast dipping. This is confirmed by graded beds that are virtually all upright and face southeast. There is no evidence of intense stratigraphic disruption or faulting on a suture-zone scale. Similar conclusions have been drawn from detailed mapping that also crossed the proposed "suture" in the Otter River area.

THE MAFIC AND ULTRAMAFIC ROCKS

The mafic-ultramafic rocks discussed here include the amphibolites and hornblendites of the basal Lynchburg Moneta Formation, which directly overlies the basement; the amphibolites associated with ultramafics in the Lynchburg Group; and the Catoctin and Slippery Creek (Evington Group) greenstones (Fig. 2).

General petrology

The mafic and felsic rocks of the Moneta are interlayered or mutually intrusive in the field, and their lithological, as well as geochemical, distributions are clearly bimodal. In terms of chemical composition and rock assemblage features, both mafic and felsic rocks in the Moneta are very similar to the late Precambrian rift volcanic rocks of the Blue Ridge Crossnore volcanic-plutonic suite of Rankin et al. (1973) and to rift volcanic rocks globally.

The mafic-ultramafic association in the Lynchburg Group Formations I, II, and III is dominated by mafic rocks (amphibolites and amphibolite schists), with a minor amount (ca. 10%) of ultramafic rocks (mostly talc-serpentine schist, amphibole-chlorite schist, and hornblendite). No ultramafic rock has been found in and above the Catoctin Formation, which is composed of greenstones-greenschists with a few locally intercalated, thin-bedded, fine feldspathic quartz arenites and schists.

Amphibolites in Lynchburg formations I, II, and III

The mafic rocks in the Lynchburg Group are dominated by relatively massive amphibolites. Amphibolite schists with the same mineral composition are seen in thin (1 to 2-m) sills and dikes. Most of the mafic rocks are interpreted to be intrusives, mostly sills, though local discordances occur in outcrop and at the map scale. Tuffs and, perhaps, lavas and volcaniclastic/epiclastic rocks also exist, especially in the lower part of the Lynchburg. The amphibolites are typically dark gray to dark brown, medium- to coarse-grained (may become fine grained near the contacts), and massive or schistose. The common mineral assemblage is plagioclase (40 to 60%) + hornblende (40 to 60%) + titanite (1 to 3%) ± magnetite (<2%) ± quartz (< 5%). Most of the hornblende crystals are poikiloblastic and display a preferred orientation.

The amphibolites believed to be volcanic are thin- to medium-layered, dark gray to dark brown, commonly laminated, and interlayered with metaarkose. A typical mineral assemblage is plagioclase (40 to 50%) + hornblende (30 to 40%) + biotite (10 to 15%) + quartz (ca. 5%) + titanite (<2%) ± magnetite.

Some mafic tuffaceous sandstones are found to be interbedded with metaarkose. They are light gray, medium-fine grained, and thin-layered and have a typical mineral assemblage of quartz (45 to 55%) + plagioclase (35 to 45%) + hornblende (5 to 10%) ± biotite ± muscovite ± garnet ± apatite ± magnetite.

Ultramafic rocks in Lynchburg formations I, II, and III

Talc-serpentine schists. These rocks are grey to greenish and commonly schistose and display a wide range of mineral compositions; however, serpentine and talc are always present. A typical mineral assemblage contains serpentine (30 to 65%) + talc (20 to 25%) + cummingtonite/tremolite (20 to 30%) + magnetite (1 to 5%) ± chlorite (<10%) ± anthophylite ± pyroxene. Most serpentines have fabric texture (antigorite), and relict pyroxenes (both orthopyroxene and clinopyroxene) occur locally.

Amphibole-chlorite schists. These are pale to dark green, fine- to medium-grained, massive to schistose rocks. A typical mineral assemblage has chlorite (40 to 60%) + amphibole (35 to 45%) + talc (1 to 5%) + magnetite/ilmenite (<3%) ± calcite ± serpentine. The amphiboles are mostly actinolite and/or tremolite, but hornblende is present in some samples. This rock type, called "Virginia Greenstone," was quarried for almost 200 yrs in the study area.

Hornblendites. These are dark brown to black, massive and typically coarse grained rocks. They all contain a mineral assemblage of hornblende (85 to 90%) + plagioclase (10 to 15%) + magnetite/ilmenite (0 to 2%). The rock commonly contains large hornblende crystals (0.5 to 2.5 cm). Most hornblendites are associated with mafic sills or dikes, and the contacts between the two may be sharp or gradational.

Different types of the ultramafic rocks (such as talc-serpentine schist and hornblendite) can be found in a single sill or dike, both along or across strike. Their contacts are mostly gradational.

Catoctin greenstones—greenschists

These rocks are green, fine-grained, and massive to bedded and generally display a well-developed schistosity. They are most commonly made up of actinolite (30 to 60%) + albite

TABLE 1a. STRATIGRAPHY AND PETROLOGY OF THE COVER SEQUENCE IN THE BLUE RIDGE OF CENTRAL VIRGINIA

Rock Unit and Age	Rock Types	Protolith Features Preserved	Current Mineral Composition	Interpretation
Evington group (Cambrian?)	Mica schist-phyllite	Locally graphitic, bedding-parallel sandy laminations common.	Quartz (40-55%)+muscovite (25-40%)±albite (5-15%)±chlorite±biotite±garnet±graphite.	Rifting-drifting sequence: distal submarine fan deposits.
	Quartz arenite	Light gray to whitish gray, medium- to coarse-grained, thick to massive bedded.	Quartz (75-95%)+plagioclase (5-15%)+muscovite (3-5%)+biotite (2-5%) ±magnetite/ilmenite±titanite.	
Catoctin Formation, Late Proterozoic (571.5 ± 5 Ma), Aleinikoff et al., 1991	Greenstone-greenschist	Massive to bedded; albite phenocrysts common, locally with amphibole/pyroxene phenocrysts; mostly lavas, some pyroclastics (laminated with 3-5% biotite) in the Altavista dome area.	Actinolite (30-60%)+albite (25-45%)+epidote (5-20%)+chlorite (2-10%)+magnetite/ilmenite (≤2%) ±quartz±titanite±biotite.	Subaqueous late-stage rifting volcanism and submarine turbidites.
	Metasandstone-mica schist	Greenish to gray, occurs as fine-grained, thin-bedded intercalations in the greenstones.	Quartz (40-65%)+plagioclase (10-25%)+muscovite (5-45%)+biotite (3-10%)+magnetite/ilmenite (trace-3%)±chlorite±granet±apatite±graphite.	
Lynchburg Group Late Proterozoic				Rifting sequence:
Lynchburg III	Fine-grained meta-arenite and mica schist	Thin- to medium-bedded, graded bedding, and bedding-parallel laminations are common; sorted to well sorted; commonly comprises the middle to upper part of upward-fining sequences.	Quartz (35-65%)+plagioclase (10-35%)+muscovite (5-50%)+biotite (3-15%)+ilmenite (trace-3%)±K-feldspar (0-15%)±garnet (0-5%)±epidote (0-7%)±titanite (0-2%)±apatite (0-1%)±graphite (0-1%).	Submarine fan overbank-interchannel turbidites.
	Coarse-grained to gravel-bearing feldspathic meta-arenite	Light gray to gray, thin to thick bedded, some beds are graded; commonly comprises the lower parts of upward-fining sequences; otherwise occurs as lense-shaped enclaves in the finer-grained rocks.	Similar composition as the sandstones, but more feldspathic (plagioclase 25-45%); gravel fraction is mainly quartz and plagioclase; some perthite and granitic rock fragments also occur.	Turbidites; submarine fan channel-fill deposits.
Lynchburg II	Graphitic schist and graphite-bearing meta-feldspathic arenite	Gray to black, mostly thin to medium bedded, locally thick bedded to massive; bedding-parallel laminations and graded bedding are common; arenites are mostly fine-grained and locally medium-coarse grained in the lower part of a bed/sequence.	Quartz (35-65%)+plagioclase (5-35%)+muscovite (3-35%)+biotite (3-15%)+graphite (1-12%)±garnet±epidote±magnetite/ilmenite±titanite±apatite±calcite±K-feldspar (some arenite are graphite free).	Anoxic basin overbank interchannel turbidites.
	Coarse-grained to gravel-bearing meta-feldspathic arenite	Light gray to gray, thick bedded to massive; mostly are lens-shaped and occur as basal parts of upward-fining sequences/beds; graded bedding and scoured basal contacts are common; lens-shaped outcrop pattern.	Quartz (45-65%)+plagioclase (20-35%)+biotite (8-10%)+muscovite (7-10%)±graphite±garnet±epidote ±magnetite/ilmenite±titanite±apatite±K-feldspar; graphite if present occurs as small graphitic schist chips.	Turbidites; submarine fan channel-fill deposits.
Lynchburg I	Feldspathic meta-arenites	Light gray to gray, mostly medium to coarse grained, with minor muddy to gravel-bearing sandstone end members; commonly massive to thick bedded, moderately to well sorted; graded bedding and basal scour surfaces are common; a few layers of volcanic-epiclastic (mafic and felsic) rocks also occur.	Quartz (40--55%)+plagioclase (35-45%)+biotite (8-15%)±perthite (0-7%)±muscovite (0-5%)±epidote (0-7%)±garnet (0-3%)±magnetite/ilmenite±titanite±apatite ±K-feldspar.	Turbidites; slope-apron high-density turbidity current deposits.

TABLE 1a. STRATIGRAPHY AND PETROLOGY OF THE COVER SEQUENCE IN THE BLUE RIDGE OF CENTRAL VIRGINIA (continued)

Rock Unit and Age	Rock Types	Protolith Features Preserved	Current Mineral Composition	Interpretation
Lynchburg Group (Late Proterozoic) (continued)				Rifting sequence (continued)
Lynchburg I	Meta-conglomeratic arkose and meta-conglomerate	Thick bedded to massive, normal graded bedding is common, locally with reversed grading; some grade laterally and upward into Lynchburg I feldspathic meta-arenites; lens-shaped outcrop pattern; some flattened clasts are the results of later deformation.	Gravel-pebble sized clasts (10-35%) are made up of quartz, perthite, plagioclase, and locally felsite; the matrix has quartz (45-55%)+plagioclase (20-30%)+perthite (5%±)+biotite (10%±) ±muscovite±epidote ±garnet ±ilmenite±titanite±apatite±K-feldspar.	Slope apron grain flow-high density turbidity current deposits.
Moneta Formation of Lynchburg Group. Late Proterozoic	Felsite (meta-volcanic to subvolcanic rocks)	Fine to very fine grained, massive to layered, relict phenocrysts of quartz, plagioclase, and K-feldspar are present; some (subvolcanic) bodies show clearly intrusive contacts, whereas others (volcanic) are interlayered with volcaniclastic-epiclastic rocks.	Quartz (20-30%)+plagioclase (15-30%)+microcline (20-35%)+biotite (5-8%)±muscovite±garnet±magnetite/ilmenite±apatite±perthite±fluorite.	Early stage rifting volcanism and alluvial fan sedimentation.
	Mafic meta-volcanic to subvolcanic rocks (amphibolites)	Dark gray to dark brown, fine to medium grained, layered to massive, some show clearly intrusive contacts with all kinds of rocks in the formation, others are laminated and interlayered with, or grade into, Moneta volcanic-epiclastic rocks.	Hornblende (50-65%)+plagioclase (25-45%)+magnetite/ilmenite (≤2%)+titanite±quartz±biotite.	
	Feldspathic meta-sandstones	Light gray, thin to thick bedded, fine to coarse grained, locally gravel-bearing or muddy (schist now), bedding-parallel laminations are common; with minor interlayered thin-bedded biotite schist.	Quartz (40-50%)+plagioclase (25-35%)+perthite (3-8%)+biotite (10-20%)±epidote±garnet±titanite±hornblende±magnetite/ilmenite±apatite.	
	Meta-conglomerates	Light gray, thick bedded to massive, poorly sorted; pebble to boulder clasts mixed with sands and mud; lack of internal textures; show lens-shaped outcrop pattern.	Clasts (30-60%) are made of quartz, feldspars, and basement augen gneiss; the coarse sand to mud matrix is arkosic in composition (similar to above).	
Grenville basement. Middle Proterozoic (≥1.1 b.y.)	Biotite augen gneiss	Light gray to light brown, medium to coarse grained, massive, predominantly biotite augen gneiss; also in the basement are minor amounts (≤5%) of quartz-biotite amphibolite schist, calcite-biotite schist, epidote-biotite schist, and metagranite.	Quartz (25-30%)+plagioclase (20%±)+perthite (10-15%)+biotite (5-15%)+K-spar (10%±)+epidote (5-10%)+titanite (≤5%)±garnet±magnetite/ilmenite.	Prerifting continental basement.

TABLE 1b. STRUCTURE AND METAMORPHISM OF THE BLUE RIDGE COVER SEQUENCE IN CENTRAL VIRGINIA

Age*	Taconian	Acadian-Alleghanian
	M_1: Prograde metamorphism; P-T increase downsection and southwestward along strike. Metamorphic facies observed include amphibolite facies (540-630 °C, 7-8 kb) — epidote amphibolite facies (480-560 °C, 7-8 kb) — greenschist facies.	M_2: Retrograde metamorphism at lower green schist facies with chlorite and/or muscovite cross-cutting preexisting M_1 fabrics.
	D_1: A pervasive foliation (S_1) defined by mica flakes and flattened quartz-feldspar aggregates in the metasedimentary rocks and by well-oriented plagioclase and amphiboles in the metamorphosed mafic rocks; almost all S_1 surfaces are parallel to the original bedding (S_1); isoclinal, outcrop-scale F_1 flow folds are present in the lower part of the sequence; no D_1 fault of regional scale is found.	D_3: The formation of Johnson Dome and transpressional movement of the Bowens Creek Fault (?) are believed to be the result of D_3 deformation; no D_3 cleavage or other microscopic structures are observed.
	D_2: Characterized by crenulation cleavage (S_2) in the more micaceous rocks of middle to upper Lynchburg Group; mesoscopic flexural folding, small-scale faulting are also common; most S_2 are subparallel to the D_1 structures.	

*Ages of metamorphism and deformation are taken from Glover et al., 1983.

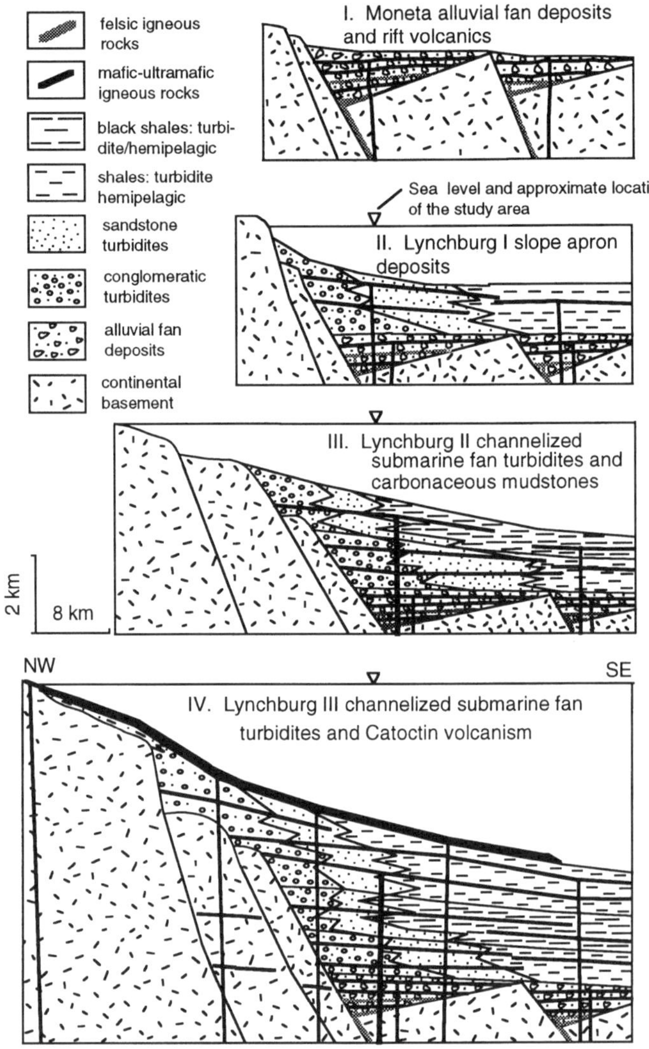

Figure 3. Basin evolution history of the Late Proterozoic cover rocks in the Blue Ridge of central Viginia. The direction of the profile NW-SE is the present-day orientation, and the scales are only approximate.

(25 to 45%) + epidote (5 to 20%) + chlorite (2 to 10%) + magnetite/ilmenite (≤2%) ± quartz ± titanite ± biotite. Plagioclase phenocrysts are common. Locally, amphibole (originally pyroxene) phenocrysts are also preserved. Some of the greenstones are laminated (with 3 to 5% biotite) and are compositionally transitional with sedimentary rocks, suggesting a volcanic-epiclastic origin. Catoctin greenstones-greenschists are believed to be submarine volcanics (lavas mostly), based on the presence of relict phenocrysts, local amygmoidal texture, and interlayering with marine sedimentary rocks.

The Catoctin has a Rb-Sr age of 570 ± 36 Ma (Badger and Sinha, 1988), and dikes associated with it have a zircon U-Pb age of 571.5 ± 5 Ma (Southworth et al., 1994; and personal communication). The Grenville basement is considered to be ca. 1 Ga (Bartholomew, 1984) Recent work on the timescale shows that the Cambrian began about 540 Ma (Bowring et al., 1993). Therefore the Lynchburg Group and Catoctin Formation are Late Proterozoic sequences.

Field relations and petrology compared with ophiolites

The stratigraphy and structure of ophiolites and their contacts with the country rocks are well known (Coleman, 1977; Panayiotou, 1980; Gass et al., 1984; Nicolas, 1989; Saunders and Norry, 1989). A complete ophiolite found on land today is commonly made up of lower ultramafic mantle tectonites, upper ultramafic cumulates, lower mafic cumulates, upper massive gabbros, sheeted dikes, basaltic pillow lavas, and deep marine pelagic sediments. This represents a normal or fast-spreading ridge sequence. Ophiolite formed from a slow spreading mid-ocean ridge may show a different rock assemblage from the normal or fast-spreading center described above. A recent study (Dick, 1989) showed that oceanic crust formed along a slow-spreading center contains a little gabbro but is dominated by serpentinite overlain by basaltic lavas. Near the center of a ridge segment between fracture zones, a more complete ophiolite stratigraphy, similar to that of fast-spreading centers, may exist. In any case, however, the ultramafic unit is at the bottom of the sequence, and the basalt lavas are at the top. Based on plate tectonic theory, ophiolite was originally formed along a midocean ridge or in a back-arc basin. The ophiolite bodies on land, therefore, must have been thrust (obducted) into position during collision, and their basal contact relations confirm this. Thus, every ophiolite body should have a tectonic (fault) contact with its underlying host rocks and perhaps at its top if the overlying sequence is not oceanic. A suture may be defined by locating the landward contacts between ophiolite and the passive continental margin or the intervening subduction mélange if that has survived the effects of collision.

Figure 4. Intrusive contact ("lit-par-lit" injection) between a mafic sill and Lynchburg I sandstone. Scale: 1:5. Railroad cut, west bank of James River opposite north end of Woodruff Island, Lynchburg, Virginia.

How the mafic-ultramafic bodies in the Lynchburg differ from those found in classic ophiolites

Most mafic-ultramafic bodies in the Blue Ridge of central Virginia are conformable sills, though dikes do occur locally. Most contacts are concordant and show no intense deformation or shearing on either macroscopic (outcrop) or microscopic (thin-section) scales. The intrusive contacts between the mafic-ultramafic rocks and sedimentary host rocks that have been observed at a number of field localities are characterized by the following features.

Fine-grained (chilled) margins. Where relict igneous texture is preserved, it is most common to find that the central part of the intrusion is medium to coarse grained, whereas the margins on either side of the sill or dike are fine grained.

Cross-cutting branches. Although the intrusions are generally parallel to bedding (sills), thin (centimeter to meter scale) branches of the intrusion are commonly found near the contact cutting across the bedding of the country rocks and then becoming parallel to it again, in a kind of "lit-par-lit" structure (Fig. 4).

Contaminated fronts. Migmatitic textures and mineral assemblages representing transitional compositions occur in many outcrops near the contacts with sedimentary rocks. These appear to be contact metamorphic effects.

Country rock xenoliths. Some of the mafic-ultramafic intrusions contain country rock xenoliths of centimeters to hundreds of meters in long dimention. The one in the Lynchburg City Quarry, for example, is about 60 m wide and at least 400 m long and occurs inside a large mafic-ultramafic intrusion. It is possible also that the xenolith is a screen that was attached to country rock. The xenolith is made up of Lynchburg II graphitic schist and is completely enclosed in mafic-ultramafic rocks. The xenolith contact is not exposed, but it can be constrained to within about 6 m locally. The graphitic schist xenolith, which is relatively sensitive to deformation, shows well-preserved bedding and laminations. No structures indicative of strong shearing or mélangelike broken rock blocks have been found.

Regional outcrop pattern. On a larger scale, the majority of the mafic-ultramafic bodies show a typical, thin but continuous sill-like outcrop pattern. In other places the sills branch and form a regional scale "lit-par-lit" structure, which is especially obvious for sills near the Lynchburg I and II boundary along the James and Otter Rivers (Fig. 2).

Mafic and ultramafic rocks occuring together do not resemble an ophiolite stratigraphy

Gradation. Through detailed field mapping it has been found that most of the ultramafic rocks grade into mafic rocks both along strike and oblique to strike. The contacts between these ultramafic and mafic rocks are so irregular and complicated that they are difficult to portray even at a map scale of 1:24,000. As a result, the mafic and ultramafic rocks are mapped (Fig. 2) as one unit, and the rock types observed in each outcrop are marked as "m" (mafic), "u" (ultramafic), or "m-u" (both).

Differentiation. Differentiation within one sill or dike is common. At several locations—for instance, at Blackwater Trail, 0.5 km north of Dearington, Lynchburg City—one can see a gradational change upward from ultramafic rocks into mafic rocks and locally from mafic into intermediate rocks. This change can be observed within a thickness of less than 100 m. The change from ultramafic to mafic rocks would occur in an ophiolite over much greater thickness, and intermediate rocks would not be present.

Location. The ultramafic rocks of the Lynchburg Group occur in many places in the center of the sill or dike, and the mafic rocks occur near the margin of the body. This could be a result of flow differentiation during magma intrusion and crystallization. In contrast, at other localities, one rock type may become an enclave within the other, and their contacts may be either sharp or gradational (within 30 cm). As observed in the WalMart site along Highway 29 South, Lynchburg City, irregularly shaped mafic enclaves of 1 to 10 m in diameter are found within the ultramafic domain, and similar sized ultramafic enclaves are found within the mafic domain (Figs. 5 and 6).

This may have been the result of multiple intrusions' occurring before the rocks were completely crystallized.

Presence of mafic rocks. Mafic dikes and sills are also found in the basement. These mafic rocks have a similar mineral assemblage, chemical signature, and deformation and metamorphic history as those in the Moneta Formation and Lynchburg Group, suggesting a similar source and origin. The basement–Lynchburg Group contact is sedimentary, not faulted, and this implies that the mafics and ultramafics in the Lynchburg traveled through the basement and are continental, not oceanic.

The suture-mélange proponents have used the westernmost known occurrence of ultramafic rock to locate the western boundary of the suture or mélange but have not presented any other evidence to support their extension of such a boundary through the Lynchburg–Otter River region. Therefore, the field relations and petrology of the mafic and ultramafic rocks described here do not in any way support either a disrupted or an intact ophiolite stratigraphy. They do, however, support a dike-sill-lava relationship in a continental rifting environment.

Figure 5. Contacts between mafic and ultramafic rocks within Lynchburg III. Cut behind WalMart store, along Rt. 29, in south Lynchburg. The mafic rocks are light gray and the ultramafic dark gray in color. Notice the irregular shapes of the mafic inclusions within the ultramafic body. Most contacts are sharp but some are gradational. The hammer is about 45 cm long.

Figure 6. Contact between mafic and ultramafic rocks within the Lynchburg III. Cut behind WalMart store, along Rt. 29, in south Lynchburg. The mafic rocks are light gray and the ultramafic dark gray. An ultramafic inclusion within a mafic body. Most of the contact is sharp but locally is gradational. The hammer is about 45 cm long.

GEOCHEMISTRY, PETROGENESIS, AND THE DISCRIMINANT DIAGRAMS

Bulk chemical analysis of 30 samples was carried out using the inductively coupled plasma (ICP) method. These samples include four amphibolites from the basement, six amphibolites and one hornblendite from the Moneta Formation, and four Catoctin and three Slippery Creek (Evington Group) greenstones (Table 2). Based on field observations, petrography, and chemistry, all of the samples analyzed are of igneous origin. The Moneta amphibolites sampled show obvious intrusive contacts, massive structure, and local relict igneous textures. Plagioclase phenocrysts are common in the Catoctin and Slippery Creek greenstones. Granite xenoliths are found in the Slippery Creek greenstones near the sampling sites along Highway 460 near Oxford Furnace, east of Lynchburg City. One of the Slippery Creek greenstones analyzed showed wavey laminations, suggesting a pyroclastic origin. In order to confirm the protolith of the sampled amphibolites, two trace element diagrams of Cr versus TiO_2 and Ni versus TiO_2 (Leake, 1964) are plotted (Fig. 7). Most of the samples plot outside the pelite (para-amphibolite) field in both diagrams, indicating an igneous origin. The Moneta ultramafic (hornblendite) sample (PW0-109-6) is closely associated with the mafic rocks at the outcrop. There should be, therefore, little doubt that here the ultramafic and mafic rocks are from the same magma source.

In order to evaluate the effects of metamorphism, a series of logarithmic oxide molecular proportion ratio (LMPR) diagrams (Beswick and Soucie, 1978) have been plotted (Fig. 8). All Moneta and Lynchburg amphibolite samples plot either within or very near the unaltered volcanic range, indicating very little chemical change due to metamorphism. A small amount of sodium and probably a little potassium may have been added to some of these rocks during metamorphism, an observation that is especially obvious in the Catoctin and Slippery Creek greenstones that plot farther toward the origin on the LMPR diagram, indicating a higher potassium content than normal mafic volcanics. It has been suggested, based on geochemical studies of amphibolites and their parent rocks, that most of the major and trace elements in mafic rocks are not mobile under amphibolite facies conditions (Sheraton, 1984; Elliott, 1973; Field and Elliott, 1974). The LMPR plots of the Moneta and Lynchburg amphibolites appear to support this conclusion, with almost all of the samples falling in the narrow fresh igneous rock fields on the diagrams.

All samples plot near the alkaline and subalkaline boundary on the SiO_2 versus ($Na_2O + K_2O$) diagram (Fig. 9), with all Slippery Creek and three Catoctin greenstones, three mafic samples in the basement, and one Lynchburg amphibolite fall-

TABLE 2a. MAJOR ELEMENT CHEMICAL ANALYSIS DATA OF THE POST-GRENVILLE MAFIC-ULTRAMAFIC ROCKS OF THE BLUE RIDGE IN CENTRAL VIRGINIA*

Sample No.	SiO_2	Al_2O_3	Fe_2O_3	FeO	CaO	MgO	Na_2O	K_2O	TiO_2	MnO	P_2O_5	M†
(Wt. %)												
PW7-288	48.54	12.91	3.94	10.61	8.49	4.58	3.11	1.34	2.94	0.23	0.42	0.366
PW7-298	44.90	12.80	4.09	11.02	10.92	6.23	2.59	0.96	2.89	0.25	0.30	0.431
PW7-297	47.52	12.57	4.01	10.80	8.66	4.82	2.41	2.06	4.39	0.26	0.62	0.374
PW7-299	50.06	13.10	3.76	10.12	9.58	5.16	2.78	0.78	2.60	0.23	0.40	0.405
PW0-109-6 50.68	16.33	2.23	5.99	10.84	7.70	3.71	0.30	1.38	0.14	0.12	0.632	
PW0-109-4 48.28	14.12	3.46	9.28	10.88	6.77	2.88	0.34	2.14	0.22	0.22	0.494	
PW0-109-2 49.92	14.12	3.11	8.39	10.38	6.97	3.68	0.36	1.67	0.20	0.20	0.526	
PW7-7-2	47.18	15.00	3.09	8.32	11.00	7.59	2.50	0.38	1.55	0.18	0.20	0.549
PW7-7-1	47.50	15.47	2.96	7.98	12.00	8.08	2.68	0.28	1.44	0.18	0.22	0.575
PW0-109-7 40.25	12.38	4.99	13.45	11.99	5.99	2.02	0.34	4.80	0.26	2.24		
PW7-338(2)	47.35	13.90	2.43	6.55	8.26	9.65	2.43	4.38	1.38	0.19	0.14	
PW7-161(1)	46.41	15.39	3.09	8.31	10.23	7.73	2.84	0.36	1.65	0.21	0.14	0.554
PW7-161(2)	48.81	15.22	2.83	7.62	11.36	6.25	3.26	0.18	1.77	0.19	0.18	0.523
PW7-161(3)	47.99	14.80	3.44	9.27	10.09	7.29	2.36	0.18	2.09	0.27	0.18	0.513
PW8-46(1) 47.40	16.57	2.75	7.41	9.84	8.95	2.63	0.12	1.22	0.15	0.06	0.618	
PW8-46(2) 48.84	17.27	2.42	6.53	9.60	7.78	3.70	0.10	1.05	0.13	0.12	0.614	
PW8-46(3) 48.16	14.87	2.65	7.13	11.24	8.47	2.50	0.06	1.27	0.16	0.10	0.613	
PW8-31	46.48	18.28	2.33	6.26	12.12	8.21	1.87	0.14	0.85	0.15	0.04	0.637
PW0-123-30	47.72	15.91	1.60	4.30	15.03	10.88	1.89	0.22	0.46	0.12	0.02	0.772
PW8-3	46.71	17.09	2.44	6.56	12.21	8.86	1.63	0.20	0.84	0.17	0.20	0.644
PW7-139	49.15	14.36	3.55	9.56	9.14	6.82	2.58	0.12	2.27	0.22	0.24	0.488
PW8-41	49.33	13.24	3.72	10.00	9.68	5.55	2.67	0.88	2.68	0.22	0.38	0.426
PW0-123-23	43.49	13.47	2.36	6.36	13.11	14.38	2.26	0.60	0.19	0.18	0.08	
PW8-327(1)	48.97	13.26	3.99	10.74	6.19	5.33	5.27	0.20	2.32	0.23	0.20	0.399
PW8-327(2)	49.76	13.40	3.50	9.43	5.85	5.35	4.31	1.92	2.78	0.28	0.42	0.431
PW0-124	46.35	12.43	4.04	10.87	9.83	5.92	2.77	0.28	3.33	0.21	0.28	0.421
PW9-13	49.98	13.23	3.51	9.44	10.19	7.33	2.48	0.14	0.67	0.21	0.10	0.509
PW0-125-2 48.04	13.43	4.07	10.96	5.85	5.58	4.50	1.00	3.09	0.30	0.44	0.405	
PW0-125-1 46.41	14.49	3.50	9.41	7.03	6.89	4.06	0.56	2.69	0.18	0.30	0.494	
PW0-126	48.04	14.24	3.62	9.74	8.48	6.14	3.95	0.38	2.69	0.17	0.44	0.457

*All chemical analyses were done by the Technical Service Laboratories at Mississauga, Ontario.
†M' is the Differentiation Index for the mafic rocks calculated from the atomic number ration of Mg/(Mg+Fe).

ing inside the alkaline field. The other samples plot in the subalkaline field. On the $K_2O + Na_2O$, Fe, MgO (AFM) diagram (Fig. 10), all but three samples plot in the tholeiite (TH) field. This implies that the post-Grenville mafic magmas have an affinity with the tholeiite series and are relatively alkaline. Generally it appears that rift volcanism tends to be somewhat alkalic, probably depending primarily on the volume of magma produced during rifting. Transitional to mildly alkalic basalt is characteristic of relatively high magma production rates (Wilson, 1989). In other words, the larger the volume of magma produced (large amount of partial melting), the less the magma alkalinity, which may explain the generally subalkaline nature of the widespread mafic-ultramafic rocks in the Blue Ridge of the Appalachians.

In order to understand whether primary magmas had been differentiated before they reached their current position, the differentiation index $M' = Mg/(Mg + Fe^{+2})$, which is also called the "Mg-number," of the analyzed samples was calculated. Also the Ni and Cr contents of these samples were examined (see Table 2a, b). It has been proposed that the M′ value, as well as Ni and Cr abundance in the magma, is insensitive to the amount of partial melting but very sensitive to differentiation (Wilson, 1989). The primary magma that is in equilibrium with the upper mantle mineralogies (olivine + orthopyroxene + clinopyroxene ± garnet ± spinel) should have high M′ values (>0.7, or 0.65 to 0.75), high Ni (>400 to 500 ppm), high Cr (>1,000 ppm), and SiO_2 less than 50%. Like most of the continental rift basaltic rocks, the post-Grenville mafic rocks in central Virginia have low Mg numbers (all but one <0.65; mostly <0.55) and low Ni (<200 ppm) and Cr (<350 ppm) content, implying that they cannot have equilibrated with normal mantle lherzolite mineralogies under water-undersaturated conditions (Cox, 1980; Prestvik and Goles, 1985). The most obvious explanation for this is that the primary magmas were not basalts but were high-Mg picrites, which subsequently underwent low-pressure fractionation to produce the observed range of eruptive and shallow intrusive magmas with evolved compositions. The primary igneous textures, where preserved, show the existence of plagioclase and

TABLE 2b. TRACE ELEMENT CHEMICAL ANALYSIS DATA OF THE POST-GRENVILLE MAFIC-ULTRAMAFIC ROCKS OF THE BLUE RIDGE IN CENTRAL VIRGINIA

Sample No.	Zr	Y	Ni	Cr	V	Sr	Hf	Th	Ta	Yb	
(ppm)											
PW7-288	220	42				270					Amphibolites in the basement
PW7-298	180	40				190					
PW7-297	280	48	30	22	350	170	6.8	2.1	1.5	3.73	
PW7-299	220	46	50	110	360	270	5.7	1.8	0.9	3.75	
PW0-109-6	95	28	120	305	192	484					Moneta amphibolites
PW0-109-4	126	44	80	135	380	199					
PW0-109-2	116	34	100	295	298	182					
PW7-7-2	87	25	130	250	298	257					
PW7-7-1	74	22	170	320	276	288					
PW0-109-7	57	67	20	40	364	240					Moneta ultramafic hornblendite
PW7-338(2)	120	28				130					Biotite-cpx gneiss (Lampropyre?)
PW7-161(1)	100	32				270					Lynchburg amphibo-lites
PW7-161(2)	120	38				180					
PW7-161(3)	130	40	80	210	370	260	2.9	0.5	0.4	3.62	
PW8-46(1)	70	22				290					
PW8-46(2)	50	20				270					
PW8-46(3)	80	26				300					
PW8-31 50	20				160						
PW0-123-30	40	12				240					
PW8-3	60	22				230					
PW7-139	149	42	80	295	428	243					
PW8-41 200	40	50	110	370	230		5.1	1.4	0.7	3.34	
PW0-123-23	50	14				120					Ultramafic schist of Wal Mart cut
PW8-327(1)	170	36	20	26	190	230	3.9	1.2	0.6	2.85	Catoctin greenstones
PW8-327(2)	200	38	10	30	360	150	5.2	1.6	0.9	3.21	
PW0-124	183	41	110	150	554	199					
PW9-13 58	25	100	160	420	118						
PW0-125-2	262	42	30	70	348	156					Slippery Creek green-stones
PW0-125-1	189	36	90	295	310	175					
PW0-126	190	38				250					

possible olivine phenocrysts in the volcanics (such as in Catoctin greenstones), suggesting that they have been involved in relatively low-pressure crystal fractionation equilibria, supporting their evolution from more MgO-rich primary magmas. The possible picritic nature of a parent magma for continental basaltic rocks, or for most of the basalts, has been known by petrologists for more than 40 years (O'Hara, 1968; Cox, 1980; Prestvik and Goles, 1985; Wilson, 1989). In other words, these basaltic rocks have evolved by fractional crystallization from more MgO-rich picritic precursors.

This assumption, if correct, implies that in areas subject to rift-basalt volcanism there may be significant additions of gabbroic and ultramafic cumulates to lower-crust and higher-level sill complexes, the total volume of which may exceed that of the surface eruptives (Wilson, 1989). Such a situation may explain the large volume of intrusives (m-u) in the Lynchburg Group and their equivalents in the Blue Ridge Appalachians.

Geochemical data have been widely used to study petrogenetic processes and to identify tectonic environments of basaltic rocks, because basaltic rocks are formed in almost every tectonic environment and are geochemically somewhat sensitive to the changes of the tectonic framework and petrogenetic history. During the last two decades, many tectonic setting discriminant diagrams have been created based on the major and trace element abundances in basaltic rocks (Pearce and Cann, 1973; Pearce et al., 1975; Pearce, 1976, 1982, 1983; Pearce and Gale, 1977; Pearce et al., 1977; Wood et al., 1979; Pearce and Norry, 1979; Shervais, 1982; Pearce et al., 1984; Fodor and Vetter, 1984; Meschede, 1986). These diagrams have been widely used by geochemists and tectonicists all over the world. Some of these discriminant diagrams have been used to support the current Blue Ridge collisional tectonic model (Hatcher et al., 1984; Conley, 1985; Stow et al., 1984), because the post-Grenville mafic rocks tend to plot in the field(s) of either MORB, or MORB/VAB (vol-

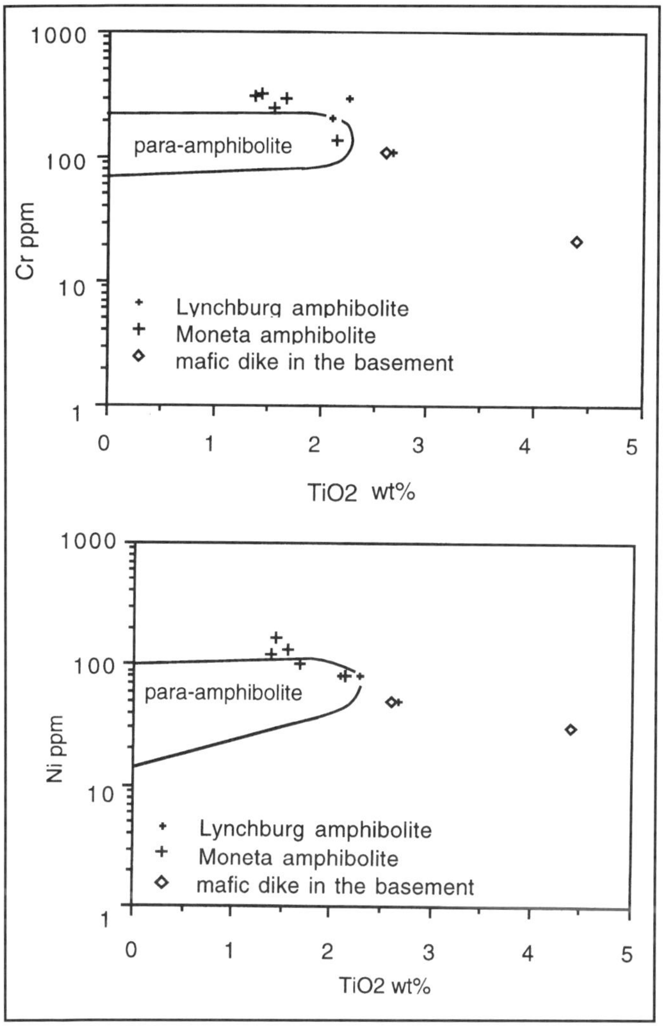

Figure 7. The Ti-Cr and Ti-Ni plot of the post-Grenville amphibolite in the Blue Ridge of central Virginia. Boundary is from Leake (1964).

canic arc basalt) rather than the field of WPB (within-plate basalt). Interpretations of these discriminant diagrams are considered to be one of the most important arguments for the ophiolite origin of the Blue Ridge mafic-ultramafic rocks.

However, our plots of the post-Grenville mafic rocks in the central Virginia Blue Ridge show that the samples fall either in more than one field on a single diagram or in different fields on different diagrams. Sample screens and qualifications are discussed in Wang and Glover (1992). Samples for this study meet those screens within the limits imposed by working with metamorphosed samples.

On the F_1-F_2 diagram of Pearce (1976), most of the post-Grenville mafic samples do plot in the WPB (within plate basalt) field; there are, however, five samples (15%, four Lynchburg amphibolite and one Catoctin greenstone) that fall into the CAB/LKT (calc alkaline basalt/low potassium tholeiite) field (Fig. 11A). Although this diagram appears to work for the Blue Ridge samples, it fails for the diabases of eastern North America (Fig. 11A and Wang and Glover, 1992) and is, therefore, ambiguous.

On the F_2-F_3 diagram (Pearce, 1976), all samples fall either in LKT or in CAB field, concentrating near the boundary (Fig. 11B). Neither field is intended to be appropriate for rift related basalts.

On the FeO*/MgO versus TiO_2 diagram (Fodor and Vetter, 1984), all mafic rocks in the basement, some Lynchburg amphibolite, and all Catoctin (except PW9-13)–Slippery Creek samples fall into the CFB (continental flood basalt) field, whereas the majority of the Moneta and most the Lynchburg amphibolites plot in the ORB (ocean ridge basalt) field (Fig. 12).

On the TiO_2-K_2O-P_2O_5 diagram (Pearce et al., 1975), all samples (except two Catoctin greenstones) fall in the oceanic basalt field as a result of a generally higher TiO_2 content for most of the samples (Fig. 13).

On the Pearce and Cann (1973) Ti/100-Zr-3Y diagram, all mafic rocks from the basement, Catoctin (except one)–Slippery Creek and one Lynchburg amphibolite fall in the WPB field, whereas all of the Moneta and Lynchburg (except one) amphibolites plot in the MORB field (Fig. 14A).

On the Pearce and Cann (1973) Ti/100-Zr-Sr/2 diagram, all (except three) samples fall in either OFB (ocean floor basalt) or LKT field, with all basement and Catoctin–Slippery Creek samples in the OFB, mainly because of their higher Sr content, whereas the Moneta and Lynchburg samples plot in both (Fig. 14B).

On both TiO_2-Zr and Zr/Y-Zr diagrams of Pearce (1980, 1982), all mafic rocks in the basement, one Lynchburg amphibolite, and all Catoctin (except one)–Slippery Creek samples fall into WPB field, whereas all of the Moneta and the rest of the Lynchburg amphibolites plot in the MORB field (Fig. 15 A, B).

On the Cr-Y diagram of Pearce (1980), all samples fall in the WPB and WPB/MORB fields (Fig. 16). In general, the Moneta and the Lynchburg amphibolites tend to have higher Cr content than those from the basement and Catoctin–Slippery Creek.

Six samples are plotted on the Hf/3-Th-Ta diagram of Wood (1980) and the Th/Yb-Ta/Yb diagram of Pearce (1983). Five out of the six samples plot very close to each other on both diagrams; these data overlap either unclassified and P-type MORB on the Hf/3-Th-Ta diagram (Fig. 17) or unclassified and WPB fields on the Th/Yb-Ta/Yb diagram (Fig. 18).

On the V-Ti/1000 diagram of Shervais (1982), most samples fall in the MORB/WPB field, with one Moneta and all Slippery Creek samples in the alkaline field and two out of three Catoctin and one out of two basement samples plotting in an unclassified field (Fig. 19).

It is obvious from the above discussion that these discriminant diagrams do not define an unambiguous tectonic setting for the mafic rocks of the central Virginia Blue Ridge, and more importantly, most of the results do not seem to be consistent with the result obtained from structure and tectonic lithofacies analysis based on field observations and petrological studies.

The limitations and misclassified results of some, but not

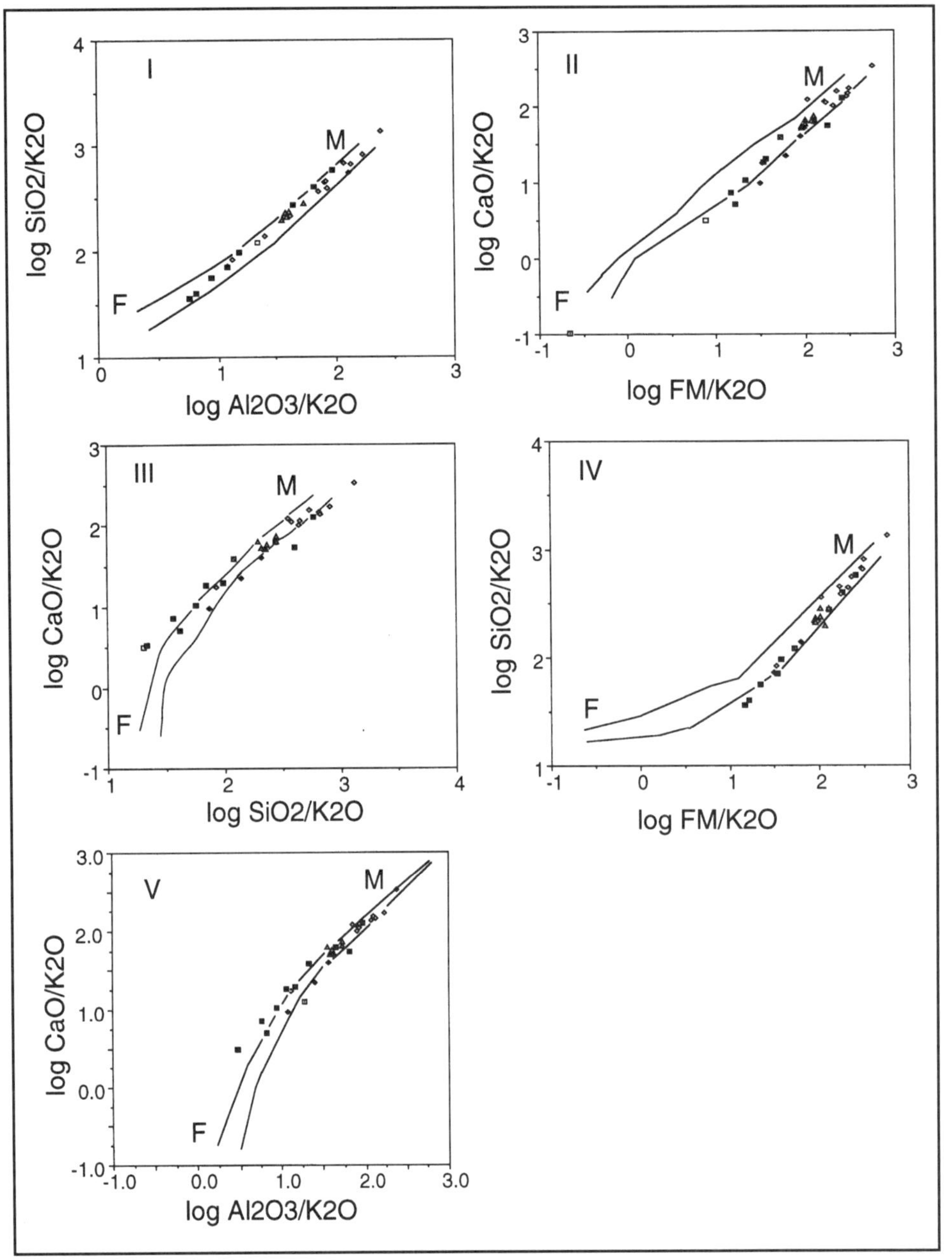

Figure 8. I–V, The LMPR (logarithmic oxide molecular proportion ratio) plot of the post-Grenville mafic-ultramafic rocks in the Blue Ridge of Virginia. Modern volcanic trends are from Beswick and Soucie (1978). M—mafic rocks; F—felsic rocks. Blue Ridge samples cannot be distinguished by rock unit at this scale.

all, of these chemical discriminant diagrams have been challenged by several other geologists (Gottfried et al., 1977, 1983; Morrison, 1978; Thompson et al., 1980; Holm, 1982; Philpotts, 1985; Duncan, 1987). Nevertheless, these challenges have generally been overlooked by tectonicists attempting to determine ancient tectonic settings.

It could be argued that the post-Grenville mafic rocks of the Blue Ridge should not be used to disqualify the discriminant diagrams, because the rocks are metamorphosed and because their own tectonic origin is in debate. It is therefore desirable to use unmetamorphosed samples from well-known tectonic settings to test the applicability of these diagrams. Nearly 200 chemical analyses of the eastern North America (ENA) Jurassic rift basalts and diabases were taken from the literature and plotted on all of the tectonic diagrams (see Figs. 11 through 19). The ENA basalts and diabases are known to be continental and were formed during the early-stage rifting of the modern Atlantic. The ENA samples, as previously demonstrated

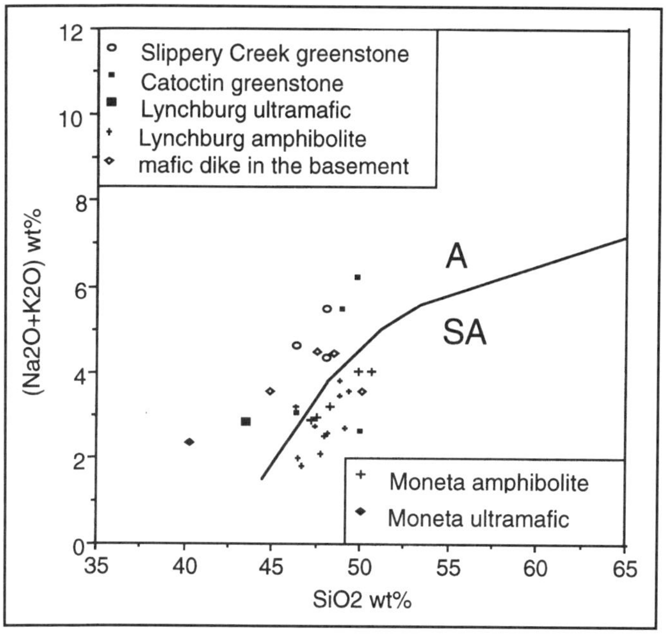

Figure 9. A/SA plot of the post-Grenville mafic-ultramafic rocks of the Blue Ridge in central Virginia. A—alkaline series; SA—subalkaline series.

for the post-Grenville mafic rocks from the Blue Ridge, tend to plot either in more than one tectonic setting in one diagram or in one setting in one diagram but different settings in the others. It is clear that most of these tectonomagmatic discriminant diagrams failed to uniquely classify the tectonic setting of ENA basalts and diabases, the majority of which plot in every field but WPB. Because continental basalts do plot in other fields, such as MORB and/or volcanic arc basalts, the discriminating ability of these diagrams thus becomes suspect. In the middle to late 1980s, more detailed studies on continental basalts were carried out, and large amounts of information about their petrogenetic processes, variable geochemical characteristics, and tectonic significance became available. The lack of representative data and incomplete understanding of the petrogenesis of continental basalts at the time when these discriminant diagrams were created, are the major reasons for the failure of the diagrams to identify continental basalt tectonic setting correctly. Moreover, because the genesis of continental basalt magma may involve a series of complicated processes that vary from place to place even in the same tectonic province, the geochemical patterns of continental basalts cover a very wide range of the compositions of isotopes, trace elements and rare earth elements (REE) as well as major and minor elements. It is therefore hard to create a universal chemical diagram that can separate continental basalts from all of the others. We propose, as have others before us, that geochemical data alone may give an incorrect tectonic classification and that a knowledge of field relations and studies of associated tectonic lithofacies are critical for definitive analysis.

The Blue Ridge mafic rocks, on the other hand, in most cases plot within or very near the ENA rift basalt–diabase field, suggesting that the two do have a similar geochemical nature and probably formed in a similar tectonic environment, that is, continental rift (see Figs. 11 through 19). The shift from the ENA field on some Ti-bearing diagrams is caused by a higher Ti content of the Blue Ridge mafic rocks, which may result from compositional differences of the magma source and/or degree of fractionation. It is clear that the Grenville basement gneisses are Ti-rich, and this is shown by the presence of titanite in almost every sample examined.

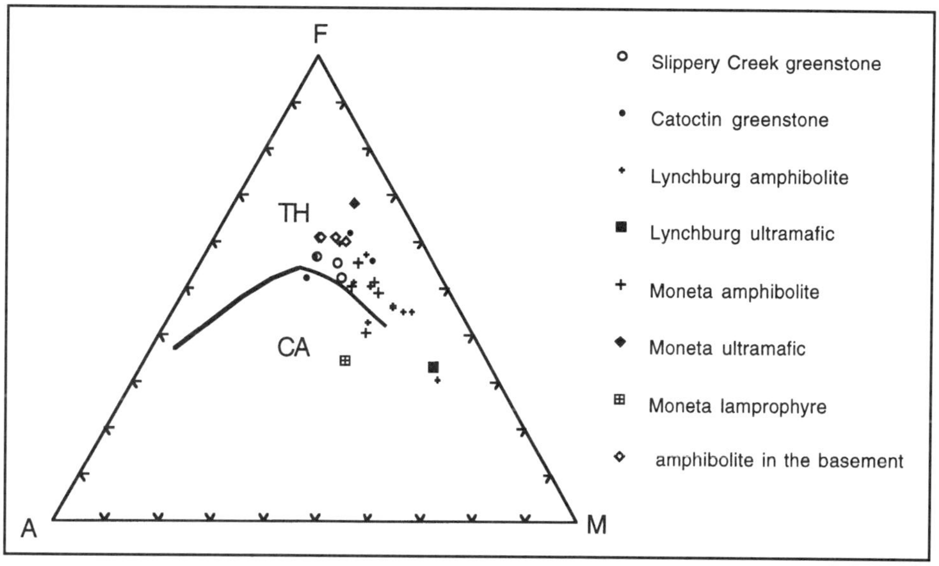

Figure 10. The AFM diagram of the post-Grenville mafic-ultramafic rocks in the Blue Ridge of central Virginia.

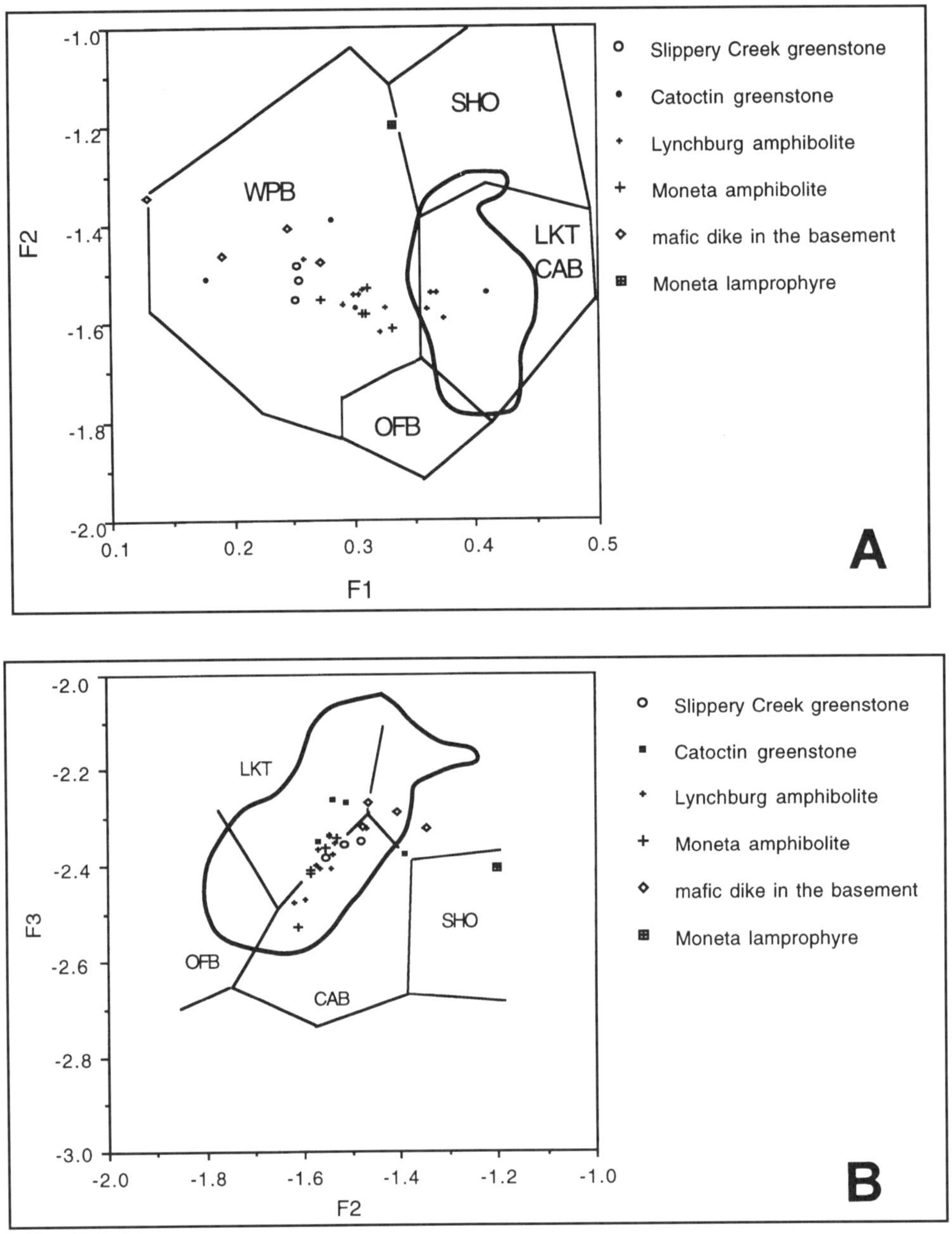

Figure 11. A, The F1-F2 plot of the post-Grenville mafic rocks in the Blue Ridge of Virginia. Boundaries are from Pearce (1976). WPB—within-plate basalt; OFB—ocean floor basalt; LKT—low potassium tholeiite; CAB—calc-alkaline basalt; SHO—shonshonite. The thick solid line field is where the eastern North America Jurassic rift mafic rocks plot. B, The F2-F3 plot of the post-Grenville mafic rocks in the Blue Ridge of central Virginia. Boundaries are from Pearce et al. (1976): LKT—low potassium tholeiite; OFB—ocean floor basalt; CAB—calc-alkaline basalt; SHO—shoshonite. The thick solid line field encloses eastern North America Jurassic rift mafic rocks.

Both MORB-normalized and primordial mantle–normalized trace element spider diagrams show that the post-Grenville mafic rocks from the central Virginia Blue Ridge have a different pattern compared with that of MORB (Fig. 20A and B). On the MORB-normalized diagram, a negative slope pattern is shown, with the absolute contents of most trace elements higher than MORB, crossing the MORB line at Yb, and the rest (Sc and Cr) lying below the MORB line. On the primordial mantle–normalized diagram, with a flat pattern, the contents of all trace elements are one to two orders higher than that of MORB, which also shows a positive trend. The variations of K, Rb, and Sr may be the result of metamorphic mobilization.

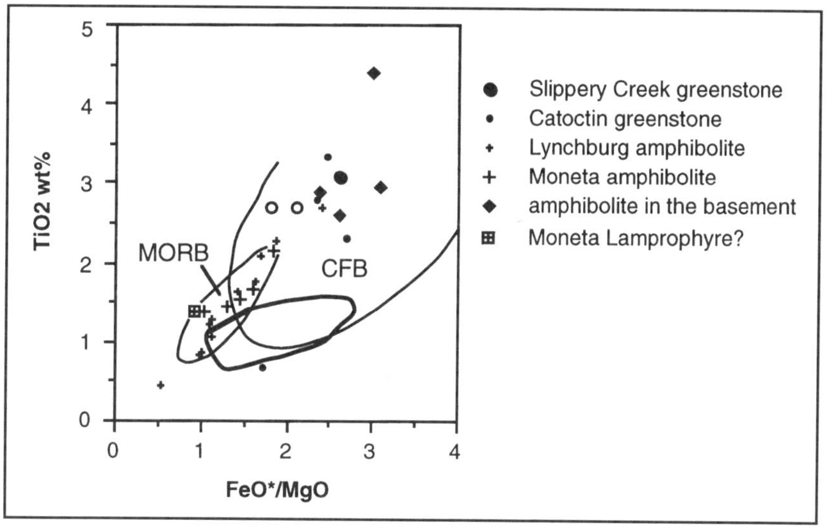

Figure 12. The FeO*/MgO vs. TiO_2 plot of the post-Grenville mafic rocks in the Blue Ridge of central Virginia. Boundaries are after Fodor and Vetter (1984). MORB—mid-ocean ridge basalt; CFB—continental flood basalt. The thick solid line encloses eastern North America Jurassic rift mafic rocks.

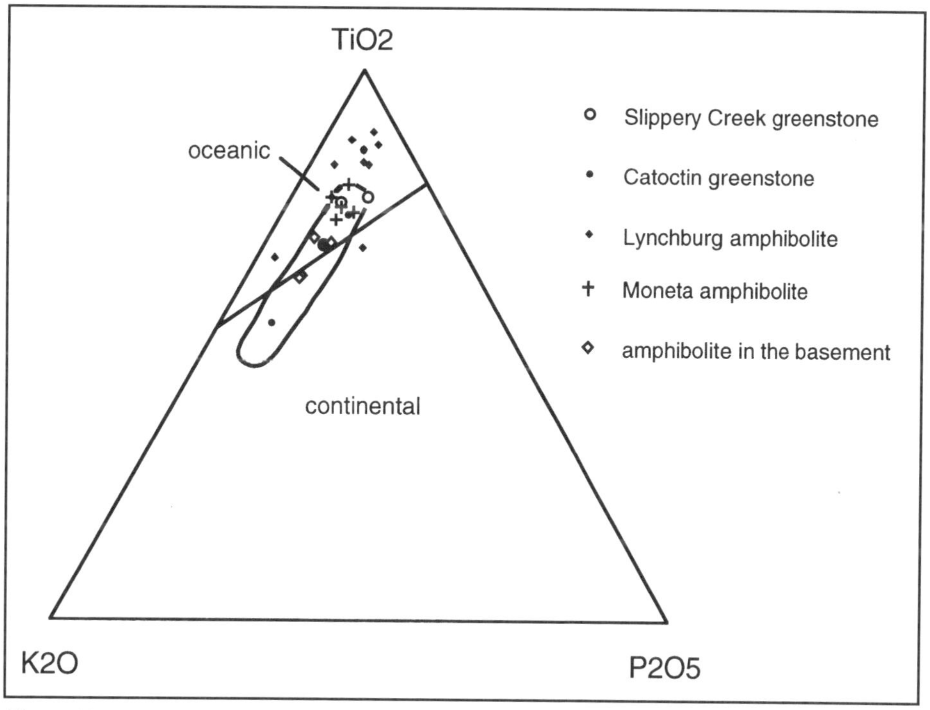

Figure 13. The TiO_2-K_2O-P_2O_5 plot of the post-Grenville mafic rocks in the Blue Ridge of central Virginia. Boundaries are after Pearce et al. (1975). The thick solid line encloses eastern North America Jurassic rift mafic rocks.

All of the post-Grenville mafic rocks in the Virginia Blue Ridge have a strongly light rare earth element (LREE) enriched pattern (Fig. 21), opposed to that of the mid-ocean ridge basalts, which show strong LREE depletion. A possible explanation for the discordant trend of PW7-161-3 is that it is located near the top margin of a thick (about 100 m) mafic-ultramafic sill and could be differentiated, as indicated by its negative Eu anomaly.

On both trace element and REE diagrams, all of the post-Grenville mafic rocks in the Blue Ridge of central Virginia show

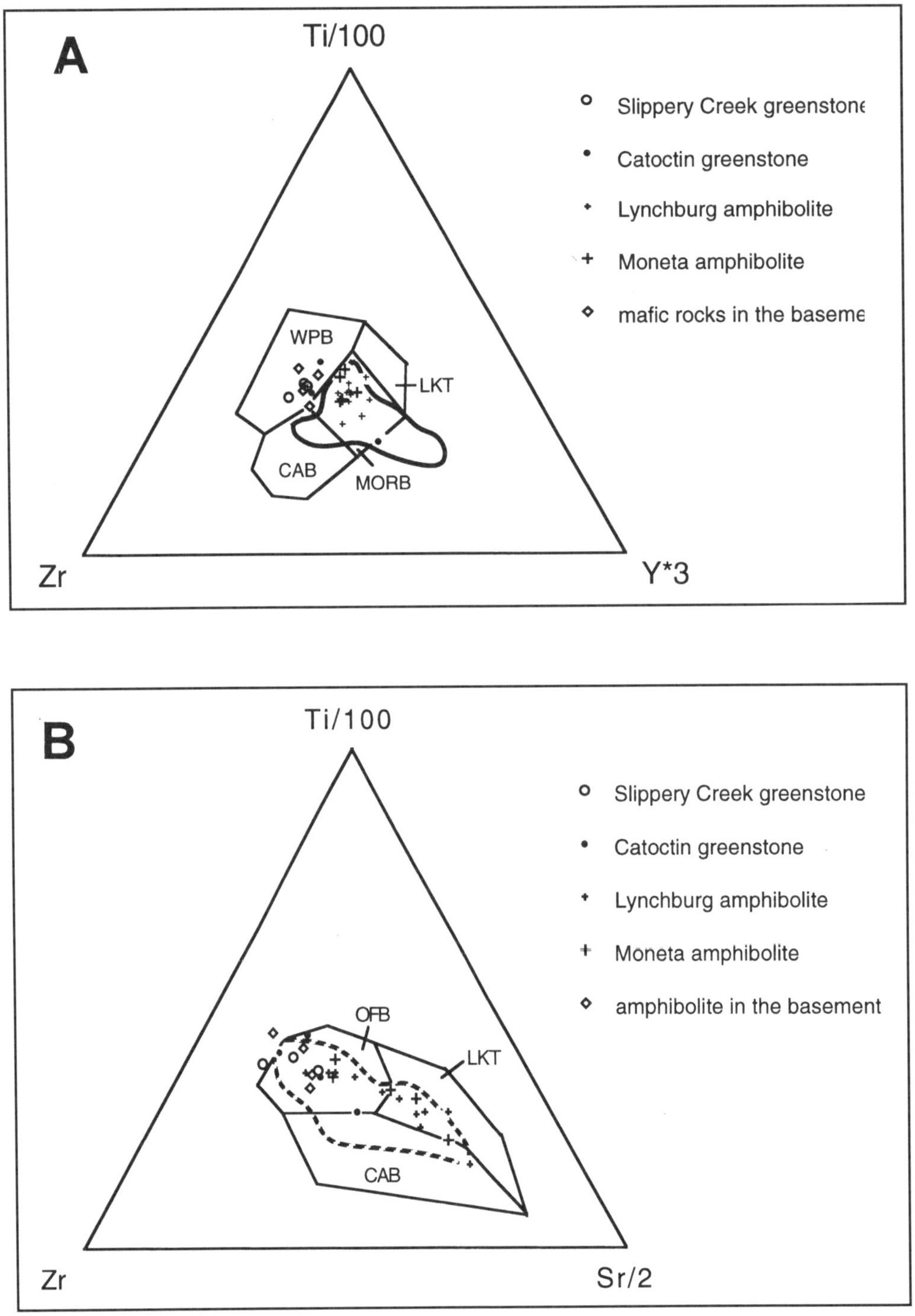

Figure 14. A, The Ti/100-Zr-Y*3 plot of the post-Grenville mafic rocks in the Blue Ridge of central Virginia. Boundaries are from Pearce and Cann (1973). LKT—low potassium tholeiite; MORB—mid-ocean ridge basalt; CAB—calc-alkaline basalt; and WPB—within plate basalt. The thick solid line encloses eastern North America Jurassic rift mafic rocks. B, The Ti/100-Zr-Sr/2 diagram of the post-Grenville mafic rocks in the Blue Ridge of central Virginia. Boundaries are after Pearce and Cann (1973). LKT—low potassium tholeite; CAB—calc-alkaline basalt; OFB—ocean floor basalt. The dashed area encloses eastern North America Jurassic rift mafic rocks.

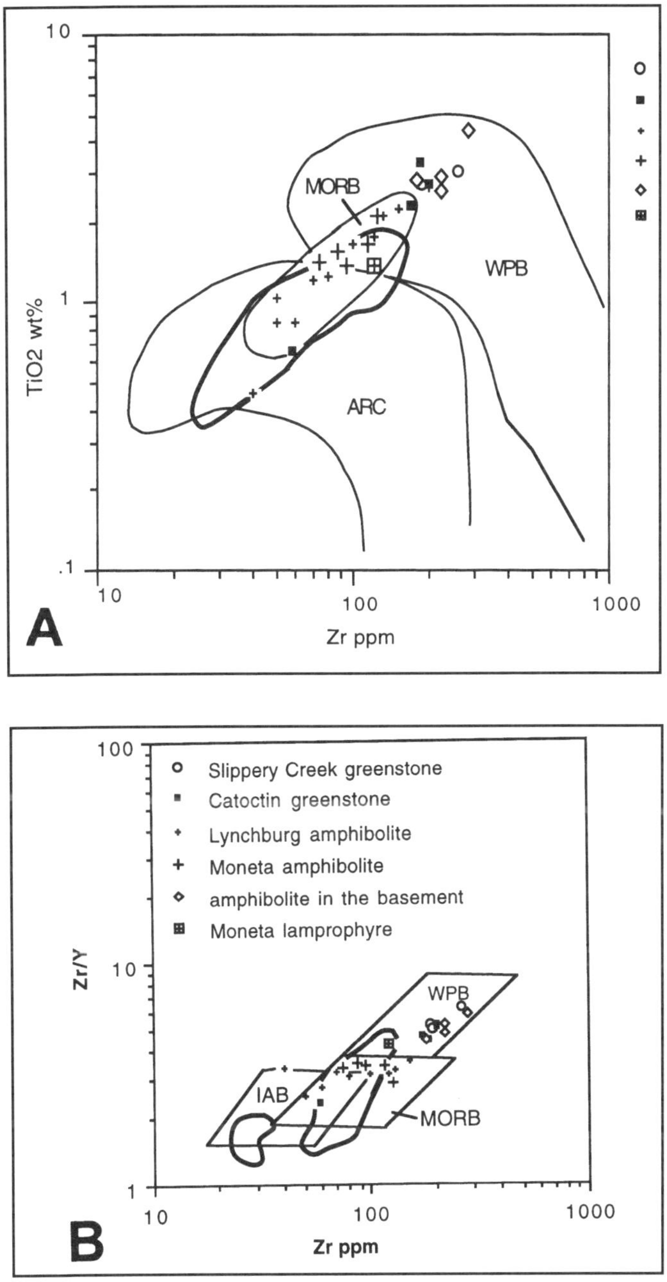

Figure 15. A, The TiO_2-Zr plot of the post-Grenville mafic rocks in the Blue Ridge of central Virginia. Boundaries are from Pearce (1982). ARC—arc rocks; MORB—mid-ocean ridge basalt; WPB—within plate basalt. The thick solid line encloses eastern North America Jurassic rift mafic rocks. B, The Zr/Y-Zr plot of the post-Grenville mafic rocks in the Blue Ridge of central Virginia. Boundaries are from Pearce (1982). WPB—within plate basalt; MORB—mid-ocean ridge basalt; IAB—island arc basalt. The thick solid lines enclose eastern North America Jurassic rift mafic rocks.

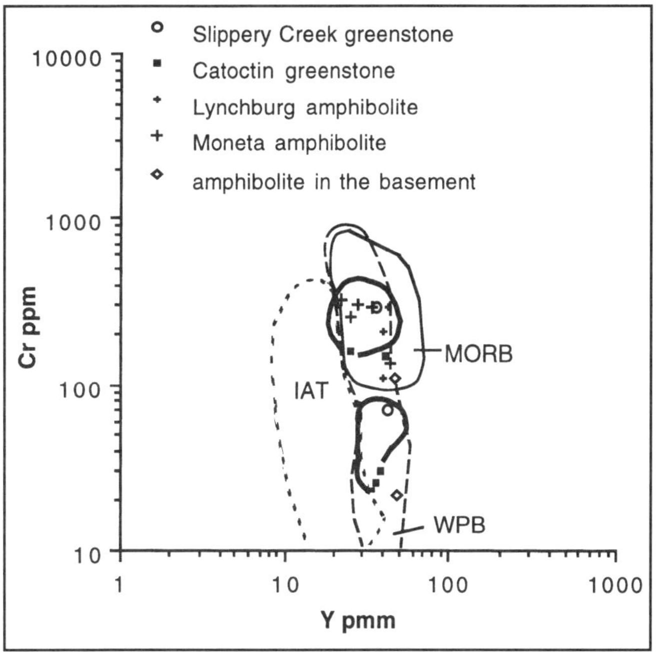

Figure 16. The Cr-Y plot of the post-Grenville mafic rocks in the Blue Ridge of central Virginia. Boundaries are from Pearce (1982). IAT—island arc tholeiite; MORB—mid-ocean ridge basalt; WPB—within plate basalt. The thick solid lines enclose eastern North America Jurassic rift mafic rocks.

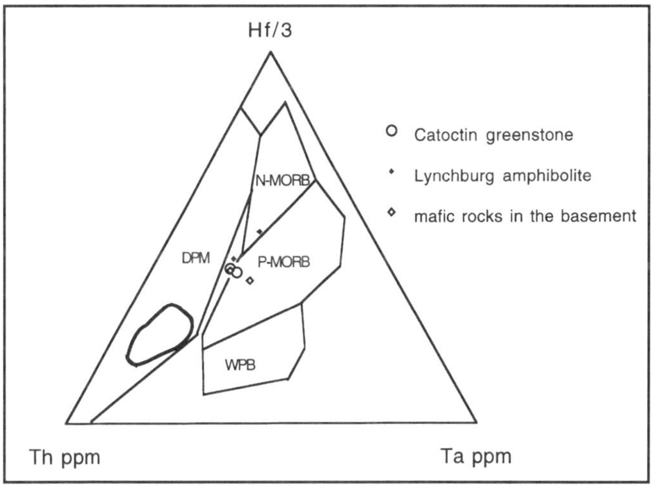

Figure 17. The Hf/3-Th-Ta diagram of the post-Grenville mafic rocks in the Blue Ridge of central Virginia. Boundaries are after Wood (1980). DPM—destructive plate margin basalts; N-MORB—normal-type mid-ocean ridge basalt; P-MORB—plume-type mid-ocean ridge basalt; WPB—within-plate basalts. The thick solid line encloses Jurassic rift mafic rocks.

a similar pattern (that is very different from MORB), suggesting a common magmatic and tectonic (nonophiolitic) origin.

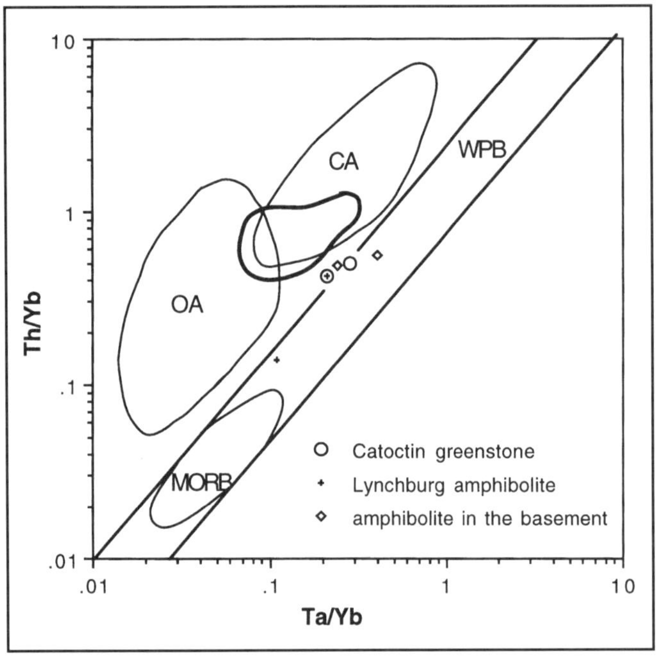

Figure 18. The Ta/Yb-Th/Yb plot of the post-Grenville mafic rocks in the Blue Ridge of central Virginia. Boundaries are from Pearce (1983). CA—continental arc; OA—oceanic arc; MORB—mid-ocean ridge basalt; WPB—within plate basalt. The thick solid line encloses eastern North America Jurassic rift mafic rocks.

CONCLUSIONS

1. The mafic/ultramafic association in the Blue Ridge cover rocks is part of an intrusive-extrusive sequence of mafic and ultramafic rocks related to the Late Proterozoic Iapetan rifting event rather an ophiolite-bearing sequence formed during the Taconian (Mid-Ordovician) or an earlier orogeny.

2. There are no terrane boundaries or sutures in the Lynchburg Group in central Virginia; the Taconic suture zone should be somewhere to the east of the Lynchburg and Evington Group.

3. A volcanic and epiclastic unit (the Moneta Formation) occurs in the lower Lynchburg Group in central Virginia. Its strongly bimodal volcanic nature and association with alluvial fan sediments suggests an early rifting environment.

4. Most tectonic setting discrimination diagrams cannot distinguish continental basaltic rocks from mid-ocean ridge basalts or island arc basalts. In these cases, field observations of the contacts both within the igneous sequence and with the clastic host rocks, the associated lithofacies assemblages and the structural patterns are of paramount importance for the tectonic setting assignment.

5. Ultramafic rocks are rarely found on or near the surface of the earth because of their high density. The abundance of supracrustal ultramafic rocks injected during the rift stage of the Late Proterozoic Iapetan margin is perhaps an anomalous, but interesting, addition to the tectonic lithofacies assemblages in rift basins. These ultramafics probably formed through fractional crystallization of a picritic-basaltic magma. This conclu-

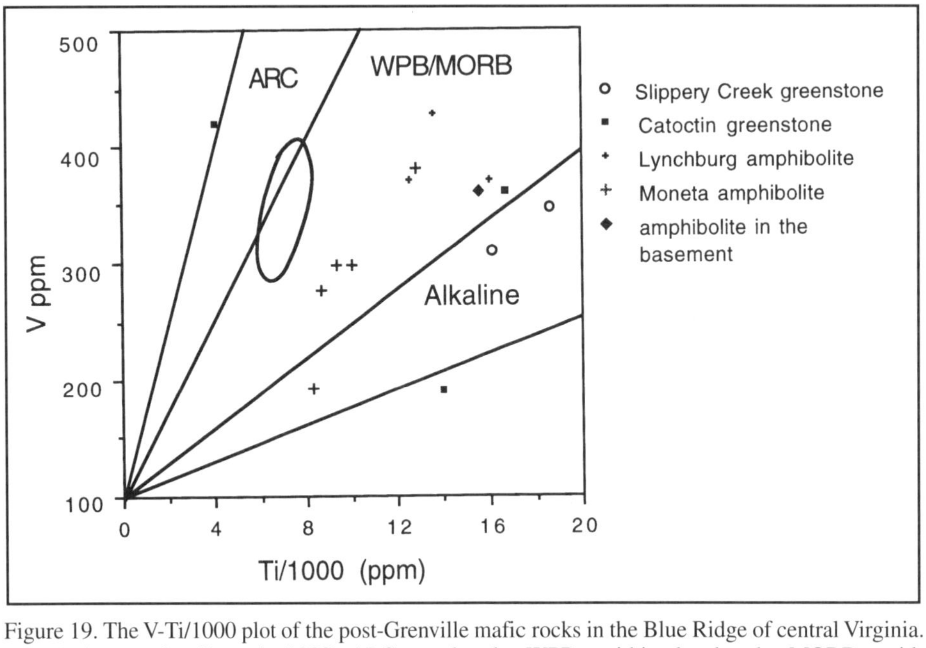

Figure 19. The V-Ti/1000 plot of the post-Grenville mafic rocks in the Blue Ridge of central Virginia. Boundaries are after Shervais (1982). ARC—arc basalts; WPB—within plate basalts; MORB—mid-ocean ridge basalts. The thick solid line encloses eastern North America Jurassic rift mafic rocks.

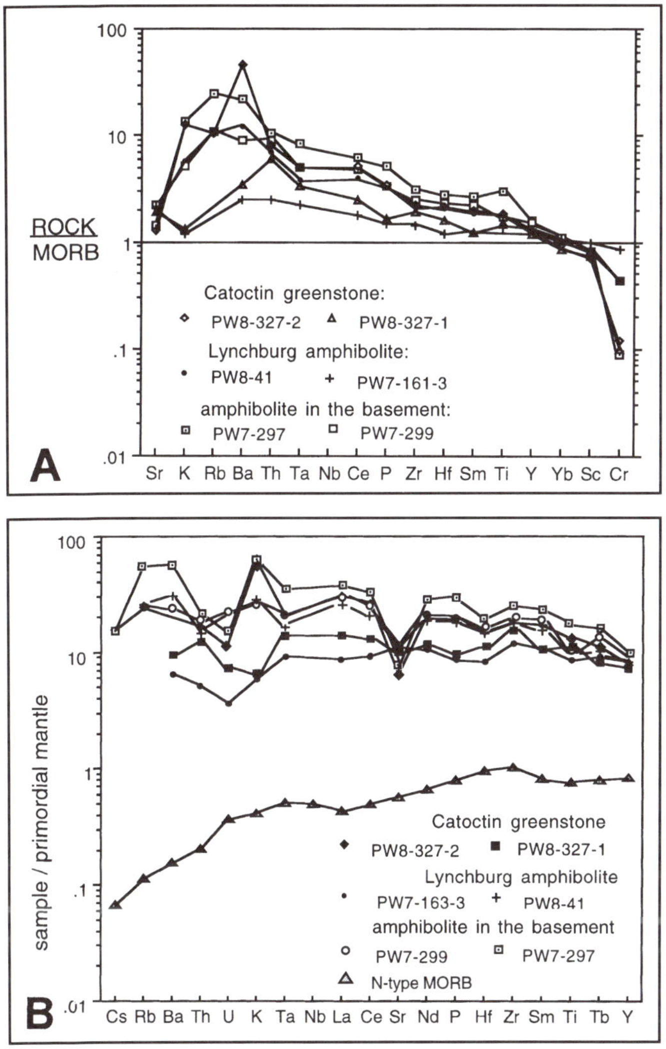

Figure 20. A, MORB-normalized trace element patterns of the post-Grenville mafic rocks from central Virginia Blue Ridge. Normalization values from Pearce (1982). B, Primordial mantle–normalized trace element patterns of the post-Grenville mafic rocks in the Blue Ridge of central Virginia and the average of N-type MORB. MORB and normalization values are from Wood et al., (1979).

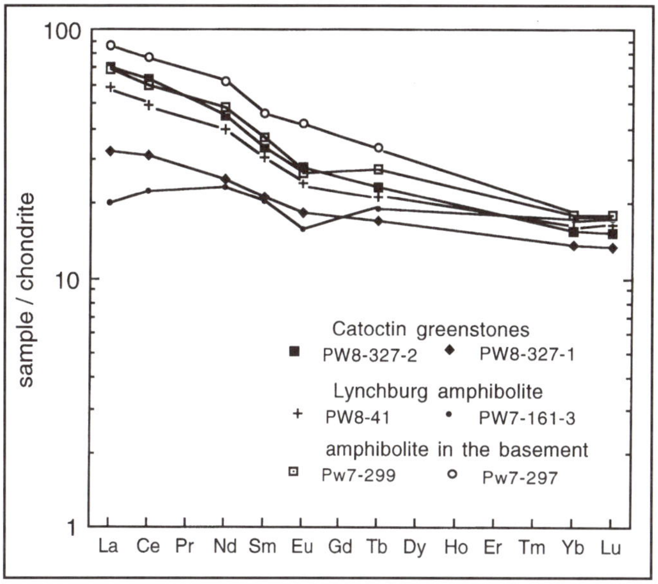

Figure 21. Chondrite–normalized rare earth element patterns for the post-Grenville mafic rocks from central Virginia Blue Ridge. Normalization values are from Boynton (1984).

sion is supported by the lower Mg-number, lower Ni and Cr content of the mafic-ultramafic rocks, and a negative Eu anomaly for some of the samples analyzed.

6. On modern continental margins, most of the rift and drift sequences are accessible only to the drill and seismic reflection exploration. Studies of ancient margins such as the one reported here still have much to tell us about the tectonic development and details of rock sequences that may be expected.

ACKNOWLEDGMENTS

We thank the reviewers, Richard D. Law, Robert J. Tracey, Kenneth A. Eriksson, James F. Tull, and Harry Y. McSween for helpful and insightful reviews.

REFERENCES CITED

Abbott, R. N., Jr., and Raymond, L. A., 1984, The Ashe metamorphic suite, northwest North Carolina: Metamorphism and observations on geologic history: American Journal of Science, v. 284, p. 350–375.

Aleinikoff, J. N., Zartman, R. E., Rankin, D. W., Lyttle, P. T., Burton, W. C., and McDowell, R. C., 1991, New U-Pb zircon ages for rhyolite of the Catoctin and Mount Rogers Formations—More evidence for two pulses of Iapetan rifting in the central and southern Appalachians [abs]: Geological Society of America Abstracts with Programs, v. 23, p. 2.

Badger, R. L., and Sinha, A. K., 1988, Age and Sr isotopic signature of the Catoctin volcanic province: Implications for subcrustal mantle evolution: Geology, v. 16, p. 692–695.

Bartholomew, M. J., 1984, The Grenville event in the Appalachian and related topics: Geological Society of America Special Paper 194, 287 p.

Beswick, A. E., and Soucie, G., 1978, A correction procedure for metamorphism in an Archean greenstone belt: Precambrian Research, v. 6, p. 235–248.

Bloomer, R. O., and Werner, H. J., 1955, Geology of the Blue Ridge region in central Virginia: Geological Society of America Bulletin, v. 66, p. 579–608.

Bowring, S. A., Grotzinger, J. P., Isachsen, C. E., Knoll, A. H., Pelechaty, S. M., and Kolosov, P., 1993, Calibrating rates of Early Cambrian evolution. Science, v. 261, p. 1293–1298.

Boynton, W. W., 1984, Cosmochemistry of the rare earth elements, *in* Henderson, P., ed., Rare earth element geochemistry: New York, Elsevier, p. 63–114.

Brown, W. R., 1958, Geology and mineral resources of the Lynchburg quadrangle, Virginia; Virginia Division of Mineral Resources Bulletin 74, 99 p.

Coleman, R. G., 1977, Ophiolites; ancient oceanic lithosphere? Berlin, Springer-Verlag, 229 p.

Conley, J. F., 1985, Geology of the southwestern Virginia Piedmont: Virginia, Department of Mines, Minerals and Energy, Division of Mineral Resources Publication 59, 33 p.

Cox, K. G., 1980, A model for flood basalt volcanism: Journal of Petrology, v. 21, p. 629–650.

Dick, H. J. B., 1989, Abyssal peridotites, very slow spreading rates and ocean ridge magmatism, *in* Saunders, A. D., and Norry, M. J., ed., Magmatism in Ocean Basins: Geological Society Special Publication 42,

p. 71–105.

Duncan, A., 1987, The Karoo igneous province—A problem area for inferring tectonic setting from basalt geochemistry: Journal of Volcanology and Geothermal Research, v. 32, p. 13–34.

Elliott, R. B., 1973, The chemistry of gabbro/amphibolite transitions in south Norway: Contributions to Mineralogy and Petrology, v. 38, p. 71–79.

Espenshade, G. H., 1954, Geology and mineral deposits of the James River–Roanoke River manganese district, Virginia: U.S. Geological Survey Bulletin 1008, 155 p.

Field, D., and Elliott, R. B., 1974, The chemistry of gabbro/amphibolite transitions in south Norway. II: Trace elements: Contributions to Mineralogy and Petrology, v. 47, p. 63–76.

Fodor, R. V., and Vetter, S. K., 1984, Rift-zone magmatism: Petrology of basaltic rocks transitional from CFB to MORB, southeastern Brazil margin: Contributions to Minerology and Petrology, v. 88, p. 307–321.

Gass, I. G., Lippard, S. J., and Shelton, A. W., eds., 1984, Ophiolites and oceanic lithosphere: Geological Society Special Publication 13, 413 p.

Glover, L., III, Speer, J. A., Russell, G., and Farrar, S. S., 1983, Ages of regional metamorphism and ductile deformation in the central and southern Appalachians: Lithos, v. 16, p. 223–245.

Gottfried, D., Annell, C. S., and Byerly, G. R., 1983, Geochemistry and tectonic significance of subsurface basalts near Charleston, South Carolina: Clubhouse Crossroads test holes #2 and #3. U.S. Geological Survey Professional Paper 1313-A, p. A1–A19.

Gottfried, D., Annell, C. S., and Schwarz, L. J., 1977, Geochemistry of subsurface basalt from the deep corehole (Clubhouse Crossroads Corehole 1) near Charleston, South Carolina—Magma type and tectonic implications: U.S. Geological Survey Professional Paper 1028-G, p. 91–113.

Hatcher, R. D., Jr., 1989, Tectonic synthesis of the U.S. Appalachians, *in* Hatcher, R. D., Jr., Thomas, W. A., and Viele, G. W., eds., The Appalachian-Ouachita Orogen in the United States: Boulder, Colorado, Geological Society of America, The Geology of North America, v. F-2, p. 511–535.

Hatcher, R. D., Jr., Hooper, R. J., Petty, S. M., and Willis, J. D., 1984, Structure and chemical petrology of three southern Appalachian mafic-ultramafic complexes and their bearing upon the tectonics of emplacement and origin of Appalachian ultramafic bodies: American Journal of Science, v. 284, p. 484–506.

Hess, H. H., 1933, Hydrothermal metamorphism of an ultrabasic intrusive at Schuyler, Virginia: American Journal of Science, Fifth Series, v. 36, p. 377–407.

Hess, H. H., 1939, Island arcs, gravity anomalies and serpentine intrusions: 17th International Geological Congress, Moscow 1937, Report 17, v. 2, p. 297–300.

Hess, H. H., 1955, Serpentines, orogeny, and epirogeny: Geological Society of America Special Paper 62, p. 391–408.

Holm, P. E., 1982, Non-recognition of continental tholeiites using Ti-Y-Zr diagram: Contributions to Mineralogy and Petrology, v. 79, p. 308–310.

Horton, J. W., Drake, A. A., and Rankin, D. W., 1989, Tectonostratigraphic terranes and their Paleozoic boundaries in the central and southern Appalachians: Geological Society of America Special Paper 230, p. 213–245.

Keppie, J. D., and Dallmeyer, R. D., compilers, 1989, 1:5,000,000 map of pre-Mesozoic terranes in circum Atlantic Phanerozoic orogens: International Geologic Correlation Programme Project 233, Geological Society of America Abstracts with Programs, v. 21, p. 23–24.

Kline, S. W., 1991, Provenance of arkosic metasediments in the Virginia Blue Ridge: Constraints on Appalachian suspect terrane models: American Journal of Science, v. 291, p. 189–198.

Leake, B. E., 1964, The chemical distinction between ortho- and para-amphibolites: Journal of Petrology, v. 5, p. 238–254.

Meschede, M., 1986, A method of discriminating between different types of mid-ocean ridge basalts and continental tholeiites with the Nb-Zr-Y diagram: Chemical Geology, v. 56, p. 207–218.

Morrison, M. A., 1978, The use of immobile trace elements to distinguish the paleo-tectonic affinities of metabasalts: Applications to the Paleocene basalts of Mull and Skye, northwest Scotland: Earth and Planetary Science Letters, v. 39, p. 407–416.

Nicolas, A., 1989, Structures of ophiolites and dynamics of oceanic lithosphere: Boston, Kluwer Academic Publishers, 367 p.

O'Hara, M. J., 1968, The bearing of phase equilibria studies in synthetic and natural systems and the origin and evolution of basic and ultrabasic rocks: Earth and Planetary Science Letters, v. 4, p. 69–133.

Panayiotou, A., ed., 1980, Ophiolites: Proceedings, International Ophiolite Syposium, Cyprus, 1979: Nicosia, Cyprus, Geological Survey Department, 781 p.

Pearce, J. A., 1976, Statistical analysis of major element patterns in basalt: Journal of Petrology, v. 17, p. 15–43.

Pearce, J. A., 1980, Geochemical evidence for the genesis and eruptive setting of lavas from Tethyan ophiolites, *in* Panayiotou, A., ed., Ophiolites: Proceedings, International Ophiolite Symposium, Cyprus, 1979: Nicosia, Cyprus, Geological Survey Department, p. 261–272.

Pearce, J. A., 1982, Trace element characteristics of lavas from destructive plate boundaries, *in* Thorpe, R. S., ed., Andesites: Chichester, United Kingdom, John Wiley & Sons, p. 525–548.

Pearce, J. A., 1983, Role of the sub-continental lithosphere in magma genesis at active continental margins, *in* Hawkesworth, C. L., and Norry, M. J., eds., Continental basalts and mantle xenoliths: Nantwich, United Kingdom, Shiva Publishing, p. 230–249.

Pearce, J. A., and Cann, J. R., 1973, Tectonic setting of basic volcanic rocks determined using trace element analysis: Earth and Planetary Science Letters, v. 19, p. 290–300.

Pearce J. A., and Gale, G. H., 1977, Identification of ore-deposition environment from trace element geochemistry of associated igneous host rocks, *in* Volcanic processes in ore genesis: Geological Society of London: Special Publication 7, p. 14–24.

Pearce, J. A., and Norry, M. J., 1979, Petrogenetic implications of Ti, Zr, Y and Nb variations in volcanic rocks: Contributions to Minerology and Petrology, v. 69, p. 33–47.

Pearce, T. H., Gorman, B. E., and Birkett, T. C., 1975, The TiO_2-K_2O-P_2O_5 diagram: A method of discriminating between oceanic and non-oceanic basalts: Earth and Planetary Science Letters, v. 24, p. 419–426.

Pearce, T. H., Gorman, B.E., and Birkett, T. C., 1977, The relationship between major element chemistry and tectonic environment of basic and intermediate volcanic rocks: Earth and Planetary Science Letters, v. 36, p. 121–132.

Pearce, J. A., Lipparel, S. J., and Roberts, S., 1984. Characteristics and tectonic significance of supra-subduction zone ophiolites, *in* Kokelaar, B. P., and Howells, M. F., eds., Marginal basin geology: Geological Society of London Special Publication 16, p. 77–94.

Philpotts, J. A., 1985, Pearce-Cann discriminant diagrams applied to Eastern North America Mesozoic diabase: U.S. Geological Survey Circular 946, p. 114–117.

Prestivik, T., and Goles, G. G., 1985, Comments on the petrogenesis and the tectonic setting of Columbia River Basalts; Earth and Planetary Science Letters, v. 72, p. 65–73.

Rankin, D. W., 1973, Stratigraphy and structure of the metamorphic belt in northwestern North Carolina and southwestern Virginia: A study from the Blue Ridge across the Brevard fault zone to the Sauratown Mountains anticlinarism: American Journal of Science, v. 273-A (The Byron N. Cooper Volume), p. 1–40.

Rankin, D. W., and nine others, 1989, Pre-orogenic terranes, *in* Hatcher, R. D., Jr., Thomas, W. A., and Viele, G. W., eds., The Appalachian-Ouachita Orogen in the United States: Boulder, Colorado, Geological Society of America, The Geology of North America, v. F-2, p. 29–42.

Saunders, A. D., and Norry, M. J., eds., 1989, Magmatism in ocean basins; Geological Society of London Special Publication 42, 398 p.

Sheraton, J. W., 1984, Chemical changes associated with high-grade metamorphism of mafic rocks in the east Antarctic Shield: Chemical Geology, v. 47, p. 135–157.

Shervais, J. W., 1982, Ti-V plots and petrogenesis of modern and ophiolitic lavas: Earth and Planetary Science Letters, v. 59, p. 108–118.

Southworth, S., Burton, W. C., Froelich, A. J., Schindler, J. S., Lyttle, P. T., and Drake, A. A., 1994, Geologic maps of Loudon County, Virginia: Geological Society of America Abstracts with Programs, v. 26, p. 63.

Stanley, R. S., and Ratcliffe, N. M., 1985, Tectonic synthesis of the Taconian orogeny in western New England: Geological Society of America Bulletin, v. 9, p. 1227–1250.

Stow, S. H., Neilson, M. J., and Neathery, T. L., 1984, Petrography, geochemistry, and tectonic significance of the amphibolites of the Alabama Piedmont: American Journal of Science, v. 284, p. 416–435.

Thompson, R. N., Morrison, M. A., Matterey, D. P., Dickin, A. P., and Moorbath, 1980, An assessment of the Th-Hf-Ta diagram as a discrimination for tectonomagnetic classification of crustal contaminations of magmas: Earth and Planetary Science Letters, v. 50, p. 1–10.

Wang, P., and Glover, L., III, 1992, A tectonics test of the most commonly used geochemical discriminant diagrams and patterns: Earth Science Reviews, v. 33, p. 111–131.

Wehr, F., 1983, Geology of the Lynchburg Group in the Calypso and Rockfish River areas, Virginia [Ph.D. thesis]: Blacksburg, Virginia Polytechnic Institute and State University, 245 p.

Wehr, F., 1985, Stratigraphy of the Lynchburg Group and Swift Run Formation, Late Proterozoic (730–570 Ma), central Virginia: Southeastern Geology, v. 25, p. 225–239.

Wehr, F., and Glover, L., III, 1985, Stratigraphy and tectonics of the Virginia-North Carolina Blue Ridge: Evolution of a Late Proterozoic–early Paleozoic hinge zone: Geological Society of America Bulletin, v. 96, p. 285–295.

Wilson, M., 1989, Igneous petrogenesis, a global tectonic approach: London, Unwin Hyman, 480 p.

Wood, D. A., 1980, The application of a Th-Hf-Ta diagram to problems of tectonomagnetic classification and to establishing the nature of coastal contamination of basalt layers of the British Tertiary Volcanic Province: Earth and Planetary Science Letters, v. 50, p. 11–30.

Wood, D. A, Joron, J. L., and Treuil, M., 1979, A re-appraisal of the use of trace elements to classify and discriminate between magma series erupted in different tectonic settings: Earth and Planetary Science Letters, v. 45, p. 326–336.

Manuscript Accepted by the Society April 30, 1996

Printed in U.S.A.

Geological Society of America
Special Paper 314
1997

Late Proterozoic cover rocks in the Blue Ridge of northern Virginia: Do they include a terrane boundary?

Grigorios D. Kasselas and Lynn Glover, III
Department of Geological Sciences, Virginia Polytechnic Institute and State University, Blacksburg, Virginia 24061-0420

ABSTRACT

The Late Proterozoic to Early Cambrian metasedimentary and metavolcanic cover rocks, which lie unconformably upon the basement on the east limb of the Blue Ridge, experienced one low-grade progressive metamorphism. Deformation was minimal and took place in two stages: D1 involved a pervasive foliation (S_1) in varying degrees of intensity throughout the sequence. Subsequent folding and faulting and development of a localized discrete cleavage (S_2) resulted from D2. The observed structures are correlated with features described by workers in adjacent areas and may be regionally extensive. The map-scale structure varies from east-dipping (D1 fold) strata directly above the eastern Blue Ridge basement, to a thrust fault(?) or tight folds at the D2 structural front 3 km farther east, and finally to open folding and gently dipping strata several kilometers still farther east. We suggest that the D2 structural front was localized along an east-dipping deformation boundary above the eastern subsurface edge of the rigid Blue Ridge Grenville basement.

Some workers have proposed tectonic models of the central Appalachians in which accreted terranes are present along the eastern Blue Ridge. In one recent tectonic model of the central Appalachians, a suture drawn along the western boundary of the belt of ultramafic rocks in the Lynchburg Group (Blue Ridge basement cover rocks) was extended northeastward through the Warrenton area, where earlier mapping had outlined an eastward incursion of metasedimentary rocks into and possibly through the Catoctin metabasalts that overlie the Lynchburg. This suture was correlated with a proposed Taconian suture in New England. Thus, rocks previously mapped as Catoctin metabasalt south of the metasedimentary rock incursion were consequently proposed to be on an exotic terrane, unrelated to the true Laurentian Catoctin. The study area of the present report straddles this proposed suture. No evidence was found to support the presence of such a terrane boundary in the Warrenton area. Detailed mapping indicates that Catoctin Formation and pre-Catoctin metasedimentary rocks and intercalated mafic and ultramafic rocks comprise a conformable stratigraphic sequence deposited upon the Late Proterozoic rifted Laurentian margin.

INTRODUCTION

Tectonic models for the central Appalachians generally follow one of two approaches. According to one approach, metamorphosed sedimentary and mafic igneous rocks that lie to the southeast of the high-grade plutonic rocks of the Blue Ridge core are interpreted as a conformable and structurally coherent assemblage of lithofacies that were deposited on the edge of ancient Laurentia (Glover, 1989; Kline, 1991). According to the second, or "terrane" approach, some of the cover rocks of the eastern

Kasselas, G. D., and Glover, L., III, 1997, Late Proterozoic cover rocks in the Blue Ridge of northern Virginia: Do they include a terrane boundary? *in* Glover, L., III, and Gates, A. E., eds., Central and Southern Appalachian Sutures: Results of the EDGE Project and Related Studies: Boulder, Colorado, Geological Society of America Special Paper 314.

Blue Ridge belong to either exotic or para-autochthonous terranes that were accreted to Laurentia during the Taconian orogeny (Hatcher, 1987; Rankin et al., 1990b).

The "terrane" model has gained many advocates in the southern and central Appalachians during the last decade. For example, Williams and Hatcher (1983) introduced the suspect Piedmont terrane (including a large part of the Blue Ridge in the Carolinas and Virginia), and Hatcher (1987) and Hatcher et al. (1989) suggested that this terrane is a rifted continental fragment of Laurentia. Figure 1 shows the location of the terrane boundary proposed by Hatcher (1987). In a similar approach Horton et al. (1989) and Rankin et al. (1990a) (or Rankin et al., 1990b) proposed the existence of an exotic disrupted terrane ("Jefferson terrane") that formed in the accretionary wedge of an eastward-dipping subduction zone on the other side of Iapetus. The terrane model authors suggested that the boundary of this terrane lies in the eastern Blue Ridge (Fig. 2) and extended it northward to the Warrenton area, in northern Virginia. There the proposed terrane boundary was considered to exit the Blue Ridge province through an incursion of metasedimentary rocks that apparently cut through the Catoctin Formation outcrop area, based on available geologic mapping. In this model, the Catoctin south of the incursion was considered to be part of the accreted terrane and fortuitously aligned with the true Laurentian Catoctin north of the incursion. The boundary of the Jefferson terrane was correlated with a proposed Taconian suture in New England.

The position of these proposed boundaries was based largely on the presence of mafic and ultramafic bodies in a rock sequence that was interpreted to be a tectonic mélange. However, these mafic and ultramafic rocks actually occur within a conformable sequence of metasedimentary rocks deposited upon the Blue Ridge basement complex (e.g., Conley, 1978; Wehr, 1983; Wehr and Glover, 1985). In contrast to the "terrane" approach, the model proposed by Wehr and Glover (1985) and Glover (1989, 1992), and supported by Kline (1991), suggested that the eastern part of the Blue Ridge geologic province is composed entirely of Laurentian rift and drift facies rocks and that the Taconian suture lies farther to the east.

The objective of this study is to resolve the structural and stratigraphic framework of the eastern Blue Ridge in the area near Warrenton. New, 7.5-minute-scale, geologic mapping data are presented and compared with those of previous work in the

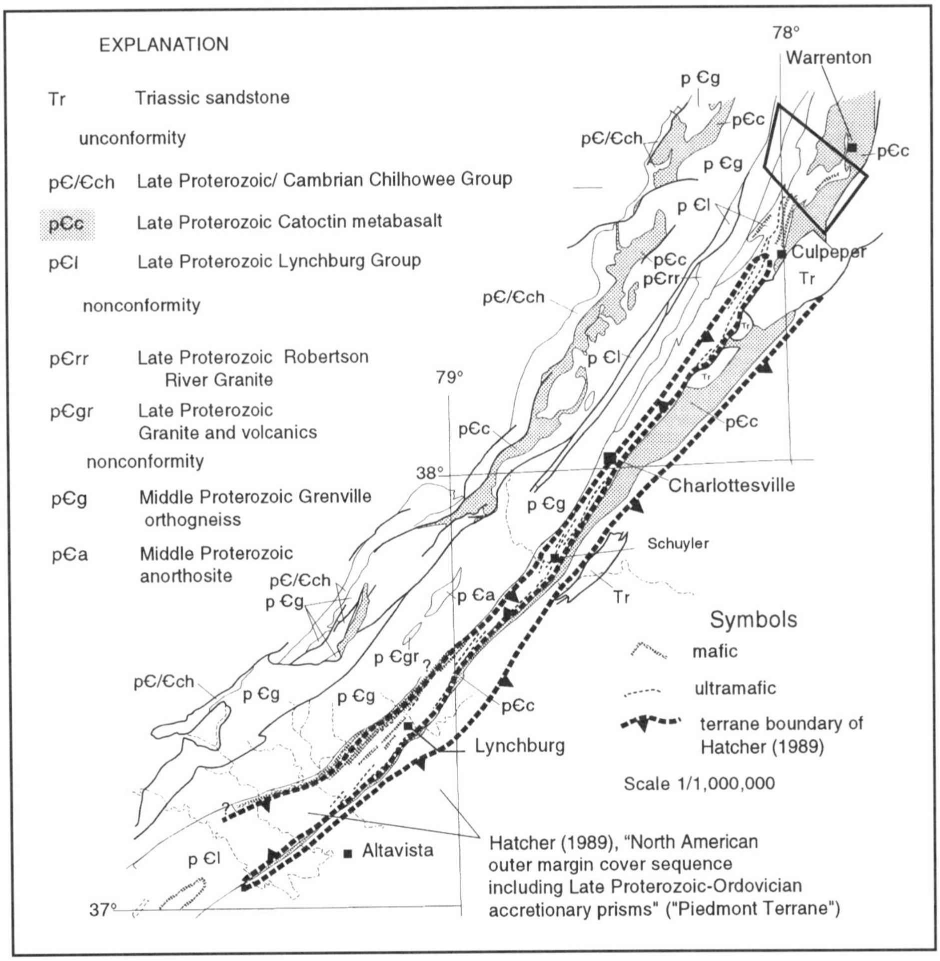

Figure 1. Virginia Blue Ridge, showing terrane boundary of Hatcher (1987) and Hatcher et al. (1989). Modified from Glover et al. (1992). Study area in heavy outline near Warrenton at northeast end of map.

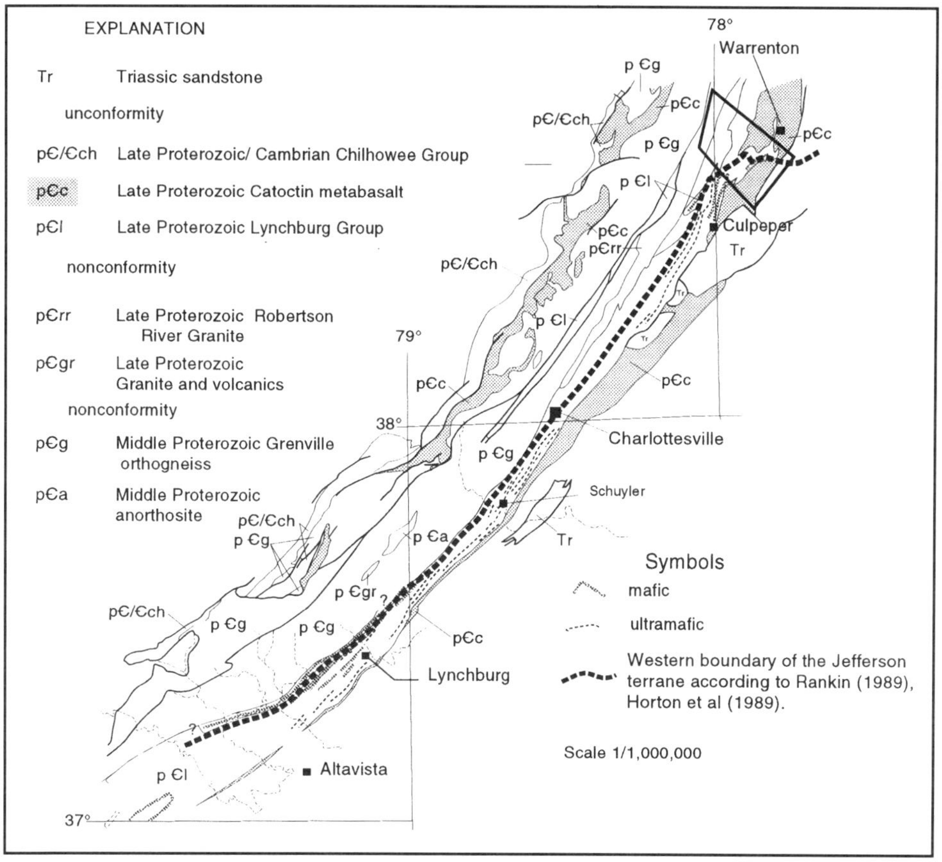

Figure 2. Virginia Blue Ridge, showing terrane boundary of Rankin et al. (1990a) and Horton et al. (1989). From Glover and others (1992). Study area in heavy outline near Warrenton at northeast end of map.

region. Subsequently, a model for the tectonic evolution of the area during the Paleozoic orogenic events is presented. The results of the study are used to test the current tectonic models for the central Appalachians.

REGIONAL STRATIGRAPHY

The crystalline basement complex of the eastern Blue Ridge province in northern Virginia is an assemblage of Middle Proterozoic (Grenvillian) gneisses intruded by the Middle Proterozoic Marshall Metagranite and by Late Proterozoic granitoids of the Robertson River Igneous Suite (or Crossnore Igneous Suite) (Bartholomew and Lewis, 1984; Clarke, 1984; Lukert and Banks, 1984; Tollo et al. 1991; Tollo and Aleinikoff, 1992).

In the study area (Fig. 3), the basement rocks consist of the Marshall Metagranite, local occurrences of small felsic intrusions probably related to the Robertson River magmatism, and a few narrow metadiabase dikes. In this study the basement rocks were not divided.

Unconformably above the basement, a thick succession of Late Proterozoic greenschist facies clastic rocks (Fauquier and Lynchburg groups) is overlain by a thick pile of metamorphosed mafic volcanic and volcaniclastic rocks of the Catoctin Formation. The clastic rocks of the Fauquier and Lynchburg Groups show a general fining upward. Coarse arkosic sandstones with minor conglomerates in the lower part of the sequence are overlain by fine-grained feldspathic and quartzose sandstone and siltstone, capped by mudstone and black schists containing isolated sandstone layers. Generalized stratigraphic columns and their correlation are shown on Figure 4. These rocks are interpreted to form a continuous retrogradational sequence that is transitional from alluvial to marine-slope to deep-water deposits (Wehr, 1985, and this study).

Voluminous metamorphosed tholeiitic basalts of the Catoctin Formation (570 ± 36 Ma, Badger and Sinha, 1988) cover a large area above the metasedimentary rocks and farther north onlap the basement. North and west of the study area the Catoctin is overlain by the rift and drift sedimentary rocks of the Early Cambrian and Late Proterozoic Chilhowee Group. Metamorphosed mafic dikes and sills, probably Catoctin feeders, intrude the basement and pre-Catoctin metasedimentary rocks.

The cover rocks on the eastern and western flanks of the Blue Ridge anticlinorium are interpreted to constitute the Late Proterozoic to Early Cambrian rift and drift sequence developed on the edge of the Laurentian continent (Wehr and Glover, 1985). The cover rocks in the study area are part of the Late

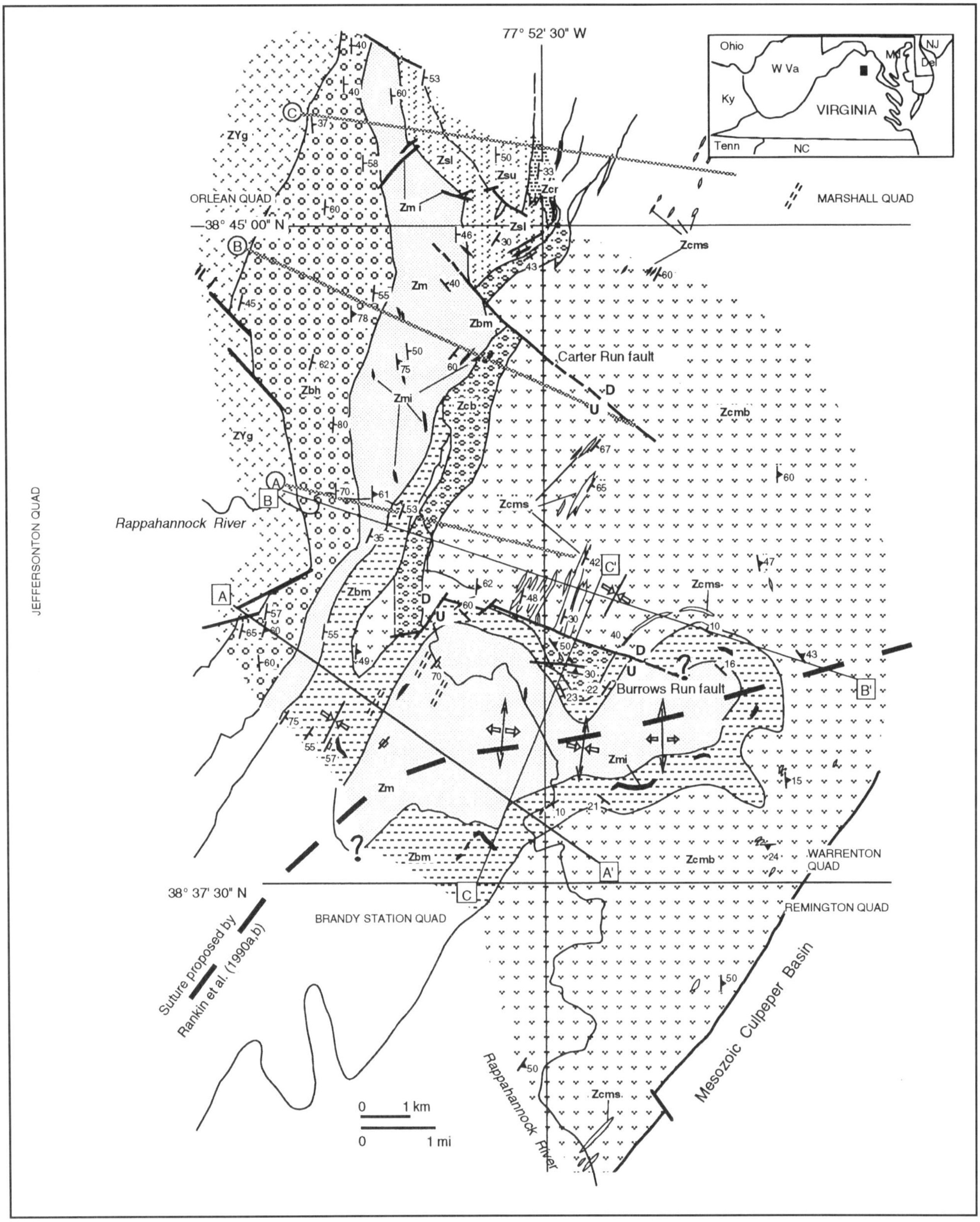

Figure 3. Generalized geologic map of the study area (patterned) near Warrenton, Virginia (explanation on the next page).

DESCRIPTION OF MAP UNITS

Catoctin Formation (Late Proterozoic)

Zcmb — Metabasalt member -- Medium to dark green, metabasalt : massive and foliated greenstone, greenschist, and epidosite.

Zcb — Metabasalt breccia member -- Light to dark green, thickly (0.5-2 m) layered, metabasalt volcanic breccia. They form crudely defined layers of "coarse"(mainly cobble size angular to subrounded clasts in a lighter colored matrix) and "fine"(scattered angular granule-size clasts in a light green aphanitic matrix.

Zcms — Metasandstone member -- Whitish to light gray, medium bedded, medium- to coarse-grained quartz and quartzofeldspathic arenite -- Quartz muscovite schist with granule size clasts of blue and white quartz.

Fauquier Group (Late Proterozoic)

Zcr — **Carter Run Formation** -- Gray to dark gray and whitish, laminated mudstones and siltstones. -- Whitish to light gray, very thick bedded to massive, dominantly medium- to coarse-grained, typically unsorted , quartz and feldspathic wackes, with local rip-up clasts of dark gray to black mudstone.

Swains Mountain Formation

Zsu — Upper Member -- Whitish to gray, massive , medium grained, quartz arenites and wackes. -- Whitish to gray, typically cross-bedded, medium- to fine-grained, quartz arenites, -- Rare interlayers of: gray, very fine sand and silt.

Zsl — Lower Member -- Gray, massive, medium-grained quartzofeldspathic wackes -- Gray, massive to thick-bedded, graded granule-conglomeratic and coarse-grained sandstones and thin interlayers (5-15 cm) of Gray mudstones and siltstones.

Zm — **Monumental Mills Formation** -- Medium to dark greenish and locally whitish, laminated to thinly bedded siltstones and mudstones and rarely fine sandstones. Sandstones commonly have scoured bases with larger grains near the base -- Greenish and gray mudstones and siltstones typically interlayered with fine to coarse, medium bedded, quartzofeldspathic wackes.

Zbh — **Bunker Hill Formation** -- Whitish to light gray coarse- to very coarse-grained feldspathic and quartz arenites with common granule conglomerates, commonly graded and/or cross-bedded. -- Light to medium gray, matrix supported granule conglomerates with rare small pebbles, with variyng ammounts of mudstone, siltstone or fine sandstone matrix, and dominantly feldspar clasts, commonly showing a crude gradation. -- Light to medium gray, medium grained, medium bedded (10-15 cm) feldspathic wackes, interlayered with thin beds (5-8 cm) of gray fine sandstone and siltstone. -- Minor occurences of dark greenish laminated and thinly bedded mudstones (thin unit : 2-3 m, near the base.

Lynchburg Group (Late Proterozoic)

Zbm — **Ball Mountain Formation** -- Dark gray to black and white, laminated, mudstones and solid black schists. -- Whitish to medium gray, thick-bedded to massive medium- to coarse-grained, quartz and quartzofeldspathic sandstones -- Uncommon medium- to dark-gray, medium- to thick-bedded, granule to pebble conglomerates

Zm — **Monumental Mills Formation** -- See Fauquier Group

Zbh — **Bunker Hill Formation** -- See Fauquier Group

........................Unconformity..

Zmi — **Mafic intrusive rocks** (Late Proterozoic ?) -- Aphanitic and fine grained metadiabase, metagabbro, and locally metapyroxenite.

ZYg — **Metamorphosed granitic rocks, undivided** (Late and Middle Proterozoic) Basement rocks, including pegmatites of the Late Proterozoic Robertson River Igneous Suite, Middle Proterozoic (Grenville ?) metagranite (augen gneiss), and Middle Proterozoic Marshall Metagranite (fine-grained and coarse grained facies).

Ⓐ Ⓑ Ⓒ Summary stratigraphic columns of Figure 4

[A] [B] [C] Structural cross sections in Figure 8

EXPLANATION OF MAP SYMBOLS

		(vertical)	(inclined)		
	minor tight fold axes			bedding	Fault :
	anticlinal axis			discrete cleavage	U / D upthrown - downthrown block
	synclinal axis			pervasive foliation	

Figure 3. Explanation of the geologic map.

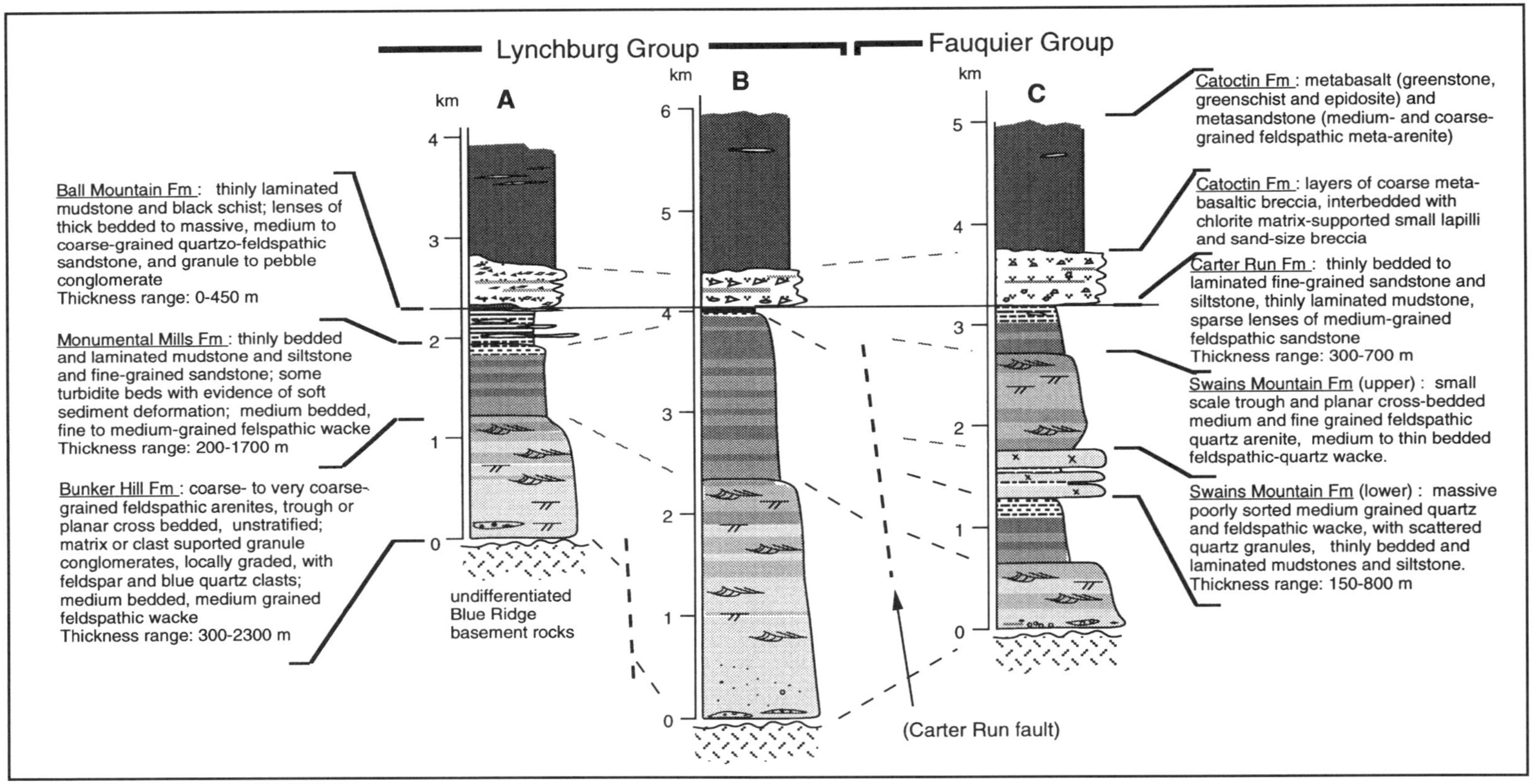

Figure 4. Generalized stratigraphic columns from three geologic map transects in the study area. Lithologic description of each unit of the clastic cover rocks (Fauquier and Lynchburg Groups) is given in terms of the protolith. The three columns are correlated on the base of the Catoctin volcanic rocks, which is interpreted to be a chronostratigraphic surface. The differences in thickness of individual units and the total clastic sequence are attributed to different subsidence histories in different parts of the rift basins, which are interpreted to be controlled by faults. (See Fig. 3 for locations.)

Proterozoic rift sequence of the southeastern Blue Ridge. Lateral facies changes observed in the clastic sequence are related to oblique orientation of the presently exposed section with respect to the axis of the rift basin. In general the rift sequence is of deeper-water origin in a southwesterly direction along strike (Wehr and Glover, 1985). Drift facies rocks do not crop out in the map area.

The observed stratigraphy (Fig. 4) and continuity of formations through the mapped area (Fig. 3) indicate that the suture of Rankin et al. (1990b) does not exist there (Fig. 3).

STRUCTURAL GEOLOGY

General description

The map-scale structure in the area varies from east-dipping (D1 fold) strata directly above the eastern Blue Ridge basement, to a thrust fault(?) or tight folds at the D2 structural front 3 km farther east, and finally to open folding and gently dipping strata several kilometers still farther east (Fig. 3; Table 1; refer to Fig. 10). Thus, the westernmost syncline divides the study area into two parts that have different structural characteristics.

The cover rocks in the study area have one pervasive foliation, S_1, that consistently strikes north-northeast and dips to the east, except in the eastern part of the study area, where the rocks were affected by D2 folding and faulting (Fig. 3) after the development of S_1. Intensity of foliation varies by rock type, but most rocks are weakly foliated. An extensional lineation is present locally in the eastern part of the area. A localized discrete cleavage associated with D2 folding occurs in the finer-grained rocks (mudstone and siltstone of the Ball Mountain Formation).

TABLE 1. SUMMARY OF STRUCTURAL AND METAMORPHIC EVENTS

	Blue Ridge Basement Complex	Blue Ridge Cover Rocks: Rift Sequence
Late Paleozoic (Alleghanian)		**D2** (Broad open and tight folds **F_2**) (Localized discrete cleavage **S_2**) (Activity of major faults)
Early Paleozoic (Cambrian or Taconic?)	Greenschist facies metamorphic overprint (Penetrative foliation) (Ductile deformation zones) (Phyllonite zones)	Greenschist facies metamorphism **M1** **D1** (Pervasive foliation) **S_1**
Late Proterozoic	Extensional faulting **D_{z2}**	
	Ductile extension? **D_{z1}** (Extensional fabric?)	
Middle Proterozoic	High grade metamorphism and deformation **D_{gr}/M_{gr}** (Grenvillian relict fabric)	

A summary of structural and metamorphic events is shown in Table 1. Based on the structural styles in the study area, four domains were established: (1) the basement complex, (2) the east-dipping cover sequence that lies above the basement (refer to domain A in Fig. 10), (3) the tightly folded part of the cover sequence (refer to domain C in Figs. 10, and (4) the open-folded cover sequence (domain B, Fig. 10). The distribution of structural data is not uniform over the area of study. Plots in Figure 10 indicate the abundance of data collected in the study area. Mapping was done at 7.5-minute scale and generalized to produce the figures in this chapter.

Bedding planes (S_0) and the pervasive foliation (S_1) of domain A intersect at low angles and dip moderately southeastward. In domain B, S_0 reflects the subhorizontal to gently domal structure of that area (Fig. 3), and the distribution of S_1 poles reflects the effect of D2 domes. Domain C shows the tight folding of S_0 and the local effects of drag adjacent to the Burrows Run fault (Figs. 3, 5).

The AB domain boundary is coincident with the possible east-dipping D2 thrust fault (mentioned later in the description of the Burrows Run fault) that would relieve a space problem in the east limb of the syncline, which contains a thinner sequence of Catoctin than the west limb. The AC domain boundary corresponds with the western boundary of D2 high-amplitude, short-wavelength folds (Figs. 3, 5).

The Burrows Run fault (Figs. 3, 5) serves as a D2 transfer fault to connect these deformation fronts.

Grenvillian structures (D_{gr}/M_{gr})

The Marshall Metagranite in the study area was deformed and metamorphosed to a high grade during the Grenvillian orogeny (e.g., Bartholomew and Lewis, 1984). Middle Proterozoic magmatism produced the Metagranite (Clarke, 1984; Espenshade, 1986). An obscure relict Middle Proterozoic foliation is locally visible in some of the basement rocks near the study area. In the Marshall quadrangle this fabric was described by Espenshade (1986) as "inconspicuous compositional layering."

Late Proterozoic structures (D_{z1} and D_{z2})

The Late Proterozoic felsic magmatic rocks located just west of the study area are considered to be anorogenic and related to an early stage of Iapetan rifting (Tollo and Aleinikoff, 1992). Elsewhere in the nearby Blue Ridge, the basement rocks have a Late Proterozoic fabric related to crustal extension (D_{z1}) during the initial rifting of the continent (Evans, 1991; Simpson and Kalaghan, 1989; Kline, 1992; Kline, 1994). Such fabrics were not identified in the study area.

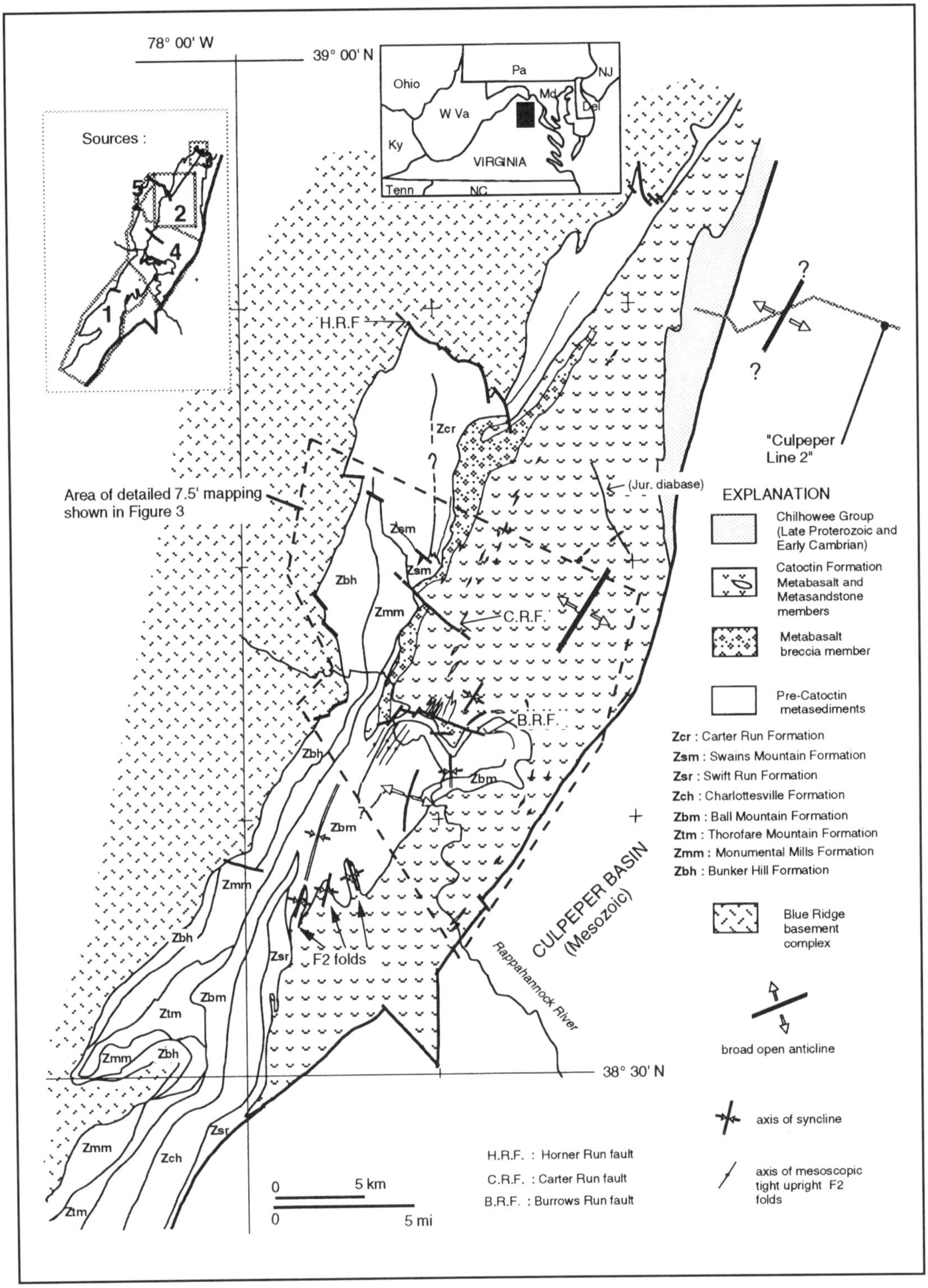

Figure 5. Generalized geologic map of the eastern Blue Ridge in the Warrenton-Culpeper area, showing the main structural features. Compiled from (1) Wehr, 1983; (2) Espenshade, 1986; (3) Kline et al., 1991; (4) this study; (5) P. Lyttle, personal communication, 1992. Other boundaries from the geologic map of Virginia (Virginia Division of Mineral Resources, 1993).

Late Proterozoic extensional tectonics resulted in brittle faulting (D_{z2}). This faulting probably occurred prior to and during deposition of the basal clastic units. The map pattern (Fig. 3) indicates that some faults affected only the basement-cover contact and appear to correspond to abrupt lateral changes in thickness of the basal stratigraphic unit. Features such as the Horner Run Fault (Fig. 5) indicate that faulting continued during the extrusion of the Catoctin Formation as well.

Ductile Paleozoic structures (D1)

D1 in the basement. A pervasive, but weakly developed, foliation is present in almost all outcrops. It strikes north-northeast and dips moderately to steeply to the east. Several phyllonitized zones (ductile deformation zones), from tens of centimeters to a few meters wide, occur in sets that are oriented parallel to the foliation. In these zones the feldspars are completely sericitized, and the resulting matrix wraps around the quartz grains. The foliation and phyllonite zones bear resemblance to the fabrics attributed by Kline (Kline, 1991; Kline, 1994) to Late Proterozoic extension, but field relations in the study area are inconclusive as to age of fabric development. The degree to which finite strain in the basement can be attributed respectively to Late Proterozoic extension and to Paleozoic D1 deformation is uncertain.

Numerous other thin (centimeter-scale) ductile deformation zones, reflecting heterogeneous accommodation of strain in the basement rocks (described by Mitra, 1979), are oriented in many different directions and are spaced from a few centimeters to a few tens of centimeters apart. According to Mitra (1979) initial deformation in homogeneous basement rocks produced conjugate shear fabrics and thin ductile deformation zones related geometrically to the regional cleavage. With continued deformation, all conjugate shear fabrics merge to become the regional (D1) foliation. These structural elements recognized locally in the basement of the study area are also similar to those described by Espenshade (1986) in the Marshall quadrangle to the north and are considered to be Paleozoic in age. In this study only reconnaissance work was carried out in the basement.

D1 in the cover rocks. A more pervasive S_1 foliation was impressed on the cover rocks. In the coarse- and medium-grained metasandstones of the Bunker Hill, Swains Mountain, Ball Mountain, and Catoctin Formations, the foliation is a weak "schistosity" (classification of Powell, 1979) or a weak "continuous cleavage" (classification of Barker, 1990). The S_1 is defined by the preferential orientation of fine grained muscovite and biotite and very small quartz grains. Muscovite, grown from feldspar, also appears to be preferentially oriented (Fig. 6a). Larger grains of quartz show undulose extinction but no other evidence of dynamic recrystallization, and some grains are fractured (Fig. 6b). Commonly, feldspar grains show evidence of cataclastic deformation and sericite fills the fractures.

Typically, fine-grained metasandstones in the upper part of the Bunker Hill and the lower part of the Monumental Mills Formations have a weak to very weak schistosity defined by biotite (Fig. 6c). Sparse sodic plagioclase grains are present. In general, the protolith sedimentary texture and grain size of the rock appear to be well preserved. In metamorphosed very fine grained sandstone, siltstone, and mudstone of the upper Monumental Mills Formation, the S_1 foliation is defined in a similar manner.

Laminated black and white metamudstones and metasiltstones of the Ball Mountain Formation have a very weak schistosity, like the fine-grained rocks in the Monumental Mills Formation. Black quartz–microcrystalline mica-graphite schist found in several locations appears to be more intensely deformed than the other very fine grained metasediments. It has a penetrative foliation defined by elongate, very fine (0.01 to 0.2 mm wide and 0.07 to 0.1 mm long) quartz and mica grains oriented at a very low angle to the bedding. A fine compositional layering is defined by recrystallized quartz-grain layers, ranging in thickness from 0.1 to 1 to 2 mm (Fig. 6d). These layers are continuous in outcrop and probably represent bedding laminae locally disrupted. Similar laminae were also observed in less-deformed, black metamudstone.

Catoctin metabasalts (greenstones and greenschists) also have one generally pervasive foliation (S_1) exhibiting various degrees of development. Greenschist is strongly foliated because of planar orientation of the minerals chlorite and actinolite. Large aggregates of epidote and quartz in the greenschist (epidosite) are massive and fractured; foliation in the greenschist wraps around the epidosite. Other greenstones have a very weak foliation, also defined by chlorite and actinolite.

A linear fabric (L) is present locally in the Ball Mountain and Catoctin Formations. In gently east-dipping black schists of the Ball Mountain, exposed along the Rappahannock River, a lineation is well defined by thin pressure fringes of quartz around small (1-mm) pyrite crystals. The pressure fringes are elliptical in top view. The foliation lies at a very low angle to bedding, and the lineation occurs on the foliation plane. In the southern part of the Marshall quadrangle a similar lineation is defined by chlorite-rich aggregates in the greenschist. This lineation, also recognized by Espenshade (1986), trends on average S 10 to 40° E and plunges 30 to 50°. This linear fabric is interpreted to be a Paleozoic lineation showing the direction of maximum elongation strain. D1 was a relatively weak deformation except in the least competent units where strain was intense.

The Late Proterozoic to Early Cambrian metasedimentary and metavolcanic cover rocks, which lie unconformably upon the basement on the east limb of the Blue Ridge, experienced one low-grade progressive metamorphism. Deformation was minimal and took place in two stages: D1 involved a pervasive foliation (S_1) in varying degrees of intensity throughout the sequence.

Paleozoic deformation D2

In the study area, the event appears in the cover sequence only in the eastern part. This deformation formed tight mesoscopic folds with localized (S_2) cleavage, broad open folds, and major faults.

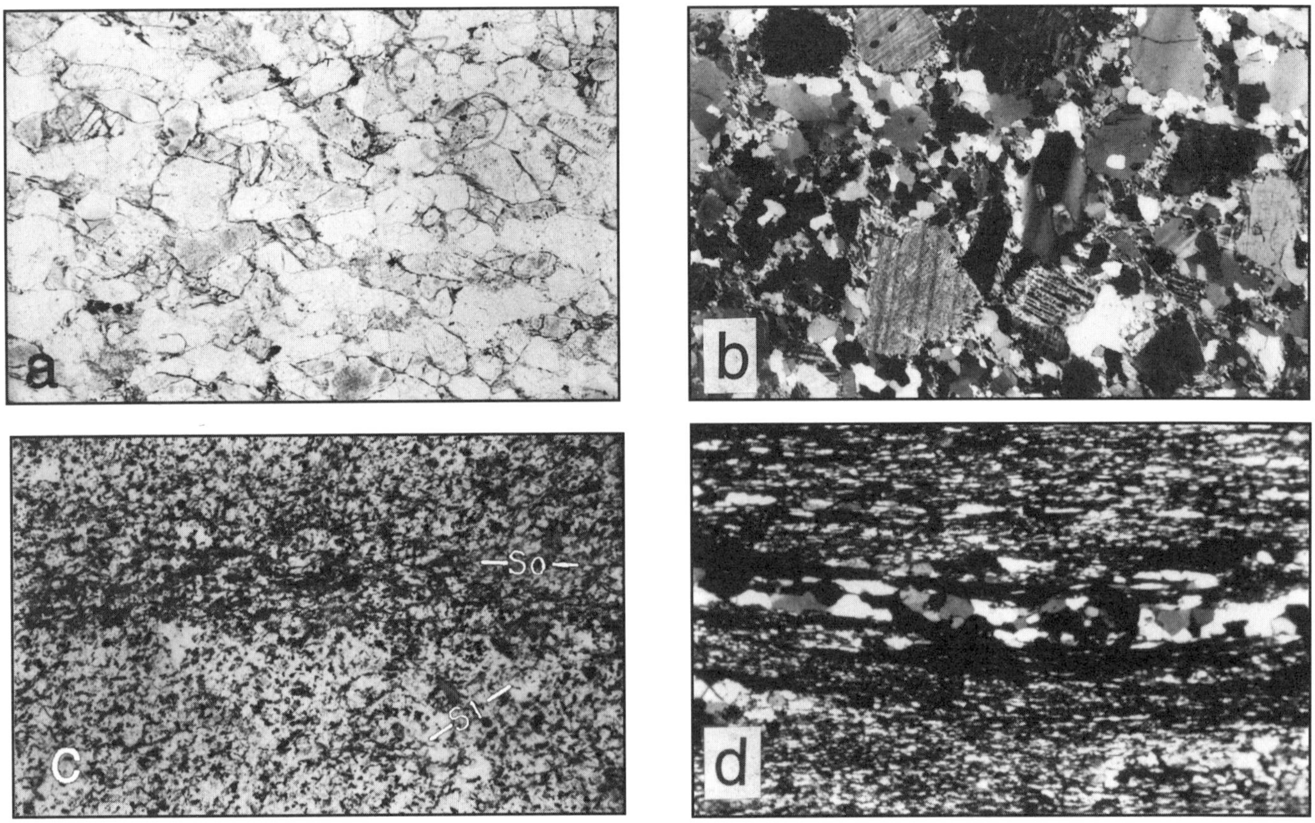

Figure 6. Deformation in different rock types. (Horizontal field of view: 6.2 mm.) a, Well-sorted feldspathic arenite of the Bunker Hill Formation. Muscovite grown from feldspars shows a weak preferred orientation, crudely defining the S_1 foliation from upper left to the lower right corner of the photo (plain light). b, Feldspathic sandstone of the Bunker Hill Formation with very weak preferred orientation of grains. Feldspars are mainly perthite and microcline and sparse plagioclase (cross-polarized light). c, Siltstone to fine sandstone in Monumental Mills Formation. The compositional layering defines the bedding (S_0), and the preferred orientation of biotite defines the weak pervasive fabric (S_1) (plain light). d, Quartz-graphite schist of the Ball Mountain Formation. S_1 is nearly parallel to bedding, which is defined by a lamina of coarser quartz crystals. In this lamina, quartz grains have been recrystallized to equant larger crystals, whereas the presence of very fine mica in the finer material probably contributed to maintaining the finer grain size.

Tight mesoscopic folds and S_2 cleavage. In the area south of the Burrows Run fault zone and east of the westernmost syncline (Figs. 3 and 5), there is a set of relatively closely spaced tight folds, ranging from outcrop scale to several tens of meters in wavelength. These folds are present in the finer-grained mudstones and siltstones of the Ball Mountain Formation and the upper part of Monumental Mills Formation. In their hinge areas a discrete crenulation cleavage (S_2) is commonly developed. In laminated mudstone-siltstone, S_2 is localized in mica-rich layers but becomes weaker or disappears in layers rich in quartz sand (Fig. 7). This cleavage cuts S_1 and strikes approximately N30° E and dips at 60 to 70° SE. Thus it formed during a later deformational event, possibly in a later orogeny. The location of tight folds is best shown on profile A-A′ (with Figs. 3, 5, and 8).

Broad open folds. In the area south of the Burrows Run Fault and east of the tight folds (Figs. 3 and 5) bedding planes and lithologic contacts define broad open folds with axes plunging 5 to 10° to the north and south and limbs that dip as much as 20 to 25°. The resulting domal structures caused the present-day map pattern (Figs. 3 and 8). Immediately south of the Burrows Run fault the metabasalt breccia member of the Catoctin Formation dips moderately to the north-northeast (Fig. 8, profile C-C′). The open folds appear to postdate S_1 in that the orientation of S_1 in the eastern part of the study area wraps around the dome (Fig. 3).

Faults. The Burrows Run fault (Figs. 3 and 5) truncates the Catoctin, Ball Mountain, and probably Monumental Mills Formations. To the west, the fault connects with a system of smaller faults. The fault could not be traced farther east.

The northern block of the Burrows Run fault is a synformal succession of Catoctin metabasalts with lenses of metasandstone (Fig. 3; Fig. 8, profile B-B′). The east limb of

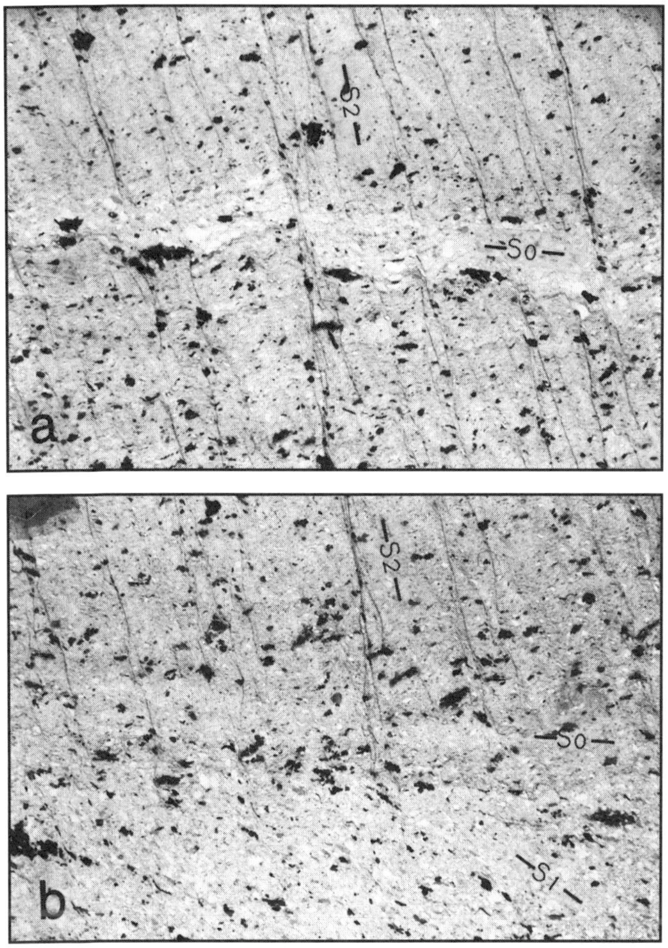

Figure 7. Discrete cleavage (S_2) in laminated meta-mudstone and meta-siltstone. a, Cleavage is less pronounced in the thin layer with more quartz grains (horizontal field of view: 3.4 mm, plain light). b, Pervasive fabric (S_1) is barely distinguishable in the silt layer (lower half) of the photograph, forming an angle of approximately 45° with S_2, which is pronounced in the mudstone layers (upper half) (horizontal field of view: 6.2 mm, plane polarized light).

the syncline contains fewer metasandstones. The west limb also has a much thicker Catoctin sequence than the east limb. The asymmetries of dip (refer to Fig. 10), thickness, and rock types suggest internal deformation in the Catoctin Formation in this area. However, exposure in the area of the syncline is poor, and the exact nature of the internal deformation is uncertain. Perhaps there is an east-dipping thrust fault. Tight folds are considered less likely because of the thickness of the metabasalts. The hypothetical east-dipping thrust is not shown on the maps and cross sections (Figs. 5, 8, and 9) but would be located approximately at the northeast-trending boundary between structural domains A and B in Figure 10.

The southern block of the Burrows Run fault has closely spaced, tight folds on the west and broad open folds of Ball Mountain and Monumental Mills metasedimentary rocks (Fig. 8, profile A-A′) on the east. Along the trend C-C′ (Fig. 8), the area immediately south of the fault shows a gentle north-northeast–dipping monocline. The change in attitude of the clastic sequence, from east-dipping to folded, occurs west of the strike of the synclinal axis north of the fault (Figs. 8 and 10).

Interpretation. The sense of displacement of the Burrows Run fault is complex. It is evident that the southern block is upthrown relative to the northern block, because most of the Catoctin Formation in the southern block was eroded away. The apparent monoclinal fold subparallel to and just south of the fault (Fig. 3) is also consistent with a relative upward motion of the southern block, and it can be interpreted as a drag fold. The magnitude of the dip-slip displacement appears to vary along the length of the fault because the sequence folded differently north and south of the fault (Figs. 3, 5, and 8).

The pervasive foliation, S_1, in the area is deformed by the fault. S_1 dips vary consistently from east-southeast north of the fault to north-northeast south of the fault (Fig. 10). Thus, the fault must postdate the deformational event D1.

The fault was probably active before, during, and after the folding because it separates two blocks that were deformed differently. Differential folding and perhaps thrusting on either side of the fault related to lateral displacement may explain the asymmetry (Nick Evans, personal communication, 1994). Tight compressional folding that appears to be associated with the fault is atypical of Mesozoic deformation in the region, and the Burrows Run fault is herein interpreted to be Paleozoic in age and to have formed during the D2 deformational event.

The Carter Run fault. The Carter Run fault truncates the Catoctin Formation with an apparent left-lateral displacement (Figs. 3 and 9a), well displayed by offset of outcrop belts of metabasalt breccia. The thickness of the breccia unit does not change across the fault, as for example near the Horner Run fault zone (Fig. 5) of Espenshade (1986). However, intercalated sandstones increase in thickness and lateral continuity south of the fault.

The position of the Carter Run fault is also related to significant changes in the stratigraphy of the pre-Catoctin clastic sequence. North of the fault, the rock types of the Monumental Mills Formation are overlain by rocks of the Swains Mountain Formation, whereas to the south of the fault the Monumental Mills is overlain by the Ball Mountain Formation, which pinches out toward the fault.

Interpretation. Lateral changes in stratigraphy lead to two suggestions. First, the Carter Run fault was initially active in pre-Catoctin time. The southern termination of the lower Swains Mountain Formation, and northward thinning of the Monumental Mills Formation, might be related to normal faulting in the Late Proterozoic, where the northern block was downthrown with respect to the southern block (Fig. 9a). However, this does not explain the absence of the Ball Mountain Formation north of the Carter Run fault. Second, because the entire sequence of formations is submarine and there is no physical evidence of subaerial erosion, Late Proterozoic sliding and submarine channeling on

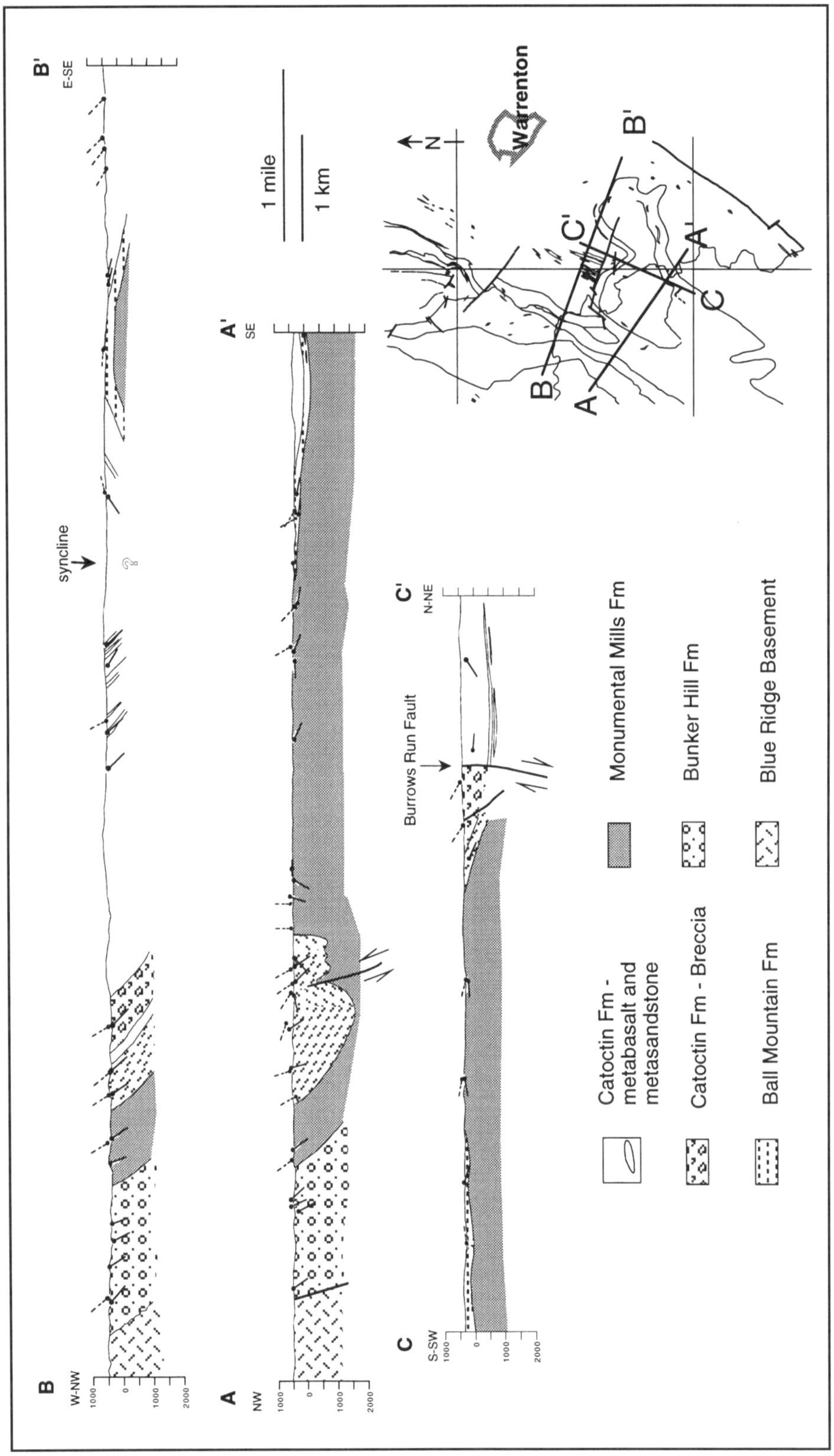

Figure 8. Cross sections of the study area. Dashed bars represent dip of the dominant foliation; solid bars represent dip of bedding surfaces. Data projected from within 0.3 km on either side of the profile. Horizontal and vertical scale are equal. Elevation is given in feet.

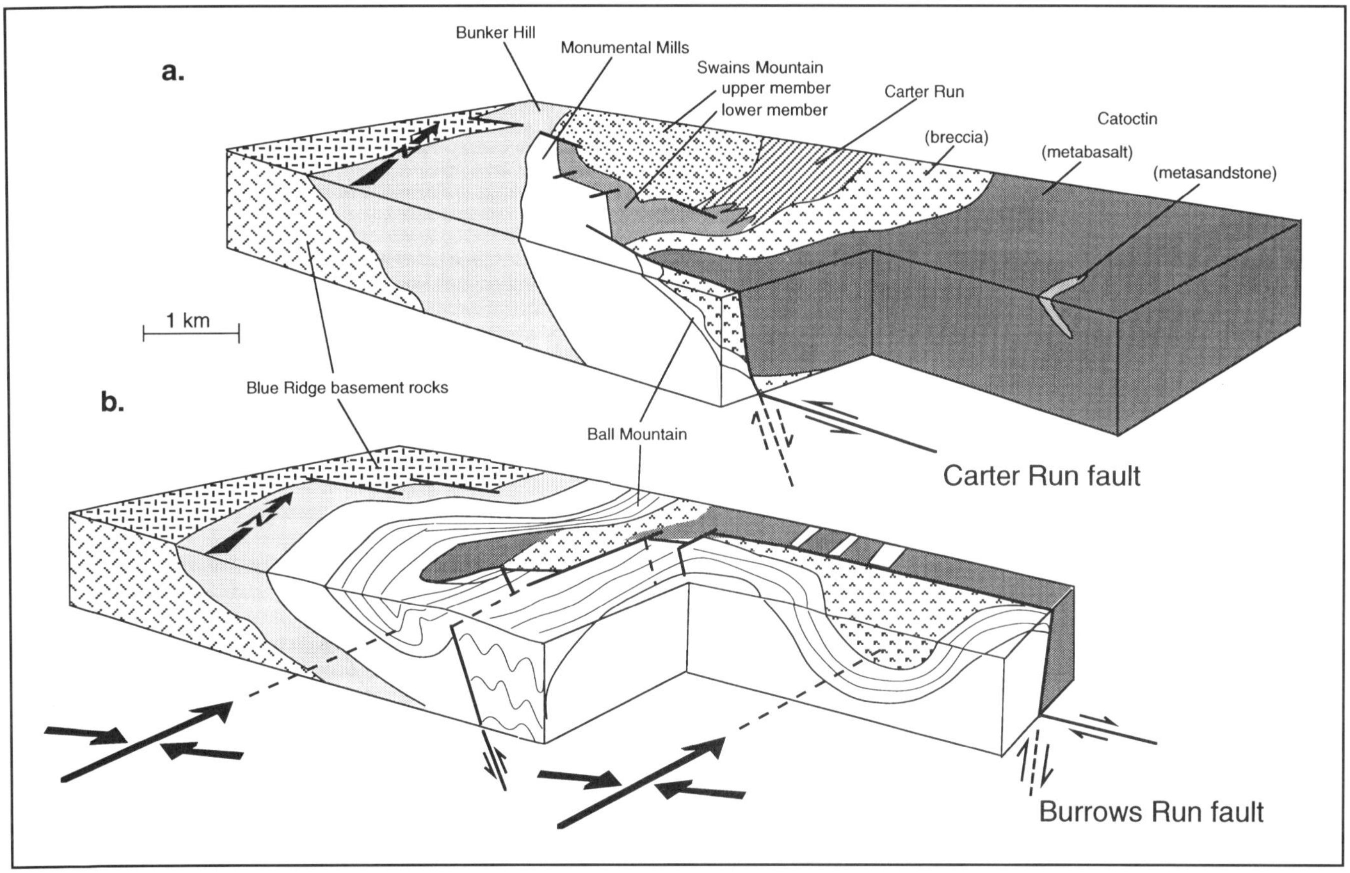

Figure 9. Block diagrams along two transects, approximately northwest-southeast, in the study area showing the present-day structure near the two major faults.

the sea floor during an episode of rifting may have cut out the Ball Mountain and part of the Monumental Mills. If so, this occurred before deposition of the Swains Mountain and Carter Run Formations.

Although the Carter Run fault does not appear to have affected the thickness and accumulation of the Catoctin volcanic breccia, the offset of the breccia as well as the western interval of Catoctin metasandstones indicates that the fault was active in late- or post-Catoctin time. This complex history of Late Proterozoic and Paleozoic(?) movement was probably related to Late Proterozoic rifting and Paleozoic D2 (?) shear accommodating, in part, the map-scale deformation during D2 folding. Mesozoic movement may have occurred on these faults, but no evidence was found.

Regional structural relations

South of the Burrows Run fault along the Rappahannock River, the belt of mesoscopic tight F_2 folds opens to merge with a broader belt of F_2 folds mapped by Wehr (1983). Perhaps the axis of the westernmost syncline may continue into the Charlottesville Formation of Wehr (1983), which appears to pinch out to the north. The Charlottesville may lie in the core of a syncline not recognized by Wehr; however, this speculation is barely justified by the poor outcrop found in the area.

As discussed earlier, the dip of rocks near the base of the Catoctin and a few bedding attitudes within the Catoctin suggest a broad anticlinal structure just north of the Burrows Run fault. This anticline may continue northeastward into the southeastern corner of the Marshall quadrangle, where its presence was suggested by Espenshade (1986). A broad open antiformal structure also appears on seismic reflection data even farther north under the Culpeper Basin ("Culpeper Line 2" in Costain and Çoruh, 1990) (line location shown in Fig. 5, this chapter). The geometry of this subsurface structure is only seen in profile; however, it appears to be on strike with the one identified on the surface.

In the Marshall quadrangle, Espenshade (1986) reported common south-southwest–plunging minor folds (shown on Fig. 3), which if projected to the southwest appear to be in a similar position just east of the westernmost syncline. Although the sequence of D2 folds north and south of the Burrows Run fault is not exactly the same, they are an expression of the same style and age of deformation. This style is characterized by tight mesoscopic folds just east of the westernmost syncline and broad open folds eastward.

Projection of the surface of Grenville basement eastward to

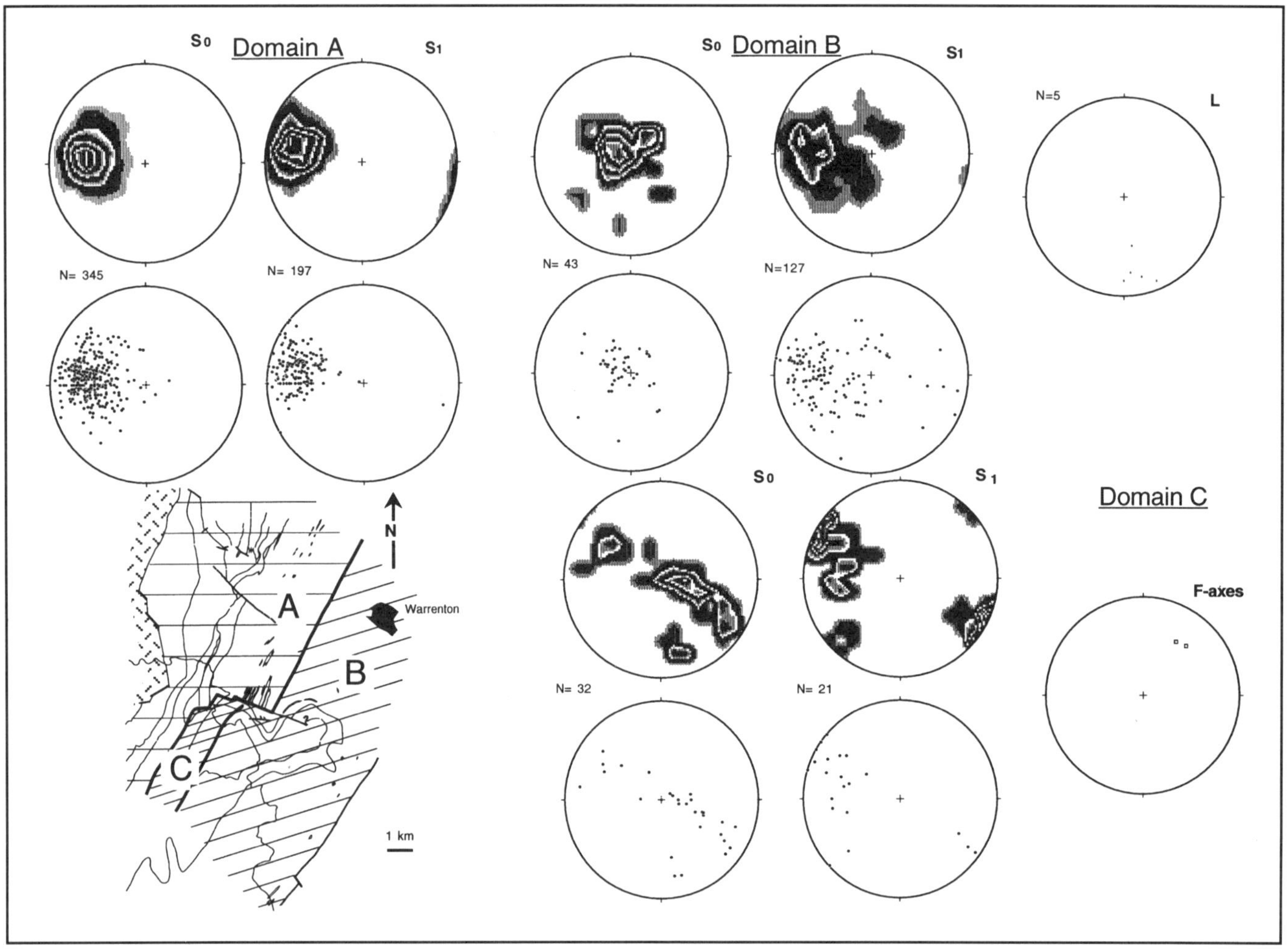

Figure 10. Scatter and contoured plots to pervasive fabric planes, and bedding planes S_0, for the three structural domains of the cover rocks in the study area. Domain A includes the east-dipping cover rocks that lie above the basement. Domain B contains the open-folded part. Domain C contains the tightly folded part. All plots are lower-hemisphere projections. Contours at 1, 2, 3, 4, 5, 6, and 7% per 1% area using "Stereonet" v. 3.1.

its intersection with the Blue Ridge décollement (seismic reflection estimate at about 3 km below the surface) (data from Costain and Çoruh, 1990) places the fault in the subsurface about 2 km east of the surface domain boundary AC. The domain boundaries AB and AC may have been localized to some extent during D2 deformation by the eastern edge of the more competent basement in the subsurface. Stratigraphic and structural data show that the mapped metasedimentary incursion (Furcron, 1939) that Rankin et al. (1990a) thought marked the position of a suture truncating Catoctin stratigraphy south of Warrenton is a cross fold with the crest eroded through the Catoctin Formation, exposing the underlying Lynchburg/Fauquier stratigraphy (Fig. 3).

METAMORPHISM

Petrographic observations on samples from the study area suggest that the cover rocks have been subjected to one phase of metamorphism, at greenschist facies (biotite grade).

The various rock types in the area record the metamorphic conditions in different ways. Coarse-grained metasedimentary rocks (coarse- and medium-grained sandstones) present in the Bunker Hill, Ball Mountain, Swains Mountain, and Catoctin Formations contain original detrital grain mineralogy, whereas the fine-grained sandstones and siltstones contain recrystallized detrital and metamorphic grains. In the coarse- and medium-grained feldspathic metasandstones, the metamorphic minerals in the matrix are muscovite, biotite, and quartz. Detrital microcline and perthite are thoroughly altered to sericite. Sodic plagioclase shows little alteration, and sparse calcium-rich plagioclase is saussuritized (e.g., Fig. 6b).

Fine-grained metasandstones and metasiltstones of the Monumental Mills Formation commonly have different bulk compositions. Variations can be observed even between adjacent thin beds. Individual compositional layers are defined by different concentrations of the various minerals (Fig. 6d). Typical assemblages include quartz + biotite + muscovite ± albite and/or quartz + epidote + chlorite ± albite ± biotite + muscovite. Abundant, large (2 to 4 mm) porphyroblasts of biotite were observed in some lithologic types of the Monumental Mills rocks (Fig. 11). These porphyroblasts contain inclusions of epidote and locally monazite, have irregular boundaries, are randomly oriented, and appear to have grown across the pervasive foliation (S_1) defined mainly by chlorite. Although these large porphyroblasts occur in metasedimentary rocks of uncommon mineralogical composition (chlorite and epidote rich), they suggest that temperature peaked after the development of the pervasive fabric. The mineral assemblage is indicative of a basic composition, and the sedimentary structures suggest a fine-grained volcaniclastic protolith. This interpretation was also favored by Furcron (1939), who described this rock as metatuff.

Metamorphosed mudstones are composed of fine-grained phyllosilicates and very fine grained quartz and sparse albite. Although the fine-grained rocks contain metamorphic minerals, the sedimentary textures are commonly well preserved.

Metamorphosed mafic intrusive rocks and lavas (greenstone) have typical greenschist facies mineral assemblages (Fig. 12a): actinolite + chlorite + epidote / clinozoisite + albite ± quartz. In many cases mafic lavas are extensively altered to greenschist (chlorite-actinolite–rich rocks) and epidosite (mostly epidote with varying amounts of quartz). An ultramafic rock was found in one locality in the study area. The rock is composed of an aggregate of chlorite surrounding crystals of green calcic amphibole (actinolite or green hornblende) (Fig. 12b), and it is tentatively interpreted as a metapyroxenite based on the preserved texture. This body appears to be concordant with the regional stratigraphy, although the actual contact is not exposed.

CONCLUSIONS REGARDING STRATIGRAPHY AND STRUCTURE

The Late Proterozoic to Early Cambrian metasedimentary and metavolcanic cover rocks (Fig. 3), which lie unconformably upon the Grenville basement on the east limb of the Blue Ridge, experienced one low-grade progressive metamorphism. Deformation was minimal and took place in two stages: D1 involved a pervasive foliation (S_1) in varying degrees of intensity throughout the sequence. Subsequent folding and faulting and development of a localized discrete cleavage (S_2) resulted from D2. The observed structures are correlated with features described by workers in adjacent areas and may be regionally extensive. The map-scale structure varies from east-dipping (D1 fold) strata directly above the eastern Blue Ridge basement, to a thrust fault(?) or tight folds at the D2 structural front 3 km farther east, and finally to open folding and gently dipping strata several kilometers still farther east. We suggest that the D2 structural front was localized along an east-dipping deformation boundary above the eastern subsurface edge of the Blue Ridge Grenville basement.

The observed stratigraphy (Fig. 4) and continuity of formations through the mapped area (Fig. 3) indicate that the suture of Rankin et al. (1990a, b) does not exist there (Fig. 3). Structural data are also consistent with this interpretation.

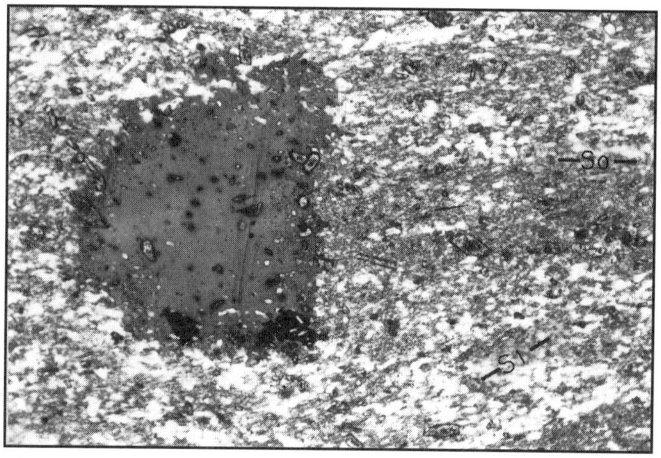

Figure 11. Biotite porphyroblasts overgrown on compositional layering (S_0) and weak pervasive fabric (S_1). Horizontal field of view: 3.4 mm, plain light.

MODEL OF TECTONIC EVOLUTION

During the last decade it has been established that the Blue Ridge province is allochthonous and was transported westward on a master décollement zone over early Paleozoic sedimentary rocks (e.g., Pratt et al., 1988). This movement was probably not the result of a single orogenic event (Glover, 1989). Wehr and Glover (1985) proposed that the crystalline core of the Blue Ridge anticlinorium was located on the hinge zone bounding attenuated continental crust of rifted Laurentia and was "decapitated" and thrust westward during the Taconian orogeny.

The age of metamorphism for the northern Virginia cover sequence, in the Warrenton area, has not been determined. However, the regional metamorphism in the higher-grade Piedmont rocks east of the Culpeper Basin is considered Taconic (Glover et al., 1983; Laird, 1988). A Cambrian age is possible based on the data presented by Glover et al. (this volume) and Gates (this volume). Low-grade rocks east of the Culpeper Basin have not been dated. Other workers have suggested that the pervasive cleavage in the cover rocks of the Blue Ridge is late Paleozoic by correlation with the cleavage in Early Devonian rocks of the Massanutten syncline (Mitra and Elliot, 1980) and by radiometric dating of muscovites (Burton et al., 1992). However, cleavage may have formed at different times in the lower parts of the eastern Blue Ridge and in the late Paleozoic strata west of the Blue Ridge. Late Paleozoic heating may also have reset mica ages.

In the area of this report D1 was synchronous with the low-grade metamorphism (M1) in the area. Metamorphism progressed and locally reached peak temperatures after the development of the pervasive foliation (S_1), because the foliation is cut by biotite porphyroblasts. Subsequently, the assem-

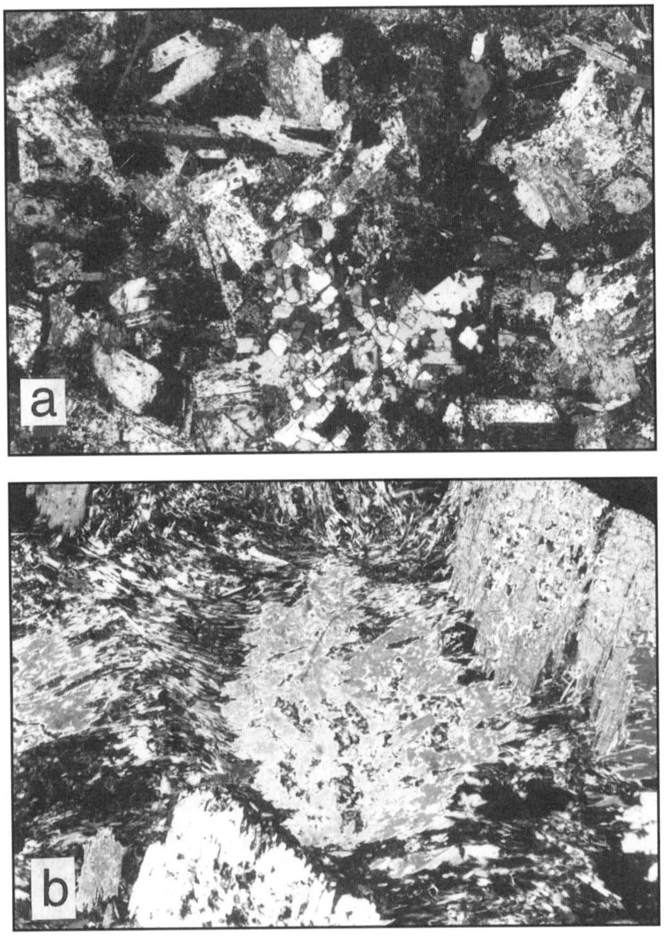

Figure 12. Metamorphosed mafic igneous rocks (cross-polarized light, horizontal field of view: 6.2 mm). a, Medium-grained metadiabase-metagabbro (1 to 2 mm). Altered plagioclase, actinolite, and chlorite (after pyroxene?) and relict hornblende and a concentration of quartz and epidote (lower center) are present. b, Metapyroxenite. Actinolite after pyroxene(?) crystals surrounded by magnesian chlorite.

blage of the decapitated crystalline basement block with its Late Proterozoic and Paleozoic cover rocks advanced westward again, probably in one major thrust sheet during D2. D2 resulted in deformation within the thrust sheet during the last stages of compression. D2 deformation of the east flank of the westernmost syncline along the eastern Blue Ridge is herein interpreted to delineate the easternmost subsurface extent of the more rigid basement rock in this thrust sheet. This interpretation is in agreement with the interpreted seismic profile ("Culpeper Line 2") by Costain and Çoruh (1990), which places the depth of the master décollement at the west end of the line at approximately 1 sec, 3-km depth, assuming an average velocity of 6 km/sec.

The location of Late Proterozoic basement highs and lows, and the boundaries between units of contrasting lithology, may have influenced the position of some faults during Paleozoic deformation.

DISCUSSION: REGIONAL APPALACHIAN TECTONIC MODELS

The hypothesis of a terrane boundary in the Warrenton area

The stratigraphic and structural framework presented in this study indicates that no terrane boundary exists in the Warrenton area. This is shown by the following: The rift stratigraphy can be traced across the proposed terrane boundary of Rankin et al. (1990a, b) (Fig. 3). The incursion of metasedimentary rocks into the Catoctin map pattern south of Warrenton (Figs. 2, 3) is explained by the open folding D2, which affected the eastern part of the rift sequence, probably in the late Paleozoic. The Catoctin is an unbroken stratigraphic unit around the area of the metasedimentary incursion (Figs. 3, 5). No discontinuity is indicated by metamorphic grade because the grade is similar for all rocks in the study area. In the study area, small mafic bodies occur within well-bedded parts of the rift stratigraphy and appear to be intrusive into the sequence. The mafic contacts are not exposed in the area, but there is no associated mélange, bounding fault, or ophiolite stratigraphy that would suggest an exotic terrane.

Terrane boundaries in the eastern Blue Ridge

The models that propose accreted terranes along the eastern Blue Ridge are based primarily on the assumption that the metamorphosed mafic and ultramafic bodies are fragments of disrupted oceanic crust caught up in an accretionary mélange and emplaced during continental collision in the Taconian or Penobscottian orogenies. However, typical ophiolite stratigraphy (ultramafics–gabbros–sheeted dikes–pillow basalts–oceanic sediments) or unequivocal parts of it has not been demonstrated anywhere in the eastern Blue Ridge. Recent detailed study of other areas in the central Virginia Blue Ridge have demonstrated the sill and dike emplacement of mafic and ultramafic rocks in the Late Proterozoic Lynchburg Group (Wang and Glover, this volume). In addition, as shown in this study and in Wang and Glover (this volume), the clastic metasedimentary rocks surrounding these bodies have a continuous, undisrupted stratigraphy. There is no evidence of intermixing of rock types as expected in a tectonic mélange. Therefore, the proposed terrane models do not explain the continuity of mapped stratigraphy across the proposed suture or the lithofacies assemblage and structure of the eastern Blue Ridge. In summary, all observations are compatible with an origin as a rift tectonic lithofacies that developed along the margin of Laurentia during the Late Proterozoic.

ACKNOWLEDGMENTS

The support for fieldwork was provided mainly by the Volunteer for Science Program of the U.S. Geological Survey and partially by a Geological Society of America Student Research Grant and a Grant-in-Aid of Research from Sigma Xi. We would like to thank Dr. Peter T. Lyttle, Dr. Kenneth A. Eriksson, and Dr. Richard D. Law for their helpful comments and reviews of this manuscript. Avery A. Drake, Steve Kline, and Nick Evans provided outstanding critical commentary.

REFERENCES CITED

Badger, R. L., and Sinha, A. K., 1988, Age and Sr isotopic signature of the Catoctin volcanic province: Implications for the subcrustal mantle evolution: Geology, v. 16, p. 692–695.

Barker, A. J., 1990, Introduction to metamorphic textures and microstructures: New York, Blackie, 162 p.

Bartholomew, M. J., and Lewis, S. E., 1984, Evolution of Grenville massifs in the Blue Ridge geological province, southern and central Appalachians, *in* Bartholomew, M. J., ed., The Grenville event in the Appalachians and related topics: Geological Society of America Special Paper 194, p. 229–254.

Burton, W. C., Kunk, M. J., and Lyttle, P. T., 1992, Age constraints on the timing of regional cleavage formation in the Blue Ridge anticlinorium, northernmost Virginia: Geological Society of America, 41st Southeastern Section Annual Meeting, Abstracts with Programs, p. 5.

Clarke, J. W., 1984, The core of the Blue Ridge anticlinorium in northern Virginia, *in* Bartholomew, M. J., ed., The Grenville event in the Appalachians and related topics: Geological Society of America Special Paper 194, p. 155–160.

Conley, J. F., 1978, Geology of the Piedmont of Virginia—Interpretations and problems, *in* Contributions to Virginia Geology—III: Virginia Division of Mineral Resources Publication 7, p. 115–149.

Costain, J. K., and Çoruh, C., 1990, Tectonic setting of Triassic half-grabens in the Appalachians: Acquisition, processing, and results, *in* Tankard, A. J., and Balkwill, H. R., eds., Extensional tectonics and stratigraphy of the North Atlantic margins, American Association of Petroleum Geologists, Memoir 46, p. 155–174.

Espenshade, G. H., 1986, Geology of the Marshal quadrangle, Fauquier County, Virginia: U.S. Geological Survey Bulletin 1560, 60 p.

Evans, N. H., 1991, Latest Precambrian to Ordovician metamorphism in the Virginia Blue Ridge: Origin of the contrasting Lovingston and Pedlar basement terranes: American Journal of Science, v. 291, p. 425–542.

Furcron, A. S., 1939, Geology and mineral resources of the Warrenton Quadrangle, Virginia: Virginia Geological Survey Bulletin, v. 54, 94 p.

Glover, L., III, 1989, Tectonics of the Virginia Blue Ridge and Piedmont, *in* Glover, L., III, Evans, N. H., Patterson, J., and Brown, W. R., eds., Tectonics of the Virginia Blue Ridge and Piedmont: International Geological Congress, 28th, Field Trip Guidebook T363, July 19–24, 1989: Washington, D.C., American Geophysical Union, p. 1–29.

Glover, L., III, Speer, J. A., Russel, G. S., and Farrar, S. S., 1983, Ages of regional metamorphism and ductile deformation in the central and southern Appalachians: Lithos, v. 16, p. 223–245.

Glover, L., III, Çoruh, C., Costain, J. K., and Bollinger, G. A., 1992, Piedmont seismic reflection study: A program integrated with tectonics to probe the cause of eastern seismicity: U.S. Nuclear Regulatory Commission report, NUREG/CR - 5731, 146 p.

Hatcher, R. D., Jr., 1987, Tectonics of the southern and central Appalachian internides: Annual Review of the Earth and Planetary Sciences, v. 15, p. 337–362.

Hatcher, R. D., Jr., 1989, Tectonic synthesis of the U.S. Appalachians, *in* Hatcher, R. D., Jr., Thomas, W. A., and Viele, G. W., eds., The Appalachian-Ouachita Orogen in the United States: Boulder, Colorado, Geological Society of America, The Geology of North America, v. F-2, p. 511–535.

Horton, J. W., Jr., Drake, A. A., Jr., and Rankin, D. W., 1989, Tectonostratigraphic terranes and their Paleozoic boundaries in the central and southern Appalachians, *in* Dallmeyer, R. D., ed., Terranes in the circum-Atlantic Paleozoic orogens, Geological Society of America Special Paper 230, p. 213–245.

Kline, S. W., 1991, Provenance of arkosic metasediments in the Virginia Blue Ridge: Constraints on Appalachian suspect terrane models: American Journal of Science, v. 291, p. 189–198.

Kline, S. W., 1992, Reaction softening of continental crust as a mechanism in the Late Proterozoic rifting of the North American craton: Geological Society of America Abstracts with Programs, v. 24, p. 328.

Kline, S. W., 1994, Late Proterozoic extensional deformation of Blue Ridge basement in Virginia: Supporting evidence from simple field relations: Journal of Geology, v. 102, p. 349–359.

Kline, S. W., Lyttle, P. T., and Schindler, J. S., 1991, Late Proterozoic sedimentation and tectonics in northern Virginia, *in* Schultz, A., and Compton-Gooding, E., eds., Geologic evolution of the eastern United States: Joint meeting of northeast and southeast of Geological Society of America, Field Trip Guidebook, p. 263–294.

Laird, J., 1988, Arenig to Wenlock age metamorphism in the Appalachians, *in* Harris, A. L. and Fettes, D. J., eds., The Caledonian-Appalachian Orogen: Geological Society of London Special Publication 38, p. 311–345.

Lukert, M. T., and Banks, P. O., 1984, Geology and age of the Robertson River pluton, *in* Bartholomew, M. J., ed., The Grenville event in the Appalachians and related topics: Geological Society of America Special Paper 194, p. 161–166.

Mitra, G., 1979, Ductile deformation zones in Blue Ridge basement rocks and estimation of finite strains: Geological Society of America Bulletin, v. 90, p. 935–951.

Mitra, G., and Elliott, D., 1980, Deformation of basement in the Blue Ridge and the development of the South Mountain cleavage, *in* Proceedings, Caledonides in the U.S.A., I.G.C.P. Project 27—Caledonide orogen: Blacksburg, Virginia Polytechnic Institute and State University, Department of Geological Science Memior 2, p. 307–311.

Powell, C. McA., 1979, A morphological classification of rock cleavage: Tectonophysics, v. 58, p. 21–34.

Pratt, T. L., Çoruh, C., Costain, J. K., and Glover, L., III, 1988, A geophysical study of the Earth's crust in central Virginia: Implication for Appalachian crustal structure: Journal of Geophysical Research, v. 93, p. 6649–6667.

Rankin, D. W., and nine others, 1990a, Pre-orogenic terranes, *in* Hatcher, R. D., Jr., Thomas, W. A., and Viele, G. W., eds., The Appalachian-Ouachita orogen in the United States, Geological Society of America, The Geology of North America, v. F-2, p. 7–100.

Rankin, D. W., Drake, A. A., Jr., and Ratcliffe, N. M., 1990b, Geologic map of the U.S. Appalachians showing the Laurentian margin and the Taconic orogen, *in* Hatcher, R. D., Jr., Thomas, W. A., and Viele, G. W., eds., The Appalachian-Ouachita Orogen in the United States: Boulder, Colorado, Geological Society of America, The Geology of North America, v. F-2, Plate 2.

Simpson, C., and Kalaghan, T., 1989, Late Precambrian crustal extension preserved in Fries fault zone mylonites, southern Appalachians: Geology, v. 17, p. 148–151.

Tollo, R. P., and Aleinikoff, J. N., 1992, The Robertson River Igneous Suite, Virginia Blue Ridge: A case study in multiple-stage magmatism associated with the early stages of Iapetan rifting: Geological Society of America Abstracts with Programs, v. 24, p. 70.

Tollo, R. P., Aleinikoff, J. K., and Gray, K. J., 1991, New U-Pb zircon isotopic data from the Robertson River Igneous Suite, Virginia Blue Ridge: Implication for the duration of Late Proterozoic anorogenic magmatism: Geological Society of America Abstracts with Programs, v. 23, p. 139.

Virginia Division of Mineral Resources, 1993, Geologic map of Virginia: Charlottesville, Virginia Division of Mineral Resources, scale 1:500,000.

Wehr, F., 1983, Geology of the Lynchburg Group in the Culpeper and Rockfish River areas, Virginia [Ph.D. thesis]: Blacksburg, Virginia Polytechnic Institute and State University, 245 p.

Wehr, F., 1985, Stratigraphy of the Lynchburg Group and Swift Run Formation, Late Proterozoic (730–570 Ma), central Virginia: Southeastern Geology, v. 25, p. 225–239.

Wehr, F., and Glover, L., III, 1985, Stratigraphy and tectonics of the Virginia–North Carolina Blue Ridge : Evolution of a late Proterozoic–early Paleozoic hinge zone: Geological Society of America Bulletin, v. 96, p. 285–295.

Williams, H., and Hatcher, R. D., Jr., 1983, Appalachian suspect terranes, *in* Hatcher, R. D., Jr., Williams, H., and Zietz, I., eds., Contributions to the tectonics and geophysics of mountain chains: Geological Society of America Memoir 158, p. 33–55.

MANUSCRIPT ACCEPTED BY THE SOCIETY APRIL 30, 1996

Printed in U.S.A

Geological Society of America
Special Paper 314
1997

Paleozoic collisions, Mesozoic rifting, and structure of the Middle Atlantic states continental margin: An 'EDGE' Project report

Lynn Glover, III
Department of Geological Sciences, Virginia Polytechnic Institute and State University, Blacksburg, Virginia 24061-0420
Robert E. Sheridan
Department of Geology, Rutgers University, New Brunswick, New Jersey 08903
W. Steven Holbrook and John Ewing
Department of Geology and Geophysics, Woods Hole Oceanographic Institution, Woods Hole, Massachusetts 02543
Manik Talwani
Department of Geology and Geophysics, Rice University, Houston, Texas 77005, and
Houston Advanced Research Center, The Woodlands, Texas 77381
Robert B. Hawman
Department of Geology, University of Georgia, Athens, Georgia 30602
Ping Wang
Department of Geological Sciences, Virginia Polytechnic Institute and State University, Blacksburg, Virginia 24061-0420

ABSTRACT

Carolinian terrane magmatic arc rocks of Late Proterozoic to Early Cambrian age extend to the present continental edge in the Mid-Atlantic states. They were emplaced upon the Cambrian Laurentian passive margin during the Early Cambrian Carolinian collision. The Carolinian plate was subsequently dismembered by orogen-parallel dextral transform, transpressional and transtensional types of shearing and finally by Mesozoic extension. In this chapter we do not subdivide all of the terranes that may lie outboard of the Laurentian margin. Instead we attempt to define the apparently diachronous Taconic/Carolinian suture and the Laurentian margin between the northern and southern Appalachians. The suture appears to have been a low-angle structure that, south of New England, crops out in the central Piedmont and was repeated at the pre-Atlantic Coastal Plain surface by faulting and folding during the Paleozoic. Possible locations of the suture under the coastal plain are the Salisbury anomaly east of Richmond and between the low-grade Carolinian volcanics and higher-grade gneiss under the Delmarva Peninsula.

We present two new models for the Cambrian and Ordovician tectonic evolution of the Mid-Atlantic margin: (1) the eastern Carolinian basement model (ECB), in which Laurentian rocks probably do not extend down dip very far east of the Salisbury anomaly in the subsurface, and (2) the eastern Goochland basement model (EGB), in which the high-grade gneisses under Delmarva are Laurentian, and the Carolinian suture dips from eastern Chesapeake Bay into the M (Mohorovičić) discontinuity along seismic reflection profiles 801, 802, and 28 offshore at the continental edge. Dating of the basement gneiss under the eastern Chesapeake and Delaware Bay areas might result in the most economical solution to the ambiguity between the models. In both the ECB and EGB models there are two variants, depending on

Glover, L., III, Sheridan, R. E., Holbrook, W. S., Ewing, J., Talwani, M., Hawman, R. B., and Wang, P., 1997, Paleozoic collisions, Mesozoic rifting, and structure of the Middle Atlantic states continental margin: An 'EDGE' Project report, *in* Glover, L., III, and Gates, A. E., eds., Central and Southern Appalachian Sutures: Results of the EDGE Project and Related Studies: Boulder, Colorado, Geological Society of America Special Paper 314.

whether the listric normal fault that dips eastward from the western Piedmont (lower-plate margin) or the present Moho itself is the plate delamination surface, producing an upper-plate margin. Lower-plate margin variants of each model are concordant with a strong structural control by the Carolinian suture on the locus of rifting at the continental edge, whereas upper-plate variants of the models are less concordant. Abundant Jurassic dikes in the Piedmont and Jurassic volcanics at the continental margin are theoretically more consistent with an upper-plate margin. The EGB model produces a trace of the suture at M depths that is more conformable with the regional convex westward salient of the pre-Triassic structure of the Mid-Atlantic region, suggesting that either the upper- or lower-plate variant of the EGB model may be the more nearly correct.

Although it is possible that the Carolinian suture influenced the locus of continental separation in the mid-Atlantic region, in New England the continental edge formed within the Avalonian plate and does not appear to follow suture boundaries.

INTRODUCTION

Sutures are fundamental plate tectonic elements of orogens, yet a generation after the recognition of sea-floor spreading (Hess, 1962) and plate tectonics (Morgan, 1968) there is no consensus on the location of the first suture formed in the evolution of the central and southern Appalachians. Largely because of the mafic-ultramafic association in the eastern Blue Ridge, a number of authors view part of the post-Grenville Lynchburg sequence (Fig. 1) as collisional ophiolitic mélange thrust upon Laurentia, thus creating a terrane boundary (suture) of Late Proterozoic to Ordovician age. Representative recent papers supporting this view are Conley (1985), Hatcher (1989), Rankin et al. (1989), Horton et al. (1991) and Stanley and Ratcliffe (1985). Glover (1989) suggested that the central Piedmont Taconic suture is the only suture in the exposed central and southern Appalachian metamorphic terrain. Wang and Glover (1992, and this volume) believe that most Pearce and Cann–type geochemical tectonic discriminant diagrams that have been used to infer an oceanic origin and terrane boundary location for western Piedmont and Blue Ridge basalt/ultramafic complexes fail because they cannot discriminate various continental basalts from oceanic types.

Sheridan et al. (1993) concluded that the continental edge off of Chesapeake Bay probably occurs near the intersection of the "Taconic" (Carolinian) suture and the Moho. This chapter attempts to evaluate and place this finding in a regional context. Two suture models, consistent with the current database are presented herein as a guide to further refinement. There are also upper-plate and lower-plate variants (Wernicke and Tilke, 1989; Lister et al., 1991) of each suture model. Implications of the models for Appalachian tectonics and the Mesozoic development of the North American margin point up the importance of further work.

We make the following major points in this chapter: Carolinian terrane (Gates, this volume) magmatic arc rocks of Late Proterozoic and Early Cambrian age extend to the present continental edge. They were emplaced upon the Laurentian margin of the central and southern Appalachians during the latest Proterozoic or Early Cambrian Carolinian collision (see below). The Carolinian plate (Secor et al., 1983) was subsequently dismembered by later Paleozoic orogen-parallel transform, transtensional and transpressional shearing, and finally by Mesozoic extension. In southern New England (Fig. 1) a different plate bearing Ordovician volcanics may have collided with the Laurentian margin in that region (Zen et al., 1989). The scope of this chapter is limited to defining the Carolinian suture and exploring the possible eastern extent of the Cambrian Laurentian margin. We do not attempt to subdivide or account for the several terranes that could lie outboard of Laurentia in New England. Lower-plate variants of each suture model are consistent with a strong structural control by the Carolinian suture on the locus of rifting at the continental edge, whereas upper-plate variants are less obviously related to inherited structure. In all models it is probable that the Alleghanian suture was east of the modern continental margin off of the Mid-Atlantic states.

In this chapter we discuss the geology of the terranes in their stacking order along seismic reflection profile I-64 from the Blue Ridge to oceanic crust. The I-64 traverse is representative of the geologic framework of the Blue Ridge and Piedmont Provinces at this latitude (Glover and Klitgord, 1995). Next we present the geology of the Mid-Atlantic basement under the Coastal Plain and then to oceanic crust along profile 803 off Chesapeake Bay (Figs. 1, 2). Finally we trace the changes in this geologic framework as one proceeds northward into the New England Appalachians. Most data used in the analysis are published elsewhere. Line drawings of seismic reflection profiles are used in this chapter.

Our purpose is to summarize recent new data and present an overview primarily of the Late Proterozoic to Ordovician tectonics along the Laurentian margin in the middle Atlantic states. We then discuss the possible influence of this geologic framework on the rift-stage evolution of the modern middle Atlantic continental margin.

THE LAURENTIAN MARGIN

Studies of Blue Ridge and western Piedmont sequences established that the Laurentian margin was born in Late Proterozoic rifting of the Grenville continent (Wehr and Glover, 1985; Simpson and Eriksson, 1989; Bond et al., 1984) and became a passive margin in Early Cambrian time. Along the I-64 profile in Virginia (Figs. 1, 2) the Grenville basement was metamorphosed about 1 Ga and consists of granulite gneisses partly retrograded to amphibolite and greenschist facies (Evans, 1991). The Lynchburg Group begins with fluvial arkoses and conglomerates deposited on Grenville basement and is capped by the Catoctin Formation. The basal Lynchburg grades upward into increasingly more mature and finer-grained sedimentary rocks through the Catoctin Formation and into the overlying Evington Group. The Catoctin is 571.5 ± 5 Ma. by zircon Pb-U dating (Aleinikoff et al., 1991; Southworth et al., 1994; Southworth, personal communication, 1994). Metabasalts of the Catoctin conformably overlie the Lynchburg Group and are in turn conformably overlain by the Evington Group (Glover, 1989). It is important to note here that the recent date of 544 Ma for the begining of the Cambrian (Bowring et al., 1993) implies that all of the Catoctin and Lynchburg are of Late Proterozoic age. The Evington could range into the Cambrian.

Basaltic sills, dikes, and lavas occur throughout the post-Grenville sequence and below the Catoctin; ultramafic rocks, probably basalt differentiates, are volumetrically rare to locally common, compared to mafic rocks. The whole sequence comprises a tectonic lithofacies assemblage characteristic of rifting, beginning in shallow water and ending in deep-water continental margin environments (Glover, 1989). The mineral composition of the sequence is consistent with derivation from the Blue Ridge Grenville basement gneisses (Wehr, 1985; Wehr and Glover, 1985; Kline, 1991; Wang, 1991; Gates, 1986) but not with magmatic arc rocks. Therefore we consider these Blue Ridge and western Piedmont rocks to be part of the Laurentian rifted margin cover. It is possible that drift-stage rocks are included in the Evington, but if so, they are not abundant and were probably removed by thrusting and erosion during collision with the magmatic arc, Carolinia. The Peters Creek and Octararo rocks of the Pennsylvania Piedmont (Gates and Valentino, 1991) and equivalents in Maryland also contain intercalated basalts and appear compositionally to be similar to upper Lynchburg and Evington Group, respectively. Similar sequences in New England—the Pinnacle, Hoosac, and Rowe Formations of northwestern Vermont and west-central Massachusetts (Stanley and Ratcliffe, 1985)—may also belong to this Laurentian rift facies setting.

Many authors place the Taconic (or "early Taconic Penobscott") suture within the Blue Ridge Lynchburg cover sequence (Fig. 3) of the central and southern Appalachians (Rankin et al., 1989; Horton et al., 1989; Hatcher, 1989). A common body of evidence and interpretation is cited to justify this suture or terrane boundary. (1) Large and small bodies of metabasalt and ultramafic rocks occur in metasediment of the Grenville cover sequence. Locally they occur together and in other places separately, although unequivocal ophiolite stratigraphy has not been found. These bodies are interpreted to be a disrupted ophiolite in a tectonic mélange. (2) Chemical tectonic discriminant diagrams commonly indicate mid-ocean ridge basalt (MORB) and locally ocean island (OIB) or island arc basalt (IAB) origins for these mafic rocks. Thus, these authors place the "Taconic suture" or a terrane boundary along the western margin of ultramafic rocks in the Lynchburg cover sequence. Recently Wang and Glover (1992) have confirmed and extended earlier arguments that these diagrams fail to discriminate continental basalts. When they plotted analyses of Jurassic rift basalts from the eastern continental United States, virtually all of the diagrams, including those previously used to determine the Blue Ridge mafic-ultramafic association, yielded identifications of MORB, OIB, and IAB rather than the correct answer: WP (within-plate) rift basalts. Further, Glover et al. (1991) have confirmed and extended earlier reports that the Virginia Blue Ridge segment of the mafic-ultramafic association consists of dikes of metabasalt, sills of basalt and differentiated ultramafic rocks, and basalt lavas injected into a stratigraphically coherent section of continental rift sediments of the Late Proterozoic Lynchburg Group. In no case is the mafic-ultramafic suite associated with mélange. Additionally, mapping recently completed (Kasselas and Glover, this volume) shows that the Catoctin is continuous southwestward along strike from Warrenton, Virginia, to the Mesozoic basin and that the mafic-ultramafic association continues northward within the Lynchburg-Fauquier sequence below the Catoctin lavas. Farther north the Catoctin overlaps the Fauquier onto Grenville basement in northern Virginia (Fig. 3). Ultramafic rocks of the association appear to end just south of the latitude of Warrenton (Kasselas and Glover, this volume). These field relations do not support a suture or terrane boundary in this part of the Virginia Blue Ridge. We must look outboard to the east in Piedmont rocks for the suture.

MÉLANGE

East of the Laurentian rift- and drift(?)-stage rocks mélanges occur structurally and probably stratigraphically above the Laurentian margin deposits (Figs. 1, 2) (Brown and Pavlides, 1981; Drake and Morgan, 1981; Muller and Chapin, 1984; Muller et al., 1989; Glover, 1989; D. W. Valentino, personal communication, 1992). Unlike the sill, dike, and lava forms of the mafic-ultramafic suite in the Lynchburg, these mélanges contain truly fragmental masses of mafic and ultramafic rocks as well as fragments of quartz, gneiss, and schist in an unstratified feldspathic sandstone matrix. Excellent descriptions of these mélanges exist in the literature (Hopson, 1964; Brown and Pavlides, 1981; Drake and Morgan, 1981). The position of the mélange between the Laurentian margin deposits and the magmatic arc rocks on the east (Fig. 1; refer to Fig. 5) implies a linkage of the formation of the mélange to the collision of a magmatic arc and Laurentia. The mélange is therefore considered to mark the suture between the terranes Laurentia and Carolinia.

CAROLINIAN TERRANE, CHOPAWAMSIC BLOCK

Magmatic arc rocks (basalts, andesites and rhyolites) of the ca. 550-Ma Chopawamsic Formation (Pavlides, 1981) are

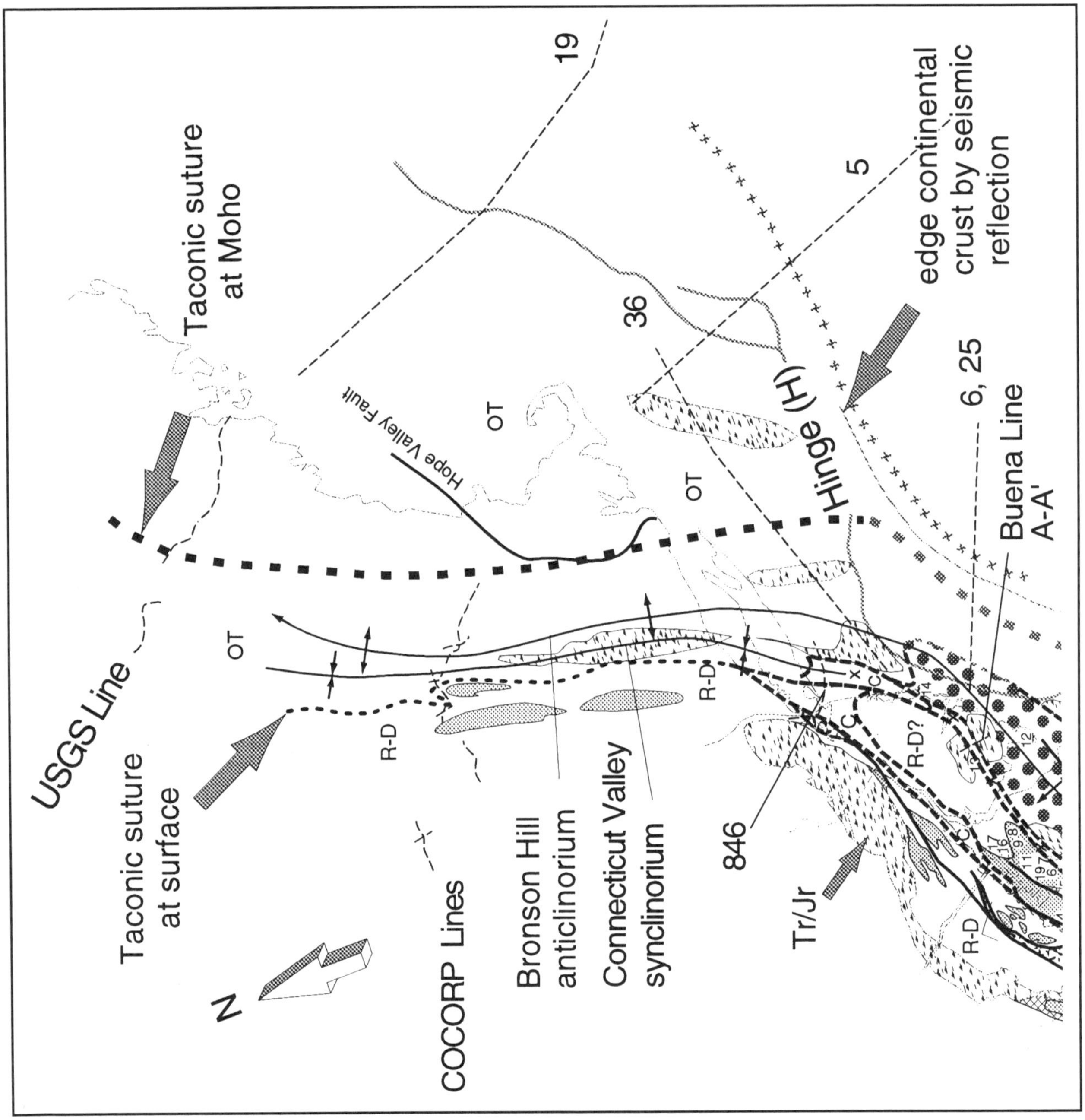
Taconic suture
at Moho
19
5
edge continental
crust by seismic
reflection
36
6, 25
Hinge (H)
Hope Valley Fault
OT
OT
OT
Buena Line
A-A'
USGS Line
Taconic suture
at surface
R-D
R-D
R-D
R-D?
COCORP Lines
Bronson Hill
anticlinorium
Connecticut Valley
synclinorium
846
Tr/Jr
N

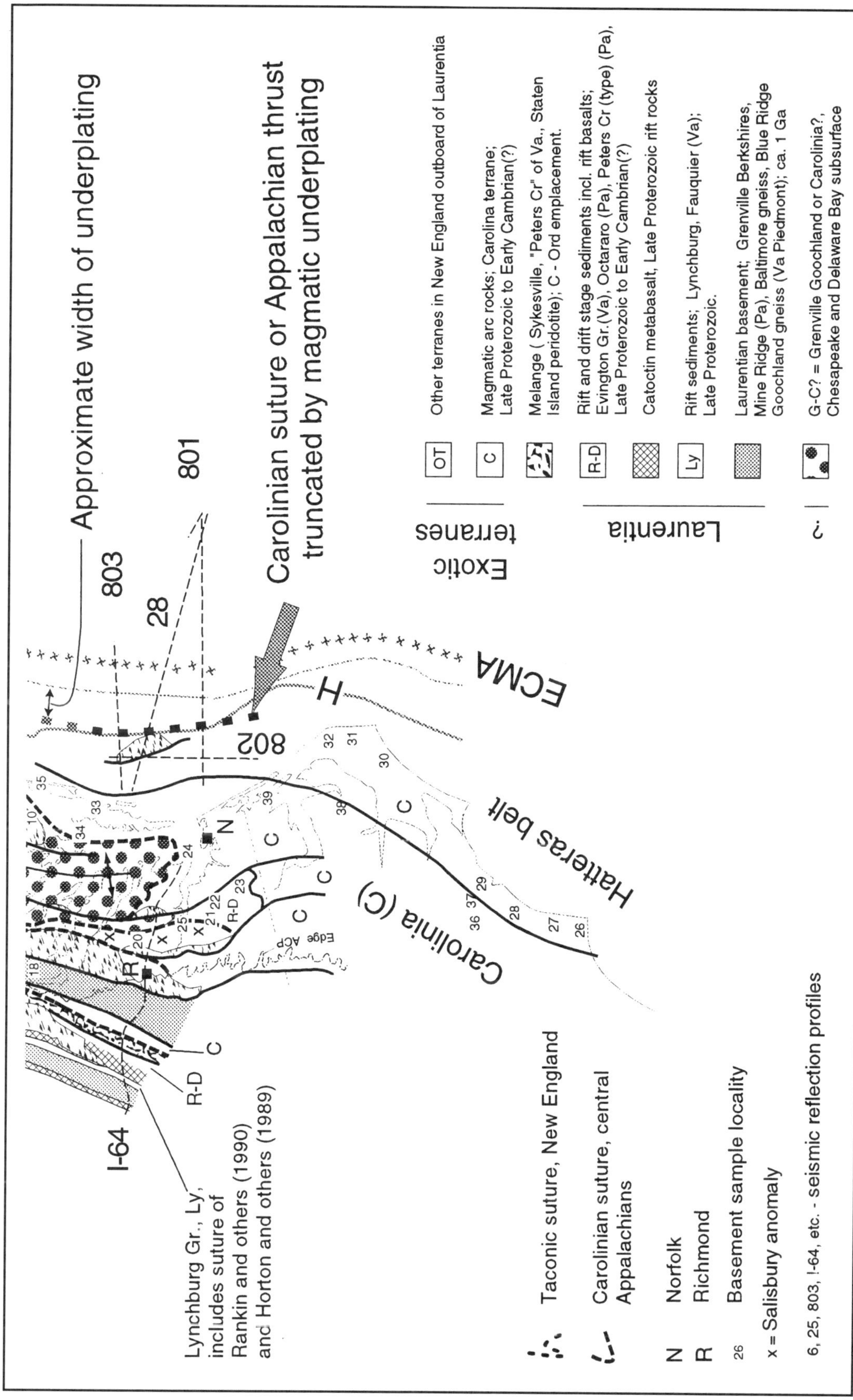

Figure 1 (on this and facing page). The Laurentian margin and Taconic and Carolinian sutures in the central and New England Appalachians; pre-Jurassic surface. ECMA = east coast magnetic anomaly.

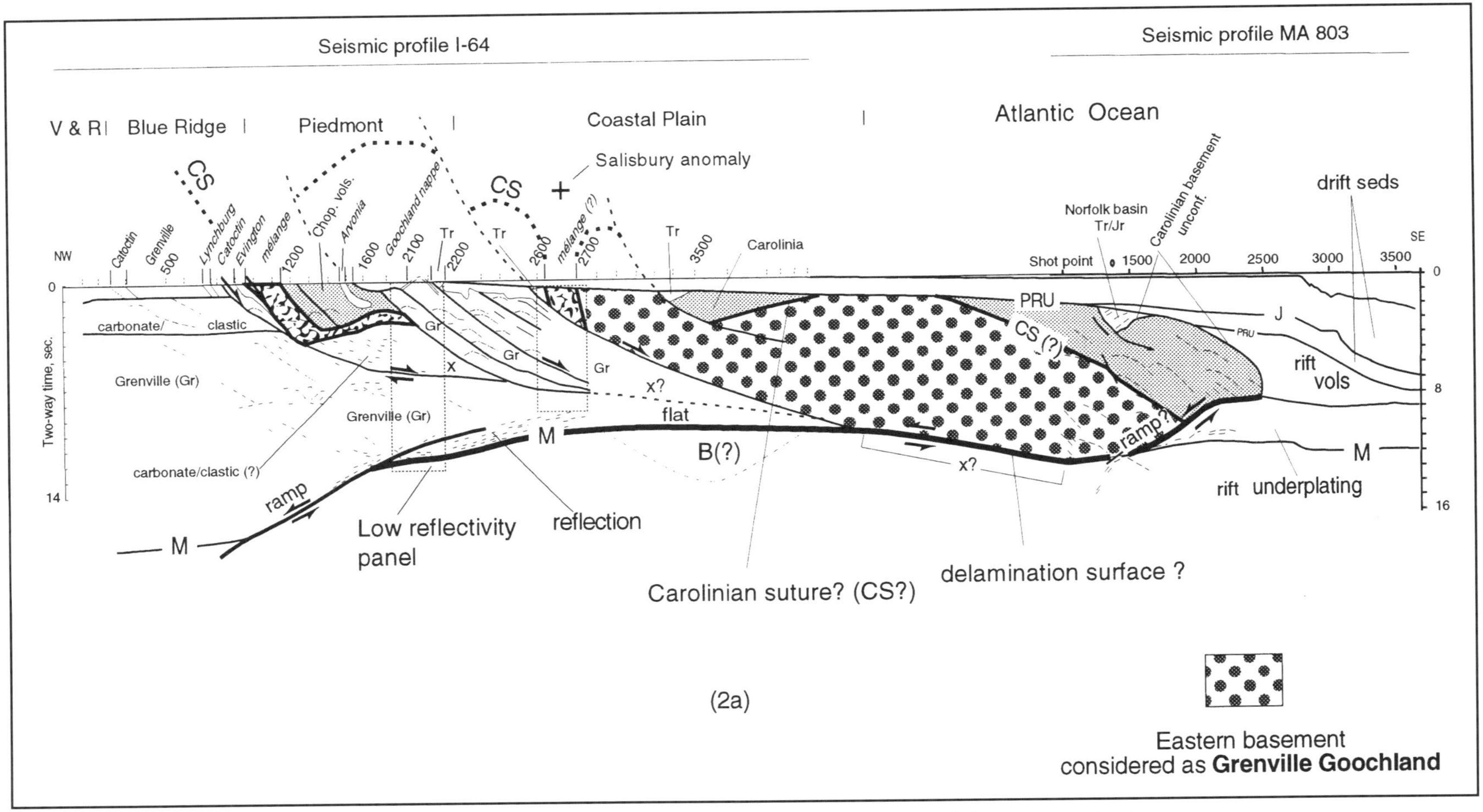

Figure 2a. Eastern Goochland basement model, upper-plate version. Carolinian suture unrelated to locus of continental separation. This alternative interpretation (and those in Figs. 2b, c, d) of I-64 profile is based on Glover et al., 1992b. All faults shown in the profile moved as thrust and/or transpressional faults in the Paleozoic. Mesozoic extensional movement is shown by arrows in cases where old contractional faults were reactivated. Lower crustal bulge B(?) may not exist (Pratt et al., 1988). J = Jurassic/Triassic boundary. M = Mohorovičié discontinuity. PRU = Postrift unconformity. x = possible delamination surface discussed in text. Mélange and Chopawamsic volcanics patterned on figure.

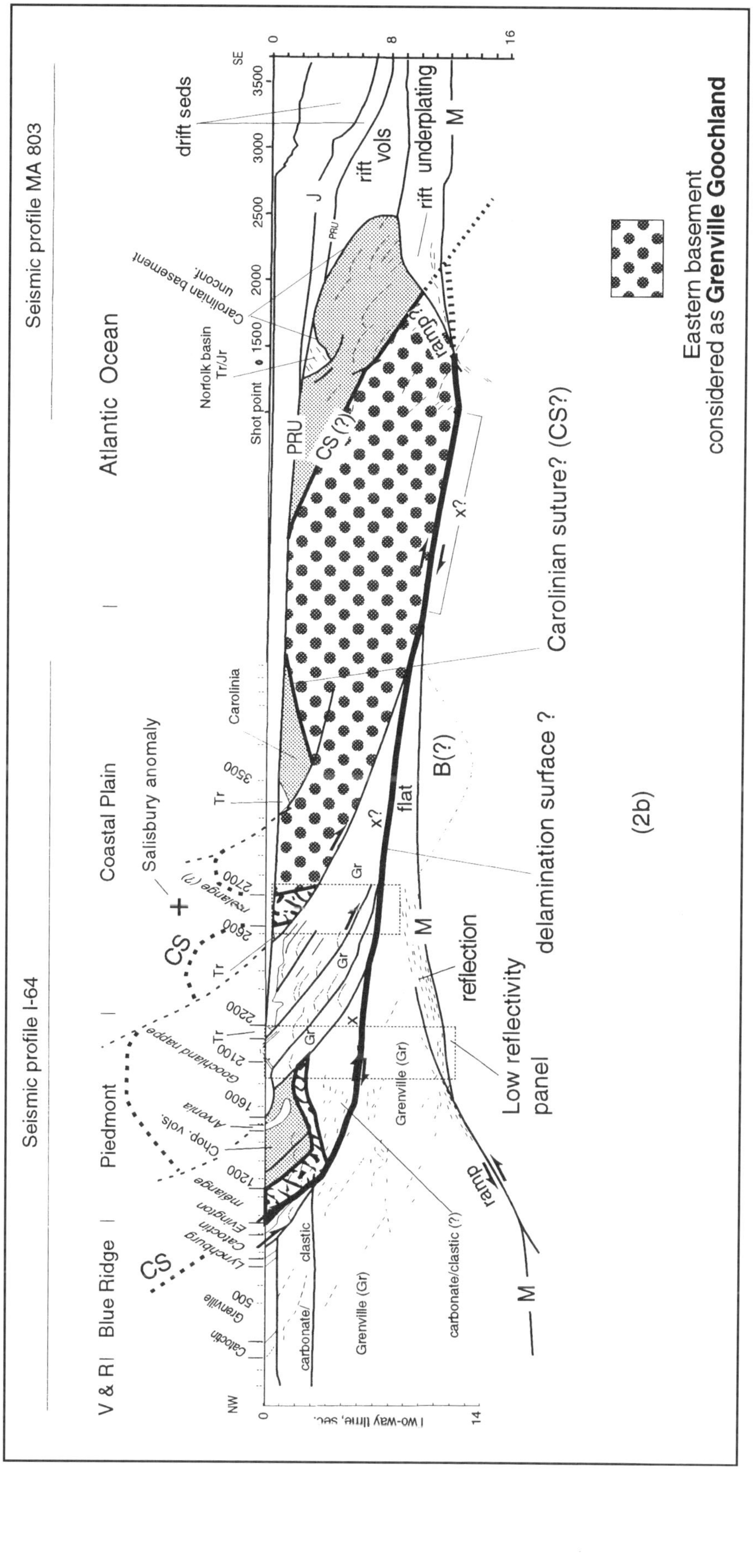

Figure 2b. Eastern Goochland basement model, lower-plate version. Carolinian suture related to locus of continental separation. Patterns and symbols same as Figure 2a.

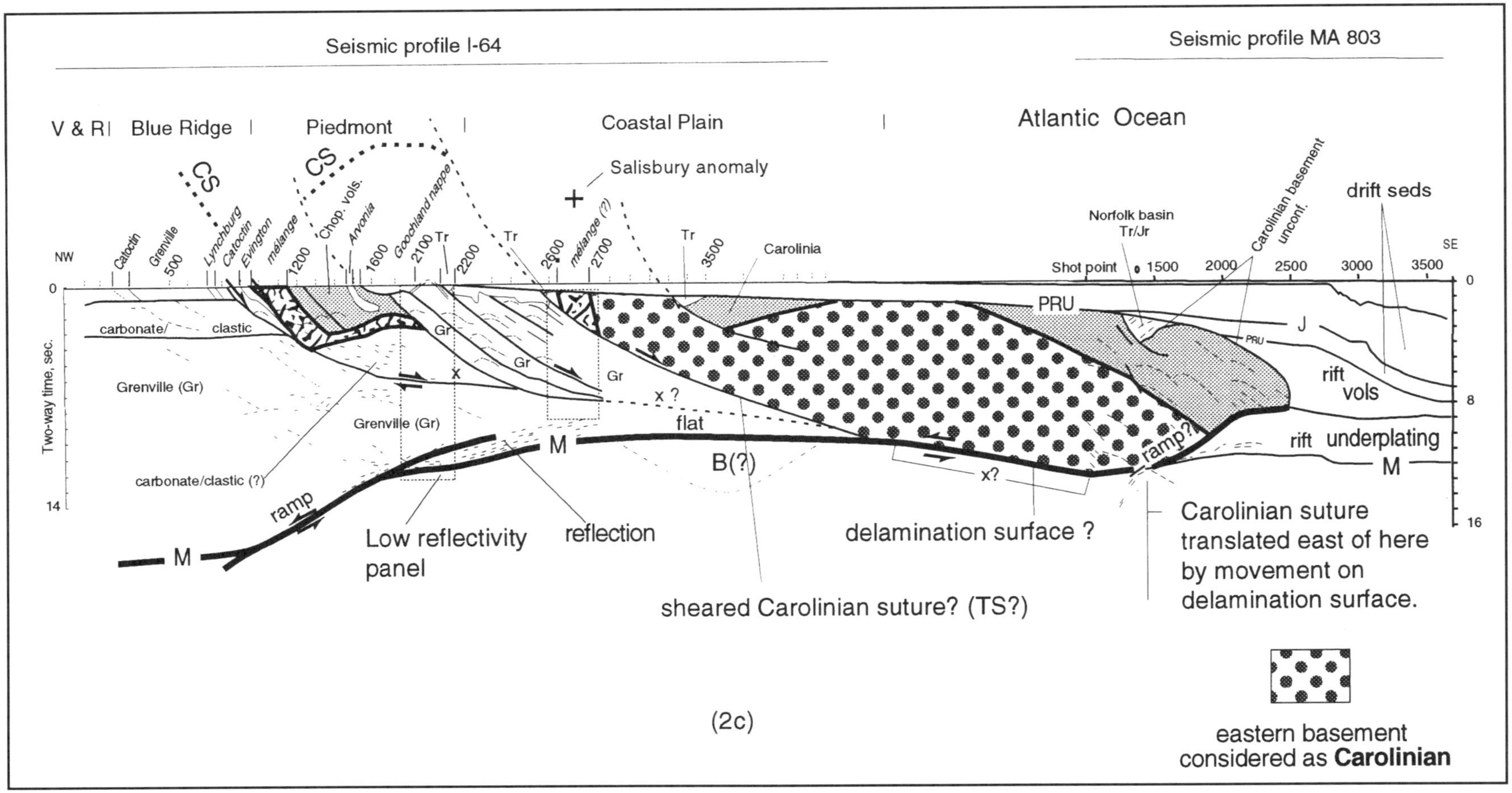

Figure 2c. Eastern Carolinian basement model, upper-plate version. Locus of continental separation unrelated to Carolinian suture structure. Patterns and symbols same as Figure 2a.

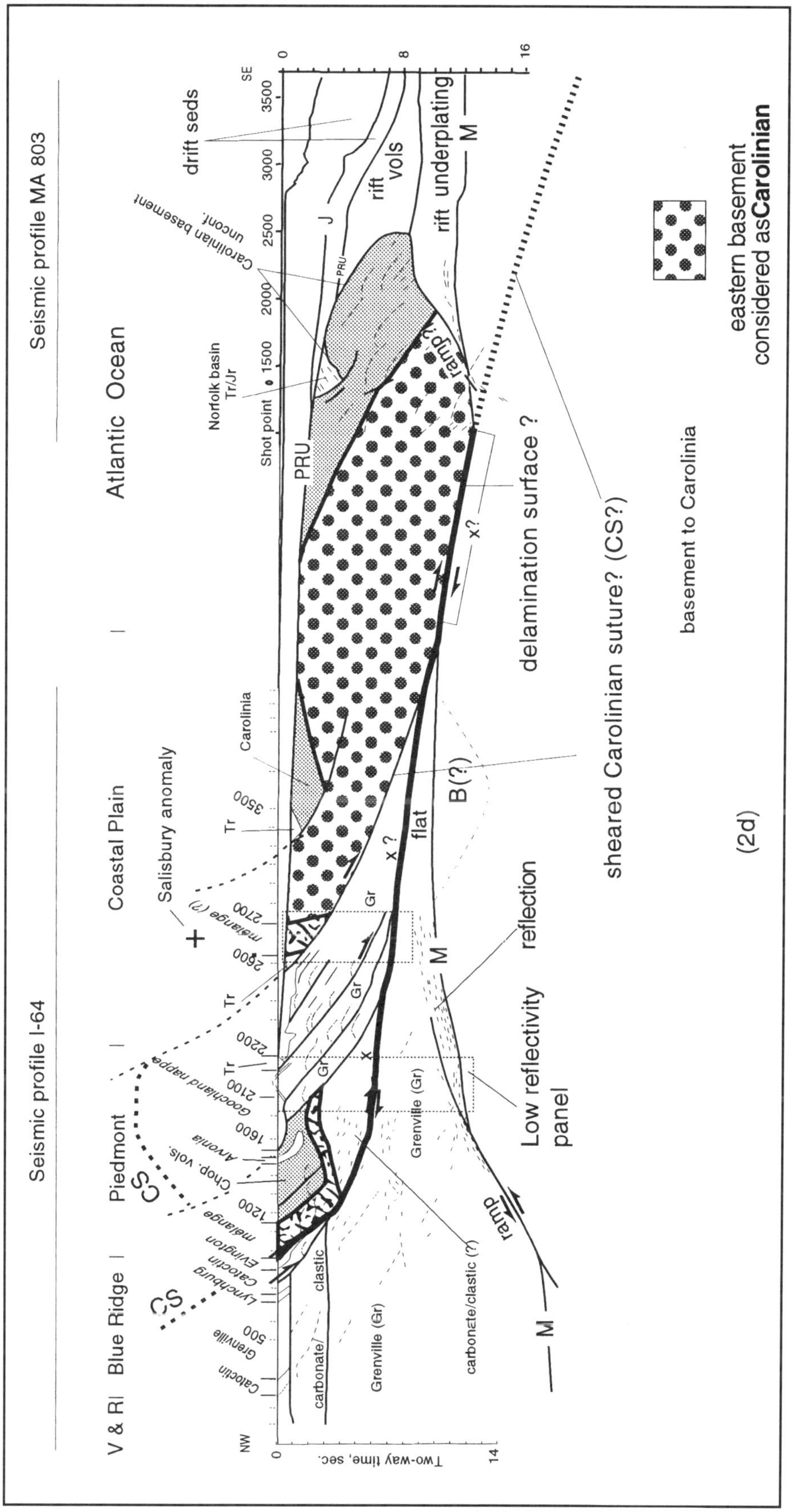

Figure 2d. Eastern Carolinian basement model, lower-plate version. Locus of continental separation related to Carolinian suture structure. Patterns and symbols same as Figure 2a.

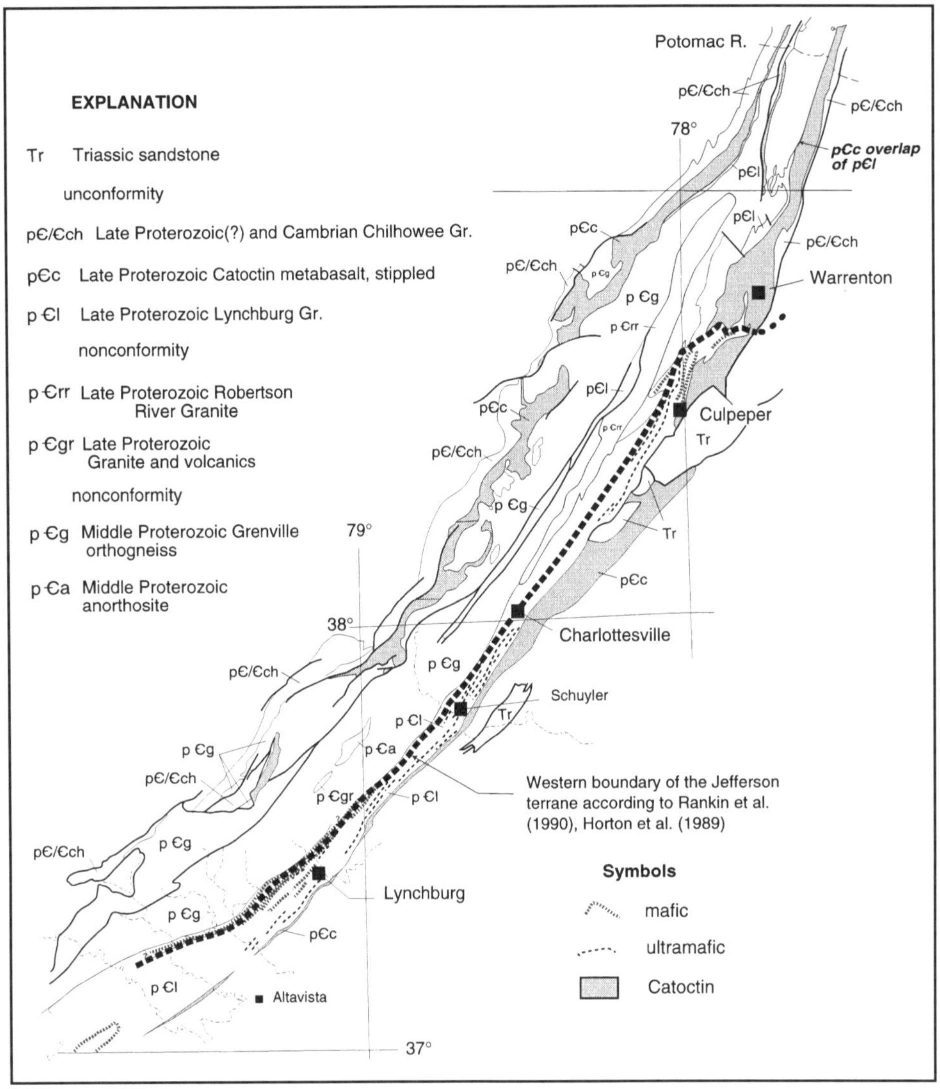

Figure 3. Virginia Blue Ridge showing Taconic terrane boundary of Rankin et al. (1989) and Horton et al. (1989)

thrust westward upon the mélange along the I-64 traverse (Figs. 1, 2). These rocks have been traced into the western edge of the Charlotte metamorphic belt (refer to Fig. 7, terrane 1) in central Virginia and are part of the Carolinian terrane that also includes the Carolina slate belt and magmatic arc rocks under the Atlantic Coastal Plain (Harris and Glover, 1988; Glover, 1989; Secor et al., 1989; Baird, 1991). Thus, the west edge of the magmatic arc and the mélange, where present, mark the first suture outboard of Laurentia. Work in progress (Glover et al., 1992a; Gates, this volume) shows that 520 to 500 Ma (Middle to Late Cambrian) plutons, such as the Occoquan Granite (Drake, 1987), Melrose Granite (Gates, 1981, this volume; Sinha et al., 1989), and Leatherwood Granite (Conley and Henika, 1973; Sinha et al., 1989), intruded the Laurentian passive margin, mélange, and the magmatic arc terrane. Each belt and terrane had already been metamorphosed and deformed prior to intrusion. Therefore, collision probably began in Early Cambrian time, although there is no sedimentary evidence for orogeny in the Valley and Ridge sequence until Middle Ordovician time. Carolinian arc rocks that are basement to the Atlantic Coastal Plain are dealt with further below.

The term "Carolinian terrane" (rather than the northern Appalachian term "Avalonian terrane") is currently preferred for these rocks in the central and southern Appalachians because they cannot be shown to be contiguous with the Avalonian and appear to have collided with Laurentia much earlier than the Avalonian (Secor et al., 1989).

LAURENTIAN MARGIN, GOOCHLAND NAPPES

Proceeding eastward along I-64 (Fig. 2) the Goochland nappes bring Laurentian Grenville basement (Glover et al. 1982, Farrar, 1984, Glover, 1989) over the arc rocks along a right lateral transpressional fault of Alleghanian age (Pratt et al., 1988; Gates and Glover, 1989). This basement nappe is in a faulted antiformal structure that plunges southward in North Carolina beneath allochthonous Carolinian arc rocks of the Carolina slate and Raleigh belts (Farrar, 1985).

In the past decade several authors (Bradley, 1982; Gates et al., 1986; Gates et al., 1988; Glover, 1989; Valentino et al., 1994) have shown the importance of late Paleozoic orogen-parallel dextral transpressional faulting in the Appalachians. The movement, at least 150 km on the Pleasant Grove–Huntington Valley mylonitic shear zone alone (Valentino et al., 1994) (Fig. 1), was distributed across the exposed Appalachians primarily east of the Blue Ridge and presumably occurred also in Appalachian rocks under the Atlantic Coastal Plain. Because the Goochland nappes lie in a structural block east of the Pleasant Grove–Huntington Valley mylonitic shear zone, they must have moved southward at least 150 km during this event.

Lefort and Max (1991) propose that there is no evidence that the outcropping Goochland belt of the eastern Piedmont (Fig. 1; p. 218) ". . . has always been, or ever was part of the Grenvillian North American (i.e., Laurentian) mainland instead of a detached fragment." However, the Goochland has a Grenville age of 1034 ± 94 Ma by Rb-Sr whole rock (Glover et al., 1982) and an unpublished age of "about 1 Ga" (Pb/U zircon analysis on State Farm by Krishna Sinha, personal communication, 1988). The Goochland was metamorphosed to granulite facies prior to Paleozoic retrogression (Farrar, 1984) and has within it a massive anorthosite body of a type largely restricted to the Precambrian. This combination of factors is unique to the Grenville in the Appalachians. Furthermore, the Goochland belt in southern Virginia and North Carolina (Fig. 2, refer to Fig. 5) is surrounded on the west, south, and east by a ca. 700- to 550-Ma cover of Carolinian terrane tectonically emplaced upon it in Late Cambrian to Ordovician time. The whole sequence was domed and thrust through its cover by dextral transcurrent deformation in Alleghanian time (Glover, 1989) to bring about the present architecture. Nothing in the details of the tectonic history (Glover, 1989) suggests a Goochland microcontinent separated at any stage from Laurentia. If the Goochland were a microcontinent, the ultramafic-mafic–bearing mélange might pass beneath it as a rootless slice, but there is no evidence for this.

We propose that the Goochland broke through its Carolinian cover by dextral transpressional faulting in the Alleghanian orogeny and has always been part of Laurentia. The Goochland west of the Salisbury appears from the data to be Laurentian Grenville in origin. The Goochland(?) east of the Salisbury anomaly may be Grenvillian or Carolinian, as discussed below.

BASEMENT BENEATH THE ATLANTIC COASTAL PLAIN

At the nonconformity below the Atlantic Coastal Plain (section along I-64 profile in Figs. 1 and 2) the Carolinian and Laurentian terranes occur duplicated by folding and thrust faulting.

Grenville basement and associated rift facies under the coastal plain northeast of its Piedmont outcrop along I-64

Distribution of Grenville basement under the Atlantic Coastal Plain west of the Salisbury anomaly in Figure 1 is modified from Thomas et al., 1989; Hansen and Edwards, 1986; and compilation from original sources. West of the anomaly and north northeast from I-64 to the head of Chesapeake Bay the outcropping Grenville Goochland nappes strike into basement characterized primarily by plagioclase-quartz-biotite (± accessory microcline) gneiss (Fig. 1, localities 1–9, 11). According to Hansen and Edwards (1986), Clifford Hopson found samples from locality 4 (Fig. 1) to be similar to Grenville basement in the Baltimore gneiss domes in which he had worked for many years (Hopson, 1964). Associated with the gneiss are chloritoid-bearing chlorite-quartz-plagioclase-muscovite phyllite (locality 18), serpentine with epidote actinolite schist (locality 19), and plagioclase-tremolite metagabbro (locality 15). The schist-mafic-ultramafic assemblage is typical of the Late Proterozoic Lynchburg/Evington rift assemblage of the eastern Blue Ridge in Virginia and might also be found in the Peters Creek rift sequence of the western Pennsylvania Piedmont. It is also similar to the mélange (Sykesville type) except for the phyllite, which is absent in the feldspar-rich mélange matrix. These gneiss and phyllite basement samples along strike from outcropping Grenville Goochland gneiss are here considered to represent Grenville gneiss associated with smaller bodies of the Laurentian rift sequence.

Farther northeast of this dominantly gneissic basement (Fig. 1), rocks at localities 16 and 17, near the head of Chesapeake Bay, mark the beginning of a muscovite-biotite-feldspar-quartz schist belt with minor graphitic schist that most investigators have compared to the "Wissahickon," a pelitic unit that probably also belongs to the Laurentian rift/drift sequence. Gneissic rocks, possibly Grenvillian, also occur in the belt.

Salisbury anomaly: A manifestation of mafic-ultramafic bearing mélange?

The Salisbury gravity and magnetic anomaly divides the relatively well known terranes cropping out in the Piedmont on the west from the poorly known terrane(s) under the coastal plain to the east. Major questions raised by this review are whether the Grenville Goochland also reappears east of the Salisbury anomaly and what the cause of the anomaly is.

The eastern boundary of the Grenville Goochland basement nappe is along the Salisbury anomaly (Thomas et al., 1989), a positive anomaly that is underlain in Virginia by gabbro retrograded to greenschist with relict ophitic texture (Fig. 1, locality 20) and saussuritized and sericitized hornblende quartz diorite (Fig. 1, locality 25).

Delaney et al. (1966) were the first to model the Salisbury anomaly. They suggested several models, including Mesozoic basic intrusives and pre-Mesozoic serpentine bodies. Pratt et al.

(1988) and Glover (1989) considered this anomaly to mark the "Taconic" suture under the Atlantic Coastal Plain (cf. Fig. 2c). Gravity modeling by Pratt et al. (1988) indicated that the mafic and ultramafic rocks of the "Taconic" suture continued southeastward to reach the Moho.

Our gravity model (Fig. 4), controlled by the I-64 seismic reflection line, suggests that mafic and ultramafic rocks of the suture do not extend below 7-km depth. Using the seismic reflection data to define body points of a metagabbro under the Atlantic Coastal Plain sediments adjacent to a Triassic/Jurassic rift basin, a two-dimensional gravity model of two polygons was computed (Fig. 4). The program for gravity anomalies from subsurface two-dimensional polygons was adapted from Talwani et al. (1959). The seismic data of the I-64 line show a reflectorless zone of about 10-km width under the Salisbury anomaly (stations 2600 to 2700, Fig. 2) where metagabbro was drilled at 200 m depth in the Atlantic Coastal Plain basement (Fig. 1, locality 20). A southeast-dipping reflection extends beneath the reflectorless zone at computed depths of 4 to 7 km. This reflection is interpreted as a detachment fault at the base of the metagabbro rocks (Fig. 2).

The seismic data of the I-64 line (just west of station 3500, Fig. 2) show a gently northwest-dipping basement to depths of 1.5 s under what is interpreted to be a Triassic/Jurassic half-graben with a steep border fault on the northwest side (Pratt et al., 1988). Assuming a reasonable velocity of 4 km/s for the half-graben fill, the basement depth of the Triassic/Jurassic basin is taken to be 3 km.

With these seismic constraints on the body points, the body density contrasts were then adjusted for the positive contrast of the metagabbro (0.17) and the negative contrast of the Triassic/Jurassic sediments (–0.23) to reproduce the peak-and-trough amplitudes of the observed anomaly (+25 and –16 mgal respectively) (Fig. 4). The resulting contrasts in density, when compared to the background density of 2.70 g/cc measured for Goochland terrane Precambrian gneiss (Keller et al., 1985), yield 2.87 g/cc for the metagabbro and 2.47 g/cc for the Triassic/Jurassic sediments. These are very reasonable densities for these Atlantic Coastal Plain basement rocks, giving independent support to the seismic reflection interpretations.

Although the peak-and-trough gravity anomaly amplitudes produced by the seismically determined bodies on the I-64 line

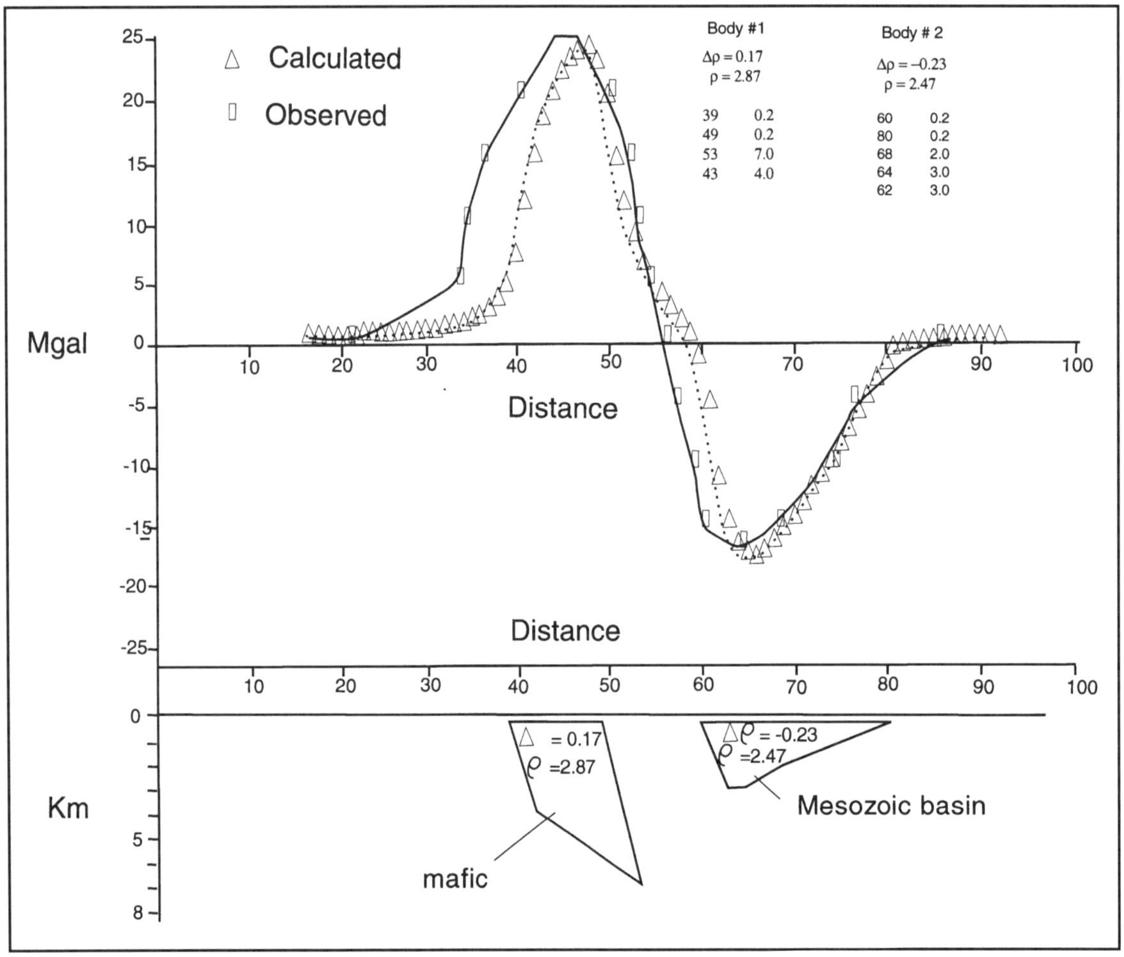

Figure 4. Gravity model of Salisbury anomaly along I-64 seismic reflection profile.

are a good fit to the observed values, the width of the observed positive anomaly is greater than that of the model, especially on the western side. The positive Bouguer anomaly contours of the Salisbury anomaly form a sinuous, linear feature characterized by lobes and saddles. The I-64 seismic line passes over the northern end of a prominent lobe that narrows markedly to the north of I-64. Consequently, the observed profile, chosen normal to the Salisbury anomaly nearest to the I-64 seismic reflection data, is influenced by the increased width of the positive anomaly lobe to the south. Accordingly, the two-dimensional model based on the I-64 data does not account for the out-of-plane positive contribution to the observed anomaly. Detailed three-dimensional modeling would be necessary to produce an exact fit, and assumptions about the body width to the north and south of I-64 would have to be made. All of this would make the model less unique. For our purposes, the two-dimensional model makes the point that shallow bodies cause the Salisbury anomaly.

Çoruh et al. (1988) called upon the strike-slip duplex model of Costain et al. (1987) and Costain and Çoruh (1990) for the structural development of the central and southern Appalachian Piedmont to explain the vertical panels of low reflectivity in a 12-km-wide panel below station 2100 along seismic profile I-64 and a wider reflectorless panel east of station 2600 on the Salisbury anomaly (Fig. 2). Reference to models by Woodcock and Fischer (1986) shows that the geometry of the Piedmont structure, as seen in the field and in the I-64 traverse (Fig. 2), does not resemble a strike-slip duplex. This is because the required vertical faults bounding vertical horsts are unknown in the region. Late Paleozoic dextral transpression occurred along *moderately eastward dipping ductile faults* (Glover et al., 1992b) such as those mapped at the surface in the Goochland nappes and traced into the lower crust on the I-64 profile west of station 2200 (Fig. 2). Therefore there were no vertical faults to be injected by the Mesozoic diabase dikes that Çoruh et al. (1988) call upon to produce the vertical panels of low to no reflectance in the I-64 profile. Lateral extension of the crust to create a 12-km-wide panel of low reflectivity implies a composite width of vertical dikes measured in kilometers. If present, it seems likely that these dikes would have reached the surface. This would be the most concentrated accumulation of Mesozoic dikes known in the Piedmont. If, as Çoruh and others believe, the dikes rose no higher than the bright reflector "F" (Çoruh et al., 1988, their Fig. 2) at about 2-sec two-way time below station 2100 (Fig. 2, this chapter), then the extensive lateral separation below that level would place the crust above it in tension so that rifts would occur there. However, the surface area in question has been well mapped at 1:24,000 scale (Bobyarchick and Glover, 1979; Poland, 1976, Reilly, 1980; Glover, unpublished) and does not contain *vertical* strike slip faults of ductile or brittle nature. The Triassic graben located over the western side of the low-reflectance panel under stations 2000 to 2250 formed as a result of brittle reactivation of the ductile Hylas fault zone, which dips moderately eastward.

Çoruh et al., 1988, p. 750, suggest that the postulated Mesozoic dike swarm under station 2100 is ". . . correlated with the positive Bouguer gravity anomaly that extends about 80 km to the *northeast*." This conclusion is not supported by the N20° to 30°W trend of the field of narrow, straight, elongate anomalies shown on the aeromagnetic map of Virginia (Zietz et al., 1977) that do correlate with the Mesozoic dikes shown on the geologic map of Virginia (Calver, 1963), published before the magnetic data were available. Further, the character of the Salisbury gravity and magnetic anomaly is a sinuous one that pinches and swells and follows Appalachian curvilinear structural trends. It is also seen to be mostly truncated by the kidney-shaped negative anomaly (Haworth et al., 1980) of the 330-Ma Petersburg granite (Wright et al., 1975), indicating that the Salisbury anomaly is *older* than 330 Ma. Such magnetic anomalies in the Piedmont owe their geometry to a complex history of Paleozoic ductile deformation and localized generation, destruction, and retention of magnetite (Robinson et al., 1982). This history contrasts with that of the characteristic Mesozoic diabase dikes in the region that trend northwest at high angles to the Salisbury anomaly as straight, narrow, high-angle dikes that were emplaced by brittle fracture in cool upper crust.

On seismic reflection line I-64 (Çoruh et al., 1988) one can also see faint dipping reflectors passing through these low-reflectance panels. Whatever the cause of the low-reflectance panels, and assuming that they are real, they are superimposed on the Paleozoic structure of the I-64 profile without deforming it. They are also most plainly seen in records using the unpublished Automatic Line Drawing display program of Cahit Çoruh (Çoruh et al., 1988)

Clearly some of the evidence is conflicting and more work needs to be done. The fact that the vertical panels of no/low reflectance are superimposed on detectable eastward dipping reflections would indicate that, if real, the structures responsible for the lower reflectance postdate the dipping reflections, that is, are post Paleozoic, perhaps Mesozoic diabase dikes. But are the blank panels real? C. Çoruh and J. K. Costain (personal communication, 1990) have not found evidence from the field records that data acquisition or processing was a problem. But it is also difficult to prove that data acquisition, processing, or geology (statics) was *not* a problem. The evidence is in part conflicting, and we suggest that the magnetic and gravity signature of the Salisbury implies that the low reflectance is caused by ductile deformation and metamorphism of Paleozoic mafic-ultramafic(?) rocks. This conclusion is also consistent with two drill hole samples on the anomaly that reveal metamorphically retrograded gabbro. Therefore we believe the Salisbury anomaly is pre-Mesozoic and that in interpreting the seismic reflection data all geological and geophysical data should be used.

Sheridan et al. (1991) conducted a seismic reflection and gravity survey over the northern end of the Salisbury anomaly in southern New Jersey and found no vertical panels of low reflectance. They concluded that the Salisbury anomaly was the result of subhorizontal mafic and ultramafic-bearing sequences that are no deeper than 8 km and are bounded on both sides by steep faults that were reactivated during the Mesozoic. In the Sheridan et al. model the Salisbury anomaly results from a synformal downwarp of suture-related mélange and associated rocks between Laurentian basement nappes, and we adopt this alternative as the eastern Goochland basement model (Fig. 1) in the middle-Atlantic states, suggesting—because of its geometry and strike—that it may be a continuation of the Connecticut Valley synclinorium. A third model (Lefort and Max, 1991) is further discussed below.

Rift/drift facies Laurentian(?) rocks adjacent to the Salisbury body southeast of Richmond

Rift/drift facies rocks may be present in southern Virginia south of I-64 (Fig. 1). Basement samples at localities 21 and 22 contain low-grade pelitic muscovite-biotite-quartz schist with accessory albite, quartzite, and metaquartz siltstone. Localities 21 and 22 are lithologies consistent with deep-water rift/drift facies rocks such as the upper Lynchburg and Evington Groups of Virginia, the Octararo Formation of the Pennsylvania Piedmont, and perhaps the Peters Creek and type Wissahickon of Pennsylvania. The lithologies at these two locations are not compatible with magmatic arc facies because of their compositional maturity. However, they are compatible with mélange matrix rocks of northern Virginia (Glover et al., 1994) and may be part of the proposed Salisbury anomaly mélange.

Amphibolite facies basement rocks east of the Salisbury anomaly: Laurentian Goochland or Carolinian?

Southeast of the Salisbury anomaly the I-64 profile has a poorly reflective character that may be caused by geological or instrumental factors (Pratt et al., 1988; Glover et al., 1992b). Because of this the eastern end of the profile is based more heavily on along-strike regional projections of basement samples using potential field data.

Three basement cores along the coast entered plagioclase-quartz-biotite ± K-spar gneiss. The Island Beach well (Fig. 1, locality 14) on the New Jersey coast cored a migmatitic plagioclase-quartz-biotite gneiss with minor microcline and granitic neosome that according to Southwick (1964) closely resembles the veined gneiss phase of the Grenville Baltimore gneiss in the domes near Baltimore. A well on Cape May (Fig. 1, locality 12) at the southern end of New Jersey also encountered a gneiss containing quartz-plagioclase-tremolite-biotite ± K-spar. Farther south in southeastern Maryland the Ohio Hammond #1 well (Fig. 1, locality 10) penetrated quartz-plagioclase-biotite gneiss with accessory microcline and hornblende. Pegmatitic veins cut a faint foliation in the rock. Table 1 and Figure 5 compare the Island Beach and WI CG 37, Ohio Hammond, cores with other rocks found in the central Appalachian Piedmont and basement under the Atlantic Coastal Plain that are compositionally similar and generally at similar amphibolite metamorphic grade.

The Island Beach, New Jersey, core presents a thorough and intensive petrographic study and is more comparable to the data available from outcrop. Because the core is migmatitic, the neosome would have to be modally merged with the paleosome to give an approximate protolith composition. There is insufficient data in the report (Southwick, 1964) to accomplish this. Nevertheless it can be seen in Figure 5 that the composition would move toward granite if minimum melt compositions (pegmatitic neosome) were added back to the paleosome.

The modal analysis of the Ohio Hammond #1 well in Maryland is based on a smaller sample and is less well constrained. This core is also reported to be pegmatitic, and the argument used above to show the direction of compositional change if the neosome is added to the paleosome (Fig.5) would apply.

The migmatitic character of the Island Beach and Ohio Hammond cores indicates high-grade metamorphism and is similar to the metamorphic history and pegmatitic character of the Maidens and other "veined gneiss" facies in the Grenville Goochland and Baltimore Gneiss belts. However, high-grade metamorphism of Carolinian rocks would also produce similar veined gneiss, although examples of it are not known to the writers. We conclude therefore that although, along with gross composition, the migmatitic character has been cited as a character of similarity between the Island Beach and the Grenvillian veined gneisses (Southwick, 1964), it is not a reliable factor in correlation of the two. The gross composition of these eastern high-grade gneisses is similar to the Grenville of the nearby Piedmont, but it is also somewhat similar to the higher-grade gneisses of Carolinia (Fig. 5). Tonalites may be rare in the Goochland Grenville, but otherwise both terranes range from tonalite through granodiorite to granite. *Therefore on the basis of available petrographic fabric and mineralogical composition it does not seem possible to make a distinction between Grenville Goochland and Carolinian rocks in the basement under the Atlantic Coastal Plain.* Because of this, two models are suggested for basement origin (in Figs. 1 and 2).

CAROLINIAN TERRANE MAGMATIC ARC: PIEDMONT AND SUBCOASTAL PLAIN GEOLOGIC FRAMEWORK

In the Carolinas magmatic arc rocks are continuous across the Piedmont and under the coastal plain from west of Charlotte, North Carolina, to Cape Hatteras (Fig. 6). In Virginia the Piedmont nappes of Goochland Grenville basement are warped into an antiformal structure that plunges southward beneath the Carolina slate belt magmatic arc rocks near Raleigh North Carolina (Fig. 6; also Farrar, 1984). The Carolinian terrane is broken by faults and interrupted by Mesozoic basins (Keppie and Dallmeyer, 1989), but there is little evidence to suggest that it comprises more than a single exotic terrane, Carolinia. Recent maps of the Atlantic Coastal Plain basement (Thomas et al., 1989; Keppie and Dallmeyer, 1989) generally agree. Horton et al. (1991), however, split Carolinia into five terranes but consider several to be possible extensions of adjacent volcanic "terranes."

TABLE 1. COMPARISON OF ISLAND BEACH, NEW JERSEY, AND WI-CG-37 (OHIO HAMMOND #1) EASTERN BASEMENT CORES WITH SIMILAR ROCKS CROPPING OUT IN THE LAURENTIAN AND CAROLINIAN TERRACES, PIEDMONT OF VIRGINIA*

	Briggs et al., 1978, Roxboro, NC, meta granite	Baird, 1989, ave of 6 Upper Hyco meta-dacite-Carolinian terrane	Poland, 1976, ave of 5 Maidens (grano-diorite gneiss) Grenville Goochland nappes	Poland, 1976, ave of 6 State Farm gn (Grenville Goochland)	Southwick, 1964, paleosome of Island Beach core
Total qtz+feldspar	91.2	86	74.2	81.99	78.4
Quartz	27.1 (29.7)	40 (47)	26.2 (35.3)	30.16 (36.8)	34.4 (43.9)
Plagioclase	34.8 (38.16)	46 (53)	35.4 (48)	35.5 (43.3)	40.4 (51.5)
K-spar	29.3 (32.13)	tr	12.6 (17)	16.33 (20)	3.6 (4.6)
Biotite	2.6 incl stilph-nomelane	10	30	13.66	18.6
Muscovite	2.4	1	0	tr	0.7
Chlorite	–	0	tr	tr	0.1
Hornblende		1	4.3	0.83	
Epidote	2.3	3	1.4	clino 1	tr
Garnet	0.2	0	5.5	3	0.1
Sphene	–	0	tr	1	tr
Scapolite		0	tr		
Calcite		0	tr		
Magnetite		tr	tr	tr	0.2
Zircon	–	0?	tr	tr	tr
Other	alanite tr				apatite 0.2
Other	apatite, fluorite tr other 1.2				0.66 tourmaline
Comments			not particularly migmatitic		similar to Maidens esp. w/k-spar of neosome added back

	Southwick, 1964, neosome of Island Beach cire	Hansen and Edwards, 1986, W1-CG-37, S MD (paleosome)	Hansen and Edwards, 1986, WI-CG-37, S MD (neosome = pegs)	Core NC26 South-port, NC, ave of 3 modes - tonolitic bi-amp-qtz-pc gneiss w/o mafic layers included	Core NC29, Camp Lejune, NC, ave of 8 Qtz monzonite, age 633 ± 61 MA
Total qtz+feldspar		73		76.97	81.3
Quartz	variable	54 (74)	probable	28.17 (37)	13.6 (17)
Plagioclase	variable	17 (23)	probable	48.7 (63)	35.8 (44)
K-spar	variable	2 (3)	reported	0.1 (0.001)	31.9 (39)
Biotite	variable	21		8.47	4.5
Muscovite		0		tr	
Chlorite		0.2	reported	tr	
Hornblende		0.3		10.17	11.8
Epidote			reported	0.8	
Garnet			reported		
Sphene		0.3			–
Scapolite					
Calcite		5			
Magnetite				–	
Zircon				–	–
Other		apatite 0.4		apatite	apatite
Other				alanite	alanite, ilmenite
Comments		similar to Maidens esp. w/k-spar of neosome added back		similar to CSB	similar to Calc-alkaline West Farrington

*All amphibolite grade with minor greenschist overprint.

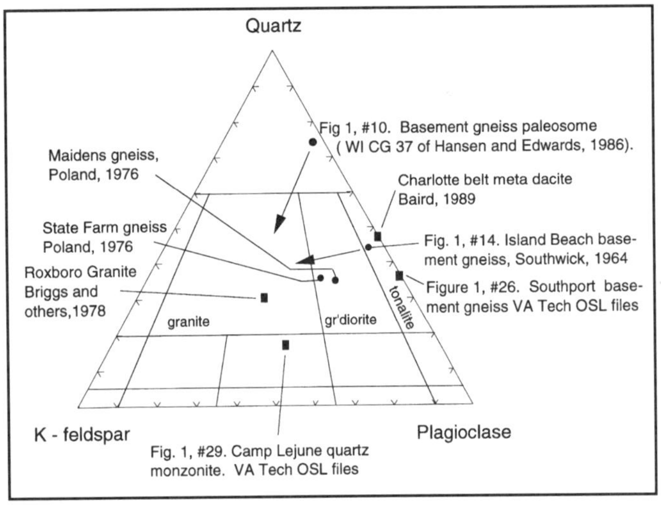

Figure 5. Comparison of selected basement cores with potentially coeval Piedmont rocks. Arrows show direction composition would move if granite minimum melt neosome were added back.

Because of the importance of determining whether other sutures may exist within these magmatic arc rocks, we comment briefly below on the principal points used by Horton et al. (1991) to split Carolinia into the multiple terranes (Fig. 7) listed herein.

The Milton Terrane (Fig. 7, terrane 1) is also the *Charlotte metamorphic terrane* of Baird (1991). Baird (1991) shows that this terrane has the same stratigraphy as the adjacent Carolina slate belt (Albemarle arc of Horton et al., 1991) from which it is separated along a metamorphic gradient and a moderately east dipping fold limb. Because of the structural orientation of the boundary, it became a stress guide and localized shearing during late Paleozoic dextral transpression. A 740-Ma age (Glover et al., 1971) from the "Milton" terrane occurs in the oldest part of the stratigraphy and therefore is not in conflict with slate belt geochronology. Baird (1991) suggested returning the Milton to the Carolinian terrane Charlotte metamorphic belt of southern Virginia and northern North Carolina and we agree.

The Chopawamsic volcanic arc terrane (Fig. 7, terrane 2) has been extended by reconnaissance and detailed mapping (Gates, 1981; Baird, 1991, and references therein) southwestward from northern Virginia into contact with the western edge of the Charlotte belt (Milton belt, terrane 1 of Fig. 7) in southern Virginia. Harris and Glover (1988) and Glover (1989) present evidence that the Chopawamsic is part of the younger slate belt sequence and lies with angular unconformity (ca. 600-Ma Virgilina deformation) upon the older sequence. Therefore the Chopawamsic is part of Carolinia. Horton et al. (1991) point out that the Chopawamsic was deformed by a "Penobscott"-age orogeny whereas other volcanic terranes were deformed at later times. However, Glover et al. (1992a) have suggested that the ca. 530-Ma (Penobscott?/Carolinian) deformation characterizes the Carolinian Charlotte belt and Chopawamsic/James Run vol-

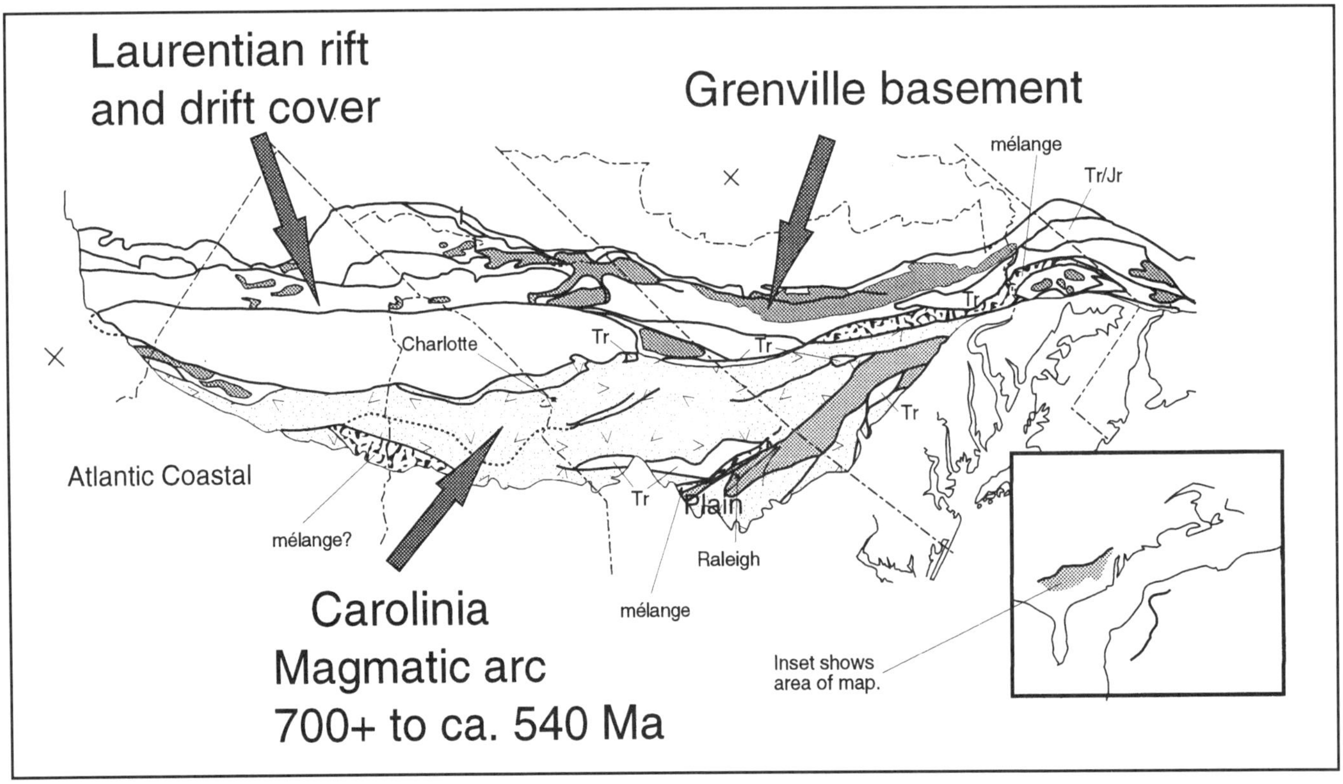

Figure 6. Terrane map, central and southern Appalachian orogen. Modified from Glover, 1989. Collision at ca. 540 Ma.

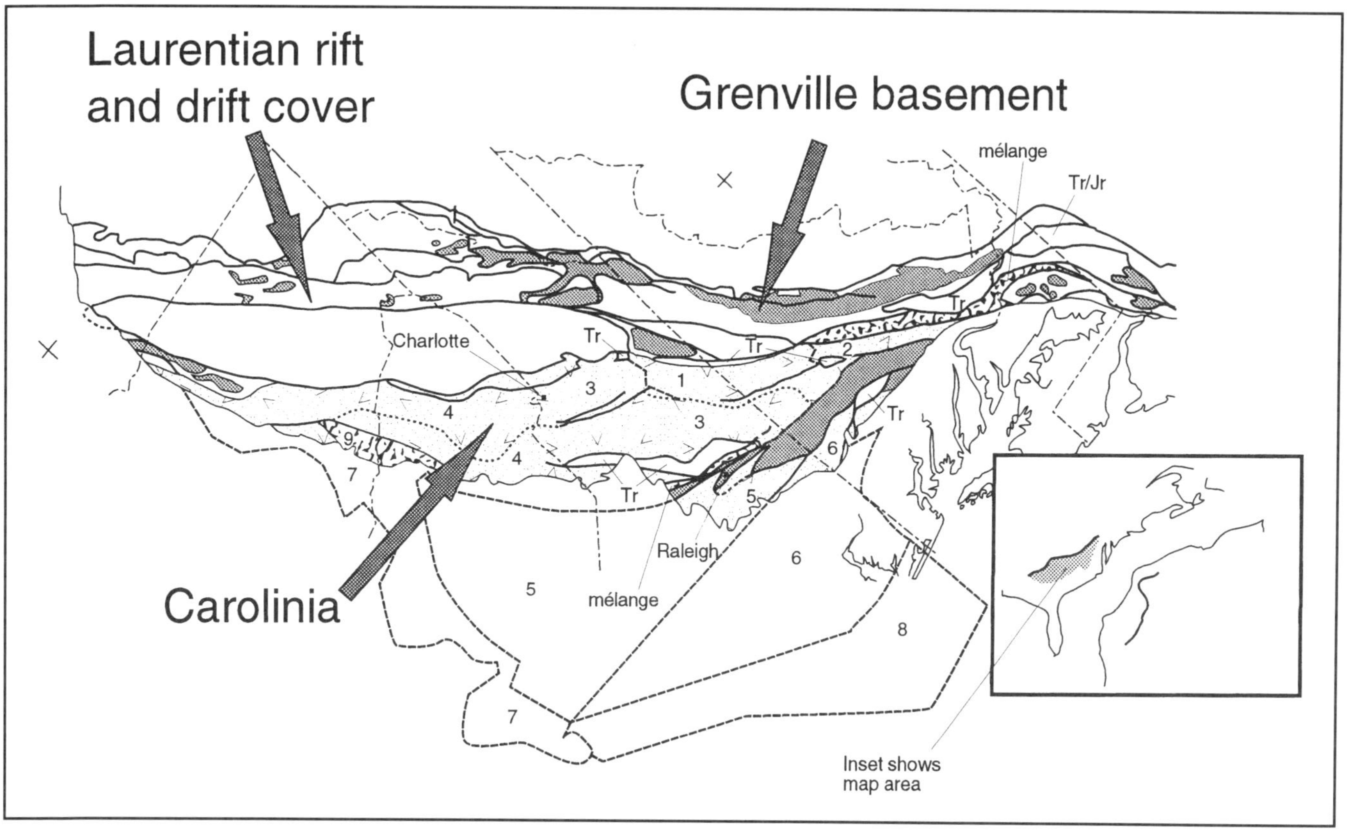

Figure 7. Magmatic arc terranes of Horton et al., 1991 (see text for discussion), central and southern Appalachian orogen. Base map modified from Glover, 1989. Terranes: 1, Milton; 2, Chopawamsic volcanic arc; 3, slate belt and Charlotte belt of North Carolina; 4, slate belt and Charlotte belt of South Carolina; 5, Spring Hope; 6, Roanoke Rapids; 7, Charleston; 8, Hatteras; 9, Savannah River.

canics from at least southern Virginia to Pennsylvania and represents the first contact of the Carolinian terrane with Laurentia.

In the Horton et al. (1991) split the Albemarle and Carolinia (restricted) terranes (Fig. 7, terranes 3 and 4, respectively) represent the slate belt and Charlotte belt of North Carolina and the slate belt and Charlotte belt of South Carolina respectively. These "terranes" strike into one another, and Horton et al. (1991) suggest that they could be the same. Subdivision, with no specified location of the boundary, was based on the possibility that stratigraphy is different (no evidence presented) and that the Carolinian (restricted) terrane was first deformed and metamorphosed at 360 to 340 Ma. This differs from Glover et al. (1983), who suggested that deformation of that part of Carolinia was "Taconic" or older. We see no reason to separate these or to remove them from the magmatic arc Carolinia.

The Spring Hope (Fig. 7, terrane 5) is separated as a terrane only because it is bounded by faults. Horton et al. (1991) suggest that it could be part of adjacent terranes. We consider it part of Carolinia.

The Roanoke Rapids terrane (Fig. 7, terrane 6) is segregated primarily because it is separated from the rest of the outcropping arc rocks by a fault. Horton et al. (1991) suggest it could be the same as Spring Hope. We include the Roanoke Rapids in Carolinia.

Mafic/ultramafic block–bearing mélange is associated with the contact between magmatic arc rocks (Albemarle, Spring Hope, and Roanoke Rapids terranes of Horton et al., 1991) and Laurentian Grenville-age Goochland basement in the Raleigh belt of the eastern Piedmont (Figs. 6 and 7). The structural position of the mélange between magmatic arc rocks of Carolinia and the Laurentian basement plus its similarity to "Taconic" mélange in northern Virginia between Carolinia and rift/drift rocks of Laurentia indicate that these are the same mélange (Glover, 1989; Horton et al., 1986).

The Charleston terrane (Fig. 7, terrane 7) was split off on the basis of its broad wavelength and high background magnetic field anomaly signature as compared to shorter-wavelength higher-amplitude anomalies of adjacent magmatic arc rocks. Magnetic character is controlled by several factors, such as composition, metamorphic grade, relative abundance of mafic versus felsic rocks, degree and style of structural deformation, and depth to source, and is thus not a sound primary basis for discriminating metamorphosed terranes. Closer inspection of the magnetic field over the Piedmont Carolinian terrane (see magnetic map of Zietz and Gilbert, 1980) reveals a variety of magnetic expressions, including the amphibolite facies Charlotte belt of Virginia and the greenschist facies Carolinia slate belt in the Albemarle district of North Carolina, both having signatures similar to the Charleston "terrane" of Horton et al. (1991). We consider the Charleston terrane also an integral part of the broader Carolinian terrane.

The Hatteras terrane (Fig. 7, terrane 8) includes basement under Pamlico Sound and under the outer banks of North Carolina. As discussed below, the plutonic and subvolcanic to volcanic nature of many of the samples from this terrane and their age span in common with the Carolina slate belt of the Piedmont indicate that th Hatteras terrane also is part of the Carolinian terrane.

The Savannah River terrane (Fig. 7, terrane 9), within the Kiokee belt, is composed of mafic/ultramafic blocks in metagraywacke matrix (Burks Mountain complex of Sacks et al., 1989). Sacks et al. (1989) consider that the complex may be derived from a unit within the Carolina slate belt or from a Precambrian basement unit beneath the Carolina slate belt. Alternatively they suggest that the complex may be a terrane exotic to both North America (Laurentia) and the Carolina slate belt. We note (as did Sacks et al., 1989) the similarity and structural position of the Burks Mountain to the Falls Lake mélange in the Raleigh belt to the north. As with the Falls Lake, we suggest that a correlation with the Cambrian mélange between Laurentia and Carolinia should also be considered. In the high-grade metamorphism case in which the mélange nature is in doubt (Burks Mountain), we note that a correlation with the Late Proterozoic to Cambrian(?) rift facies (mafic/ultramafic sills and dikes in Lynchburg/Evington Groups deep-water turbidites—Glover et al., 1992b) is possible. Therefore in the Kiokee belt Burks Mountain complex we may be looking through the Carolinian plate into the Late Proterozoic to Cambrian mélange that separates Carolinia from Laurentia at depth.

The foregoing indicates that it is reasonable to consider most of the magmatic arc terranes discussed above to be part of a single terrane Carolinia (name and limits modified from Secor et al., 1983).

The Carolinian magmatic arc is characterized by calc-alkaline volcanic rocks (Rogers, 1982) deposited in a dominantly submarine environment (Harris and Glover, 1988; Secor et al. 1989). The volcanic rocks of the terrane range in age from 740 to 700 Ma (Harris and Glover, 1988) in the older slate belt of North Carolina to ca. 550 Ma in the Chopawamsic of northern Virginia and its correlative, the upper slate belt sequence in North Carolina (Harris and Glover, 1988; Glover, 1989). Metamorphic grade (Glover et al., 1983) ranges from greenschist (Carolina slate belt of Virginia, North Carolina and South Carolina; Chopawamsic volcanic rocks of northern Virginia, eastern slate belt of North Carolina, and most of the terrane under the Atlantic Coastal Plain of the Carolinas and southern Virginia) to amphibolite facies (Charlotte belt of Virginia and the Carolinas, Raleigh belt of Virginia and North Carolina, and the Hatteras belt of coastal Carolinas basement). Carolinia was metamorphosed regionally in the Early Cambrian and/or Taconic and has late Paleozoic faulted antiformal structures exposing higher grade cores in the Charlotte belt of Virginia, and the Raleigh and Kiokee belts of the Carolinas (Glover et al., 1983; Glover, 1989). The Raleigh and Kiokee belts were also most strongly overprinted by Alleghanian medium- to high-grade metamorphism (Glover et al., 1983; Secor et al. 1986a, b), suggesting that the subsurface higher-grade Hatteras belt may have a similar structure.

About 200 wells have been drilled to basement in the coastal plain of North Carolina (because they are so abundant, most are not shown on Fig. 1; see Daniels and Leo, 1985.). These wells indicate a uniform low-grade volcaniclastic basement, only locally intruded by granite, extending from the low-grade volcanic Carolina slate belt in the eastern Piedmont to the western edge of the Hatteras belt (Daniels and Leo, 1985; Thomas et al., 1989). At the eastern boundary a few quartz-rich sandstones and muscovitic phyllites associated with carbonate beds occur (localities 36, 37, 38, and 33, Fig. 1 and Table 1). All are low greenschist facies except locality 39, which is a staurolite schist. The zone appears to be narrow. As noted above, the zone is bounded by a wide terrain of low-grade volcanics to the west in North Carolina and also has low-grade felsic tuff in the Chrisfield, Maryland, well (Fig. 1, locality 34) of the Delmarva Peninsula. The quartz-carbonate–bearing beds clearly indicate a compositionally mature stratigraphic interval of diminished volcanic activity reminiscent of the (unique?) Middle Cambrian Asbill Pond Formation at the top of the volcanic sequence in South Carolina (Rankin et al., 1989).

The Hatteras belt (Fig. 1) is a pluton-rich belt of amphibolite metamorphic grade. The western boundary is an abrupt change to greenschist facies volcaniclastic rocks and may be a fault. Basement samples having rather complete petrographic descriptions and/or age dates are given in Table 1. All of the rocks in the Hatteras belt appear to be metaigneous and range in composition from tonalite gneiss with mafic amphibolite layers (Fig. 1, locality 26) through quartz monzonite (locality 29) to granite (locality 30) to cordierite-bearing granite (locality 32). A biotite-hornblende amphibolite was encountered at locality 28 and highly altered sodic to intermediate igneous rock at locality 27. The most reliable dates appear to be Rb-Sr whole-rock ages of 583 ± 46 Ma for the granite at locality 32. A 633 ± 61-Ma age was obtained for the quartz monzonite at locality 29 and 610 ± 60 Ma for the granite at locality 30. Except for the age data, the Hatteras belt is compositionally somewhat similar to the eastern high-grade basement of the Mid-Atlantic states (Fig. 1; Table 1) and without the ages could not be easily discriminated from it.

The Hatteras suite has a compositional range appropriate for magmatic arcs on continental crust. The age of emplacement falls within the (740 to 700) to 550-Ma age range of the Carolina slate belt (Harris and Glover, 1988). The ages suggest that the Hatteras belt contains amphibolite facies older slate belt, that is, pre-600-Ma Virgilian deformation sequence of

Harris and Glover (1988). This older sequence may have been thrust westward over lower grade and younger parts of the same terrane. Additionally, these rocks indicate that Carolinia extends at least to within a few tens of kilometers of the continental margin (Fig. 1).

Gravity maps (Jachens et al., 1989) indicate that the plutonic Hatteras belt (negative anomaly) extends northward along strike offshore into the western part of seismic lines 801, 803, and 28 and over much of 802. This connection provides the firmest evidence that at least the upper continental crust along these seismic lines is a Carolinian plutonic/volcanic complex.

SEISMIC REFLECTION STUDIES AND LOCATION OF THE EASTERN MARGIN OF LAURENTIA

Geophysical studies at sea provide the principal evidence for the composition and structure of the submerged continental margin. Results from the EDGE marine seismic investigations off of the Virginia coast (Holbrook et al., 1992; Sheridan et al., 1993; Holbrook et al., 1994) provided the impetus for assembling the onshore and offshore data presented in this review. The Middle Atlantic Chesapeake Bay Corridor EDGE Project, begun in 1991, was carried out under the guidance of Lynn Glover, Robert E. Sheridan, and W. S. Holbrook. It is a contribution to the EDGE Program, a program to connect offshore geophysical studies with onshore geology and geophysics across continental edges. The EDGE program was under the leadership of Manik Talwani and associates at the Geotechnology Research Institute, Houston, Texas. The Middle Atlantic Chespeake Bay Corridor EDGE Project was funded jointly by the National Science Foundation and Texaco, Inc. The EDGE data and profiles (lines 801, 802, and 803) have been published and extensively discussed in the above-mentioned articles. Therefore only representative line 803 is discussed below.

Line 803 (Figs. 1, 2)

Line 803 is representative of three profiles (801, 802, and 803 shown in Fig. 1) in a deep seismic reflection study of the continental margin off Chesapeake Bay. Study of these profiles reveals Carolinian volcanic rocks at the continental edge possibly overlying Grenville basement (Sheridan et al., 1993). Jurassic magmatic underplating of the continental edge (Talwani et al., 1995) truncates the possible suture that was projected into the profile from its proposed position in the basement under the bay. From the foregoing an alternate interpretation is that the contact is simply a major fault or nonconformity in Carolinian terrane separating high- and low-grade volcanics. The profiles show compressional (convex upward thrust folds) and extensional (Mesozoic basin on reactivated listric fault) features. There is no horizontally layered lower crust in these profiles, and compressional structural signatures extend *discordantly* into the Moho, which might suggest that the layered lower crust was removed by low-angle normal faulting (delamination) as the Atlantic basin opened. Evidence of low-angle extensional faulting is also observed at the M (Mohorovičić) discontinuity (Fig. 2) under the Piedmont in seismic profile I-64 (Glover et al., 1992b).

Profile 803 in Figure 1 is used to extend the geological interpretation of the I-64 profile (Figs. 1 and 2), as reprocessed by Virginia Tech, to the continental margin. It is important to note here that, as shown by Austin et al. (1990), Mesozoic volcanism and magmatic underplating explain the East Coast Magnetic Anomaly (ECMA) and that this feature of the continental edge is not related to an ancient Alleghanian suture as postulated by Nelson et al. (1985). Seafloor magnetic stripes date the time of breakup and volcanism at the margin as middle Jurassic (Sheridan, 1983; Klitgord et al., 1988). All Mesozoic basins in the exposed U.S. Appalachians were initiated in the Late Triassic, and many, if not most, continued subsiding into the Early Jurassic (Manspeizer, 1988). Musser et al. (1992) and Musser (1993) indicate that subsidence of the Norfolk basin also began in the Triassic and continued into the Jurassic.

LASE (Large Aperature Seismic Experiment) lines 25 and 6 (Fig. 8)

These lines provide insufficient definition of reflectors to interpret the possible position of a suture within pre-Mesozoic crust. The Island Beach well (Fig. 1, locality 14; Fig. 8) is onshore just north of the strike of the line. As mentioned above (Table 1), this rock is similar to Grenville outcrops in the Baltimore gneiss domes.

Because Line 36, discussed later, is better understood in the light of seismic lines farther north, we here skip to the U.S. Geological Survey line in Maine before working south.

U.S. Geological Survey Quebec–Maine–Gulf of Maine (Fig. 9)

The Quebec–Maine–Gulf of Maine transect shown in Figure 9 (Spencer et al., 1989; Stewart et al., 1991, 1993) offers unresolved choices of the subsurface position of Laurentian Grenville basement (and therefore the Taconic suture). Spencer et al. (1989) traced the reflective surface of Grenville basement from the St. Lawrence platform eastward to a point about 25 km below the Maine Chain Lakes body. The Cameron–Brompton–Boundary Mountains composite terrane that lies above this (exotic terrane and successor basins in Fig. 9) was considered allochthonous and probably not Laurentian. Stewart et al. (1991) recognize a recent finding (Trzcienski et al., 1992) that the Chain Lakes is a ca. 500-Ma sedimentary mélange and not a basement block. However, higher seismic wave velocities below the mélange and above the autochthonous Grenville of the 1989 paper are interpreted by Spencer et al. to be Grenville also. This would appear to put Grenville in both the Laurentian and magmatic arc plates. Further compounding matters is the fact that the subsurface extent of upper crustal reflectors has been obscured by subcrustal flow related to Mesozoic extension below about 25 km (laminated lower crust of Fig. 9) over much of the profile. For purposes of establishing the eastern margin of the Grenville, the problem is somewhat mitigated by the following observations. Projected southeastward down dip the upper surfaces of either of the "Grenville" blocks would pierce the present Moho about 100 to 120 km east of the eastern margin of the Chain Lakes mélange. We take this to be the approximate position of the Taconic suture (eastern margin of Laurentia) with respect to the unstretched Laurentian block prior to Mesozoic extension. The eastern margin of Laurentia in New England is coextensive with the eastern margin of Laurentia in the central and southern Appalachians, but different terranes may have collided with Laurentia at different times in the two regions.

COCORP (Consortium for Continental Reflection Profiling) line

The COCORP line of Ando et al. (1984) across the Appalachians in New York, Vermont, and New Hampshire provides some control on the position of the Taconic suture (eastern margin of Laurentia) at depth (Fig. 10). Current work (e.g., Stanley and Hatch, 1988) places the Taconic suture in the Rowe Formation of the Rowe-Hawley zone. The Rowe is a quartz-muscovite-chlorite schist with lensoid bodies of mafic and ultramafic rock. In the Stanley and Hatch model the Rowe is a deep-water Laurentian margin deposit with tectonically admixed fragments of ophiolite. Structurally higher parts of the Rowe-Hawley contain basaltic and dacitic volcanic rocks, with some andesite(?) that Stanley and Hatch consider to be forearc deposits from the Bronson Hill magmatic arc to the east. Ultramafic rocks occur throughout the Rowe-Hawley but are not as common in the upper part. The age is thought to be Early Cambrian to Middle Ordovician, but there is little control. In the COCORP line (Fig. 10), the Taconic suture east of the Green Mountains corresponds approximately to the "PHT (Prime Hill thrust) surface." Ando et al. (1984) projected the surface beneath the Connecticut Valley Synclinorium into the Bronson Hill anticlinorium and then monoclinally down toward the east following the seismic structure to depth beneath the Merrimack synclinorium to a projected intercept with the Moho. This intercept is between 17 and 37 km east of their line 6, depending on whether Moho is at 30 or 35 km. This location places the eastern edge of Laurentia about 125 km east of the Laurentian hinge zone (transition from deep to shallow water where the normal continental thickness began to markedly neck down during Iapetan rifting) on the Green Mountains anticlinorium, a

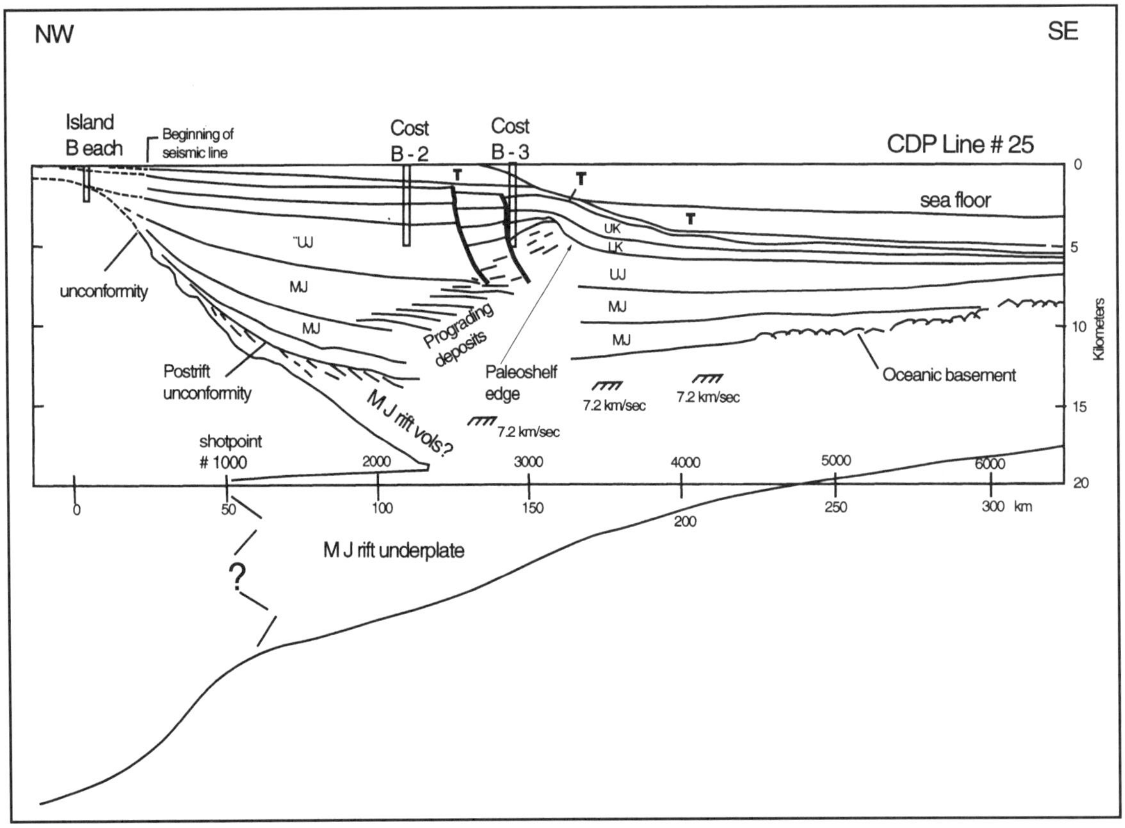

Figure 8. Combined line drawing summaries of LASE lines 25 and 6 showing rift underplate, modified from Grow et al., 1988, and Diebold et al., 1988. CDP = common depth point.

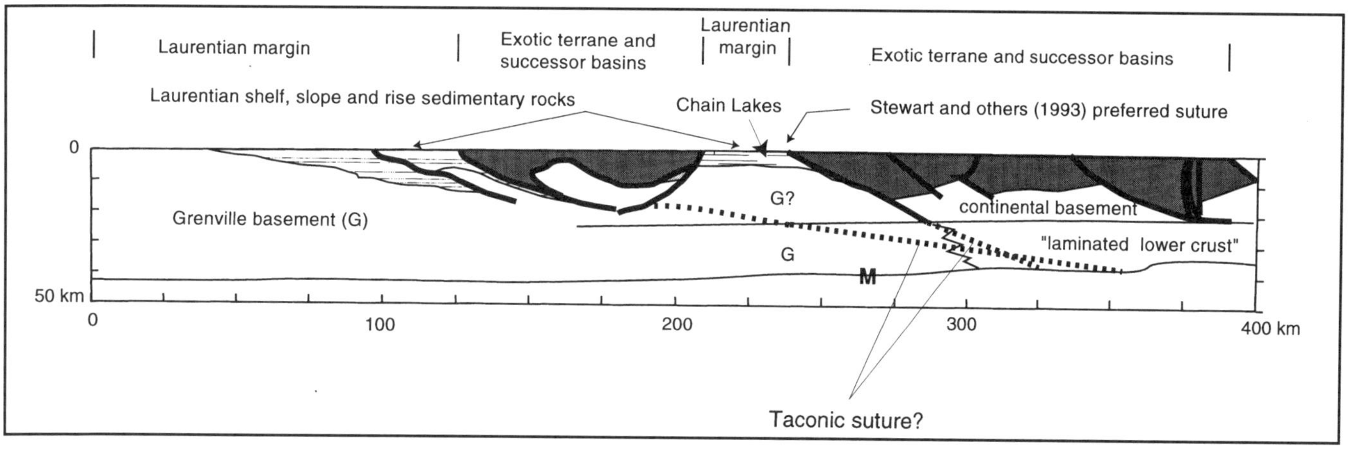

Figure 9. Western part of Quebec–Maine–Gulf of Maine transect modified from Stewart et al. (1991, 1993). Depending on crustal model adopted, Taconic suture intercepts Moho between 325 and 350 km.

reasonable figure. Analogy with the central Appalachians (Glover et al. 1991) suggests that the ultramafics and mafic rocks of the Rowe-Hawley might be rift-related dacites and basalts in which the basalts differentiated to ultramafics at high crustal levels. The whole assemblage might then have been broken and disrupted by shearing during Taconic collision. This interpretation, however, would not move the projected suture very far because of the relatively thin units involved. The reflections in this profile commonly are convex upward, consistent with folding and development of thrust folds during compression. The smooth transition of the horizontal fabric of the shear surface below the allochthonous Green Mountains into the monoclinally east dipping reflector package east of the Green Mountains is also consistent with a dominantly compressional shear fabric for most of the reflectors in this line. However, the angular discordance of the upper plate along the projected suture suggests westward rollover—an extensional phase, probably Mesozoic in age.

Hudson River to ocean Line 846

This line (Fig. 1; Fig. 11) begins in the Atlantic just north of the west end of U.S. Geological Survey Line 36, turns north, and goes up the Hudson River (Bond, 1991). The line shows moderately east dipping reflectors that persist to 8 sec where the profile crosses Cameron's Line (near marker for the Taconic suture). East of the Cameron's Line shear gently undulating reflectors characterize the terrane as they do in nearby U.S. Geological Survey Line 36. Because this terrane is on strike with rocks we interpret as Carolinian in the basement of the Salisbury Embayment, we interpret this crossing of Cameron's Line as the first of three possible structural repetitions of the Taconic suture as the New England part of the orogen swings into the central Appalachians (Fig. 1).

U.S. Geological Survey offshore line 36

The western margin of the Long Island Basin on U.S. Geological Survey Line 36 (Fig. 1; Fig. 12) is underlain by a major shear zone that dips eastward and soles into the subhorizontal reflectors associated with Moho at about 25-km depth (Phinney and Roy-Chowdhury, 1989). The Moho also shallows below this shear, confirming that it is at least in part of extensional origin. Most deep reflectors in Line 36 are characterized by being flat or concave upward in contrast to the convex upward curvature found in compression-related reflections. This line appears to be extensively reworked by deep-rifting tectonics during the Mesozoic, and the mid to lower crustal reflectors may be largely of extensional origin. Extensive experience in the Appalachians shows that virtually all Mesozoic basins formed along Paleozoic compressional faults that were reactivated in extension during initiation of Atlantic basin rifting (Ratcliffe et al., 1986). The seismic signature of this shear zone in terms of depth, dip, and rollover is similar to the projected Taconic suture on the COCORP line in Vermont and New Hampshire, although its position is somewhat east of the COCORP shear and it lacks the convex upward reflectors indicative of compressional tectonics. Thus, for somewhat different reasons, we tend to agree with Phinney (1986) that this probably is the Taconic suture on Line 36, but our conclusions do not depend upon it. To the west the deeper reflection structure is broadly horizontal with low apparent dips because Appalachian structural strike begins to turn more toward the trend of the profile. The west edge of the New York Bight basin is bounded by a nearly vertical fault, according to Phinney.

Buena line, New Jersey Salisbury anomaly

Sheridan et al. (1991) conducted a seismic reflection and gravity survey of the Salisbury anomaly in southeast New Jersey (Fig. 1) to compare with the results obtained in Virginia where the anomaly was proposed to mark a segment of the Taconic suture (Pratt et al. 1988; Glover, 1989). Velocity, gravity, and geometric constraints favor a model of nearly horizontal tectonic stacking of ultramafic, volcanic, and sedimentary rocks under the anomaly in a nearly vertical fault-bounded

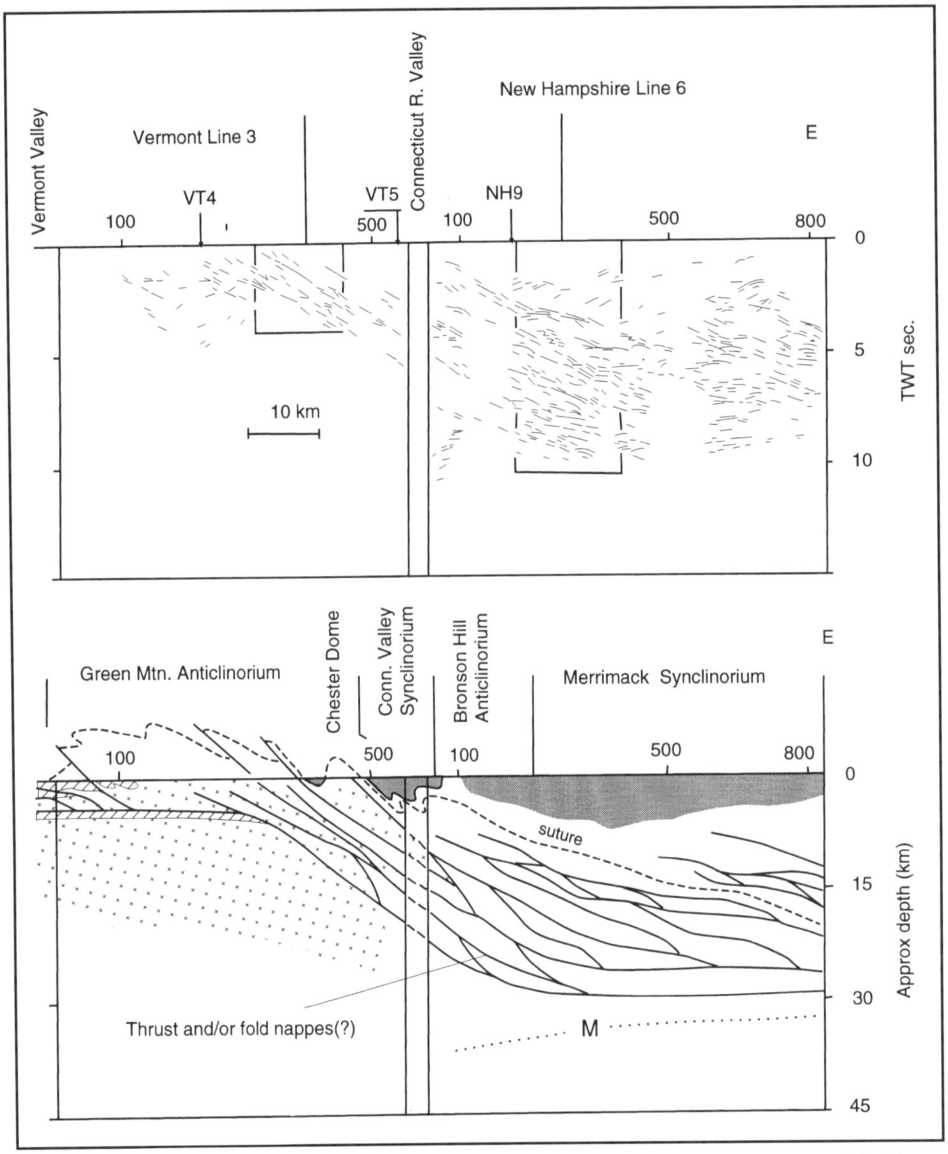

Figure 10. Slightly modified from Ando et al., 1984, to show dashed Taconic suture east of the Bronson Hill anticlinorium and M at about 32-km depth under the Merrimack Synclinorium. This depth for M is consistent with the Stewart et al., (1991) placement of M in central Maine at the base of the reflective section where Vp increases to 8.1 km/sec.

structure between Mesozoic graben. The tectonically stacked ultramafic and volcanic rocks are interpreted to be a synformal remnant of rootless Taconic suture rocks that do not penetrate below 8 km (Fig. 13). Drilling and potential field data have also outlined a Mesozoic basin along the west flank of the Salisbury anomaly in New Jersey, Vermont and northern Virginia that is coextensive with the Richmond and Taylorsville basins near Richmond (Wilkes et al., 1989).

Lefort and Max's tectonic model of the Mid-Atlantic states

The foregoing review of regional geology and geophysics bears on a model by Lefort and Max (1991), who have interpreted the Salisbury anomaly (Figs. 1, 2) as arising from a Hercynian (Alleghanian) suture between Gondwanan and "Avalonian"-Carolinian terranes. Among the lines of evidence cited are deep westward-dipping crustal reflections below and just west of the Salisbury anomaly, interpreted to be remnant from a west-dipping suture. However, Glover et al. (1992b) have shown from several adjacent crossings of the same reflections that they are part of the laminated lower crust overlain by related Mesozoic extensional structures in the upper crust. The westward dip is due to west-dipping low-angle ductile normal faulting (Fig. 2) that occurred late in the Mesozoic extensional development of these reflections (Glover et al., 1992b). It is now widely appreciated that laminated lower crust is most commonly a product of extensional shear and

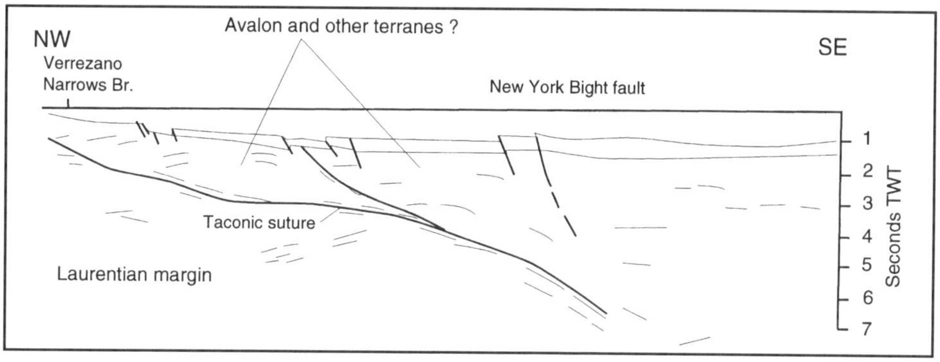

Figure 11. Seismic profile Hudson River to New York Bight. Modified from Bond (1991).

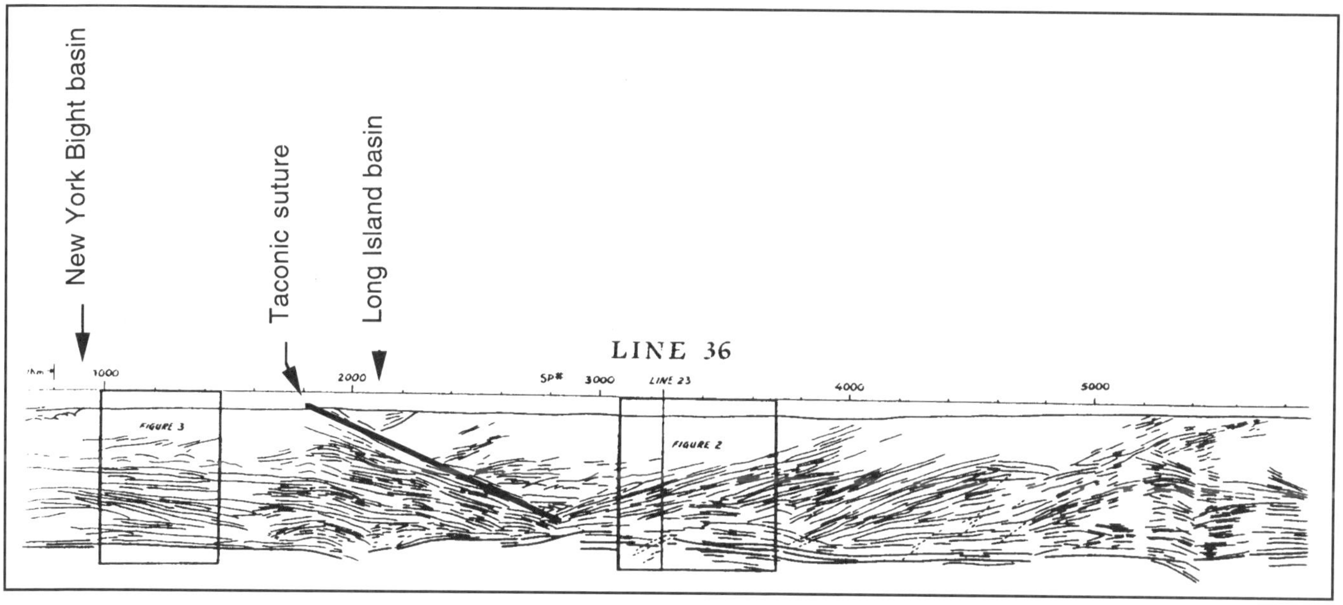

Figure 12. Line 36 on Long Island Platform. From Phinney (1986).

flowage that tends to wipe out earlier structures (Stewart et al., 1991, 1993, and references therein).

In the foregoing we have also shown that the basement cores east of the Salisbury anomaly in Virginia, Maryland, and New Jersey are similar to both the Laurentian and Carolinian terranes. Thus it is not surprising that such common rock types may also be found on the west African promontory. These observations do not require a third terrane of Gondwana affinities in the Mid-Atlantic states, as Lefort and Max propose, and we conclude that any crustal structures of the Alleghanian suture must be east of the present continental edge.

DISTRIBUTION OF THE TACONIC AND CAROLINIAN SUTURES

The distribution of the coextensive, diachronous(?), and structurally repeated Ordovician "Taconic" (New England) and Cambrian Carolinian (central Appalachian) sutures is shown on Figure 1. On the west there is a common terrane, Laurentia. On the east, different terranes seem to exist in New England and the central Appalachians. Perhaps they come together in southern Connecticut.

The easternmost positions of the suture are the most in doubt, and they define the principle differences between the EGB (eastern Goochland basement) models and the ECB (eastern Carolinian basement) models (Fig. 2). The rocks associated with the Salisbury anomaly (Figs. 1, 2) may occur as younger rocks originally downwarped or infolded into older rocks in either model. Subsequent extensional faulting provided normal movement on many of these Paleozoic compressional/transtensional faults. The isostatic gravity anomaly (Jachens et al., 1989) associated with the Salisbury appears to continue into the Connecticut Valley synclinorium of New England, suggesting a continuation of that structure into the Mid-Atlantic states (Fig. 1). Similarly, the continuation of the Bronson Hill anticlinorium into the Mid-Atlantic region is suggested by the position of the higher-grade culmination of amphibolite facies gneisses along strike in the lower Chesapeake Bay and Dela-

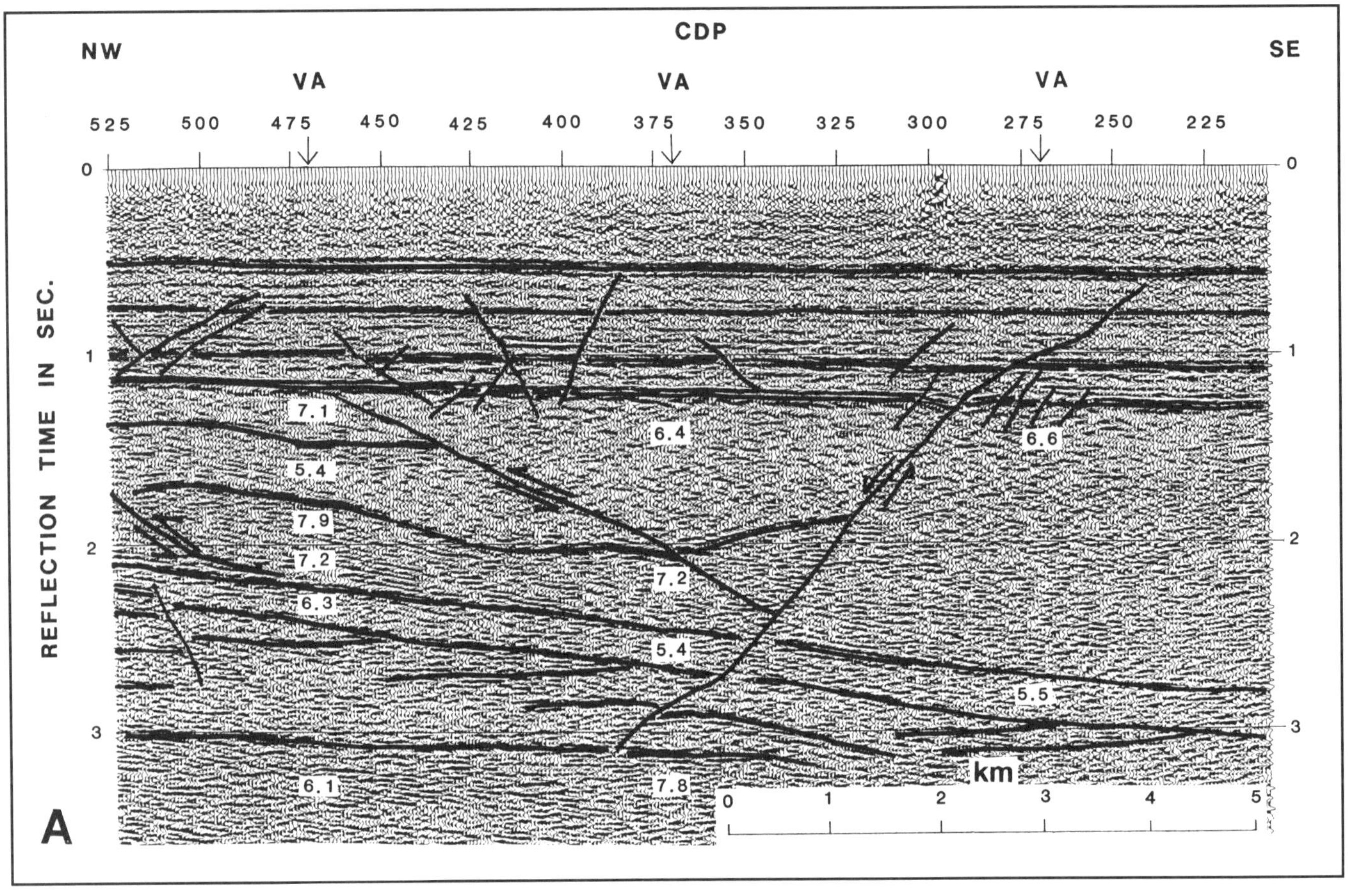

Figure 13 (on this and facing page). Buena Line A-A' of Figure 1. A, Line drawing on original seismic data of southeastern half of Buena, New Jersey, vibroseis line. Note velocity inversions at velocity analyses (VA), interpreted as the Carolinian suture where high-velocity ophiolitic rocks are thought to overlie lower velocity Laurentian margin rocks, such as at 7.2 km/sec/6.3 km/sec, 7.2 km/sec/5.4 km/sec, and 6.6 km/sec/5.5 km/sec. B, Diagrammatic cross section of Appalachian provinces and Coastal Plain showing the interpretation of the Buena vibroseis line and the synformal folding and intricate thrusting of the Carolinian suture rocks. Hatched symbol represents metavolcanics of Carolinia and ophiolitic fragments in mélange; block symbol represents Laurentian margin sedimentary rocks; dot symbol represents Triassic/Jurassic rift-basin sedimentary rocks. Both A and B from Sheridan et al., 1991.

ware Bay region (Fig. 1). These form structural belts that are conformable with the westward convex Middle Atlantic salient. In the ECB models, the Carolinian suture at the Salisbury anomaly is not repeated again to the east and must penetrate the M discontinuity well inboard of the continental edge. However, seismic and gravity models suggest that the suture rocks do not penetrate more than 8-km depth along the Salisbury anomaly, suggesting that it may not be a plate-penetrating shear zone.

STRUCTURAL IMPLICATIONS AND RELEVANCE OF THE EGB AND ECB MODELS TO THE MESOZOIC EVOLUTION OF THE MID-ATLANTIC STATES PASSIVE MARGIN

The contractional "snakehead" fold geometry exhibited by the signatures of the east-dipping reflectors in profiles 801 (Sheridan et al., 1993) and 803 (Figs. 1 and 2) indicates that they are a record of Paleozoic Appalachian west-verging thrust-and-fold structure. In profile 803 (Fig. 2) these southeast-dipping reflectors are truncated at a west-dipping reflector surface interpreted to be the top of magmatic underplating by Sheridan et al. (1993). The west-dipping reflector may have been an original "retroshear" of the Carolinian suture complex (e.g., Quinlan et al., 1993), although no other west-dipping reflectors are seen in the profile. This west-dipping reflector merges westward with the Moho where the underplating ends, yet the east-dipping reflectors continue to bear a high-angle intersection relation to the Moho. The absence of a horizontally layered lower crustal zone in profile 803 at the site of maximum continental extension and crustal thinning suggests that it was removed in the process of continental separation. Figure 2a shows how this might occur if the Mid-Atlantic margin is an upper-plate (proximal) margin and M (continental) is a delamination surface as shown. On the other hand, if fault x in

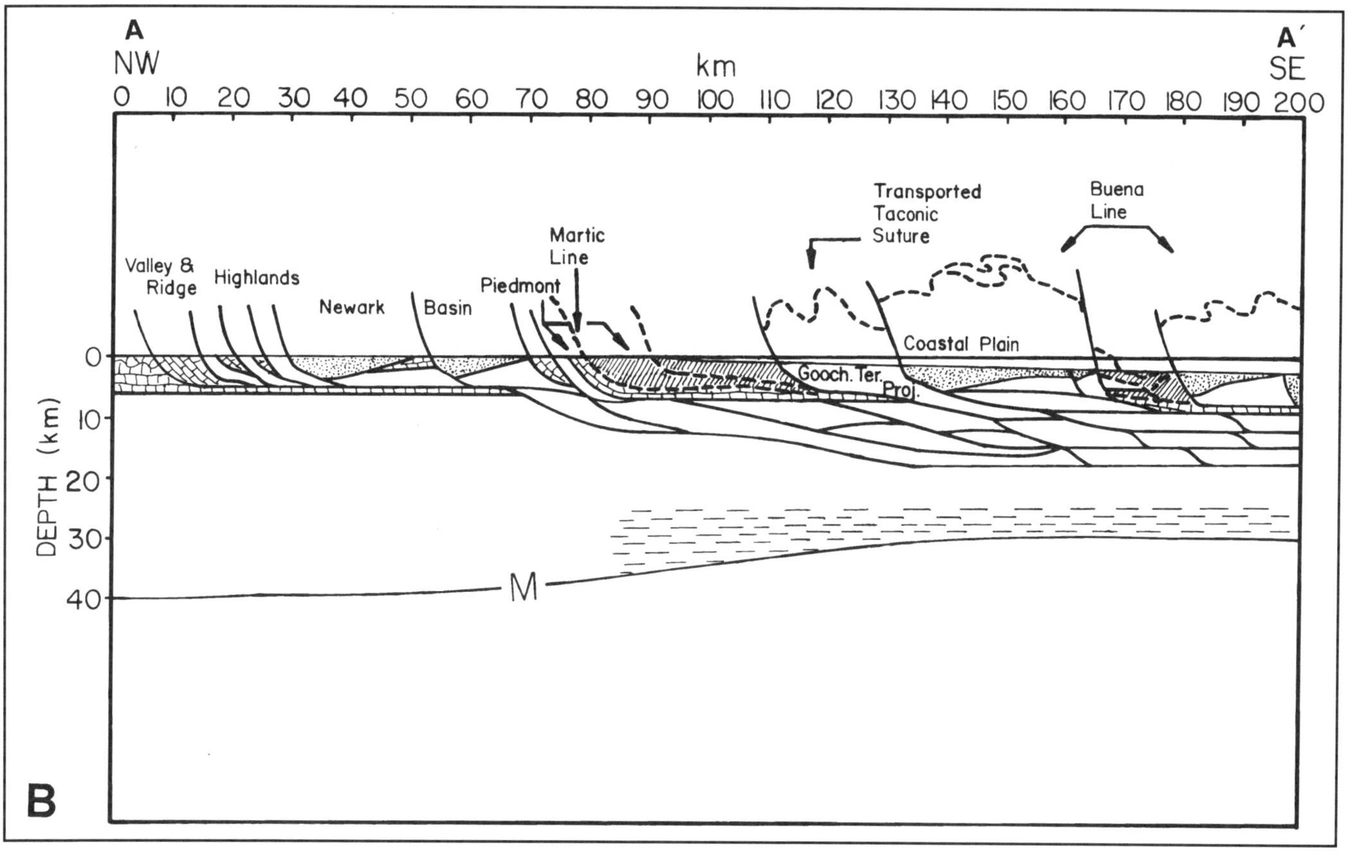

Figure 2b is the delamination surface, then the eastern ramp is an antithetic fault along which the conjugate margin broke away as underplating and volcanism filled the void. In this case we have a lower-plate margin, albeit one volumetrically dominated by the upper-plate remnant. Both models would have ramps in the more rigid upper and lower crustal zones and flats in the mid-crustal zone of greater ductility.

The basement unconformity on Carolinian rocks was first rotated counterclockwise an unknown amount about a horizontal axis during uplift and erosion as rifting began and was then rotated on a horizontal axis clockwise about 15° during the rift and drift stages. Therefore the east-dipping contractional structures are probably restored to approximately their pre-rifting dip values. The displacement of an upper plate with respect to the lower plate across a delamination surface obviously displaces structural features such as the suture unless the delamination surface coincides with it at least in part. In Figure 2a, the *upper-plate* EGB model, the Carolinian suture is truncated at a high angle by the delamination surface. In this model it is difficult to see any relation between the Carolinian suture and localization of the continental edge during rifting unless a collisional backthrust preceded the ramp (but that would be unique in the reflection profile).

In Figure 2b, a *lower-plate* EGB model is possible if the surface x? is the delamination surface and x? projected eastward into the Carolinian suture at depth. In this case the ramp at the top of the underplating bears an antithetic relation to the delamination surface. Localization of the ramp may have been induced by injection of magma (or vice versa) and displacement of the upper plate crust at that boundary.

In Figure 2c, profile 803, if the Moho and eastern ramp comprise the delamination surface, it would not coincide with the Carolinian suture at depth, and the Carolinian suture lies east of the base of the ramp, having been displaced across the delamination surface during faulting. In this case we have an upper-plate margin in which the Carolinian suture had no relation to localizing the site of continental separation. The continental edge formed along an antithetic fault, the "ramp," into which basalt was injected. The antithetic fault location is only very broadly localized by the width of the upwelling band of asthenosphere. Within that width it must break or the process cease. Rigidity is relative. The delamination surface follows a path of ramps through rigid plate and flats through the more ductile parts. There is probably a certain amount of serendipity controlling the choice of breaks from ramps to flats, as there is in shallow-level thrusting and low-angle normal faulting.

In Figure 2d, profile 803, the sheared Carolinian suture? may coincide with the easternmost x? segment of the delamination surface in a lower-plate margin. In this case the Carolinian suture had a more direct structural influence on the localization of the surface of continental separation providing that the Carolinian suture has not been displaced much in this segment by late Paleozoic orogen-parallel faults with a transform component of motion (Gates et al., 1986). Under this scenario the locus of separation would be along an antithetic fault (the "ramp") in the upper plate, allowing basalt injection and development of a volcanic-rich margin.

Lister et al. (1991) and Wernicke and Tilke (1989) have concluded that this segment of the Atlantic margin is an upper plate (proximal) margin, but their analysis lacks the consideration given here of structure in the hard rock crust.

In the EGB and ECB models, the suture itself, a fundamental lithospheric plate discontinuity, would be implicated as a localizing factor in versions of the model in which the delamination surface is listric normal dipping gently eastward from the western Piedmont. In these cases the continental edge is a lower-plate margin. In the ECB model the Carolinian suture may intersect the Moho 35 to 40 km west of the EGB model. Jurassic basalt diking of the Piedmont and abundant Jurassic volcanics at the continental edge (ECMA) are more consistent with upper-plate variants, although Harry and Sawyer (1992) show that the Jurassic diking is a precursor compatible with a lower-plate simple shear in the mantle.

In all cases the continental edge swings eastward near southern New England before turning north again. Clearly it does not follow the Carolinian suture there, nor is it conformable with Appalachian structure regionally.

CONCLUSIONS

Carolinian terrane magmatic arc rocks of Late Proterozoic to Early Cambrian age extend to the present continental edge in the Mid-Atlantic states. They were emplaced upon the Cambrian Laurentian passive margin in the central and southern Appalachians during the Carolinian collision beginning in the latest Proterozoic or Early Cambrian.

The Carolinian plate was dismembered during the Alleghanian by orogen-parallel dextral transpressional shearing and finally by Mesozoic extension (Bradley, 1982; Gates et al., 1986; Gates et al., 1988; Glover, 1989) In southern New England, current thinking (e.g., Zen et al., 1989) would suggest that a different plate bearing Ordovician volcanics collided with the Laurentian margin in that region.

In this chapter we have not attempted to subdivide all of the terranes that may lie outboard of the Laurentian margin. Instead we have defined the "Taconic" and Carolinian sutures and the Laurentian margin, respectively, between the northern and central Appalachians. The suture appears to have been a low-angle structure that, south of New England, was repeated at the pre–coastal plain surface two or three times by faulting and folding during the Paleozoic. East of the Salisbury anomaly the suture may be present between the low-grade Carolinia volcanics and higher-grade gneiss under the Delmarva Peninsula, or the higher-grade gneiss may be part of Carolinia.

Previous models fail to account for several aspects of the geological framework. Here we present two new models for the Carolinian and Taconic evolution of the Mid-Atlantic margin: (1) the eastern Carolinian basement model, in which Laurentian rocks probably do not dip very far east of the Salisbury anomaly in the subsurface, and (2) the eastern Goochland basement model, in which the high-grade gneisses under Delmarva are considered to be Laurentian, and the suture dips from eastern Chesapeake Bay into the M discontinuity along profiles 801, 802, and 28 offshore at the continental edge.

In both the ECB and EGB models there are two variants, depending on whether the listric normal faults that dip eastward from the western Piedmont lower-plate margin or the present Moho itself is the plate delamination surface, producing an upper-plate margin. Lower-plate margin variants of each model are concordant with a strong structural control by the Taconic suture on the locus of rifting at the continental edge, whereas upper-plate variants of the models suggest lesser control. Jurassic magmatism in the Piedmont and at the continental edge is consistent with upper-plate variants of the two models.

The EGB model produces a trace of the suture at M depths (Fig. 1) that is more conformable with the regional convex westward salient of the pre-Triassic structure of the Mid-Atlantic region, suggesting that either the upper- or lower-plate variants of this model may be the more nearly correct.

Dating of the basement gneiss under the eastern Chesapeake and Delaware Bay areas might result in the most economical solution to the choice between the eastern Carolinian basement and eastern Goochland basement models.

ACKNOWLEDGMENTS

This work was supported by National Science Foundation grant ES8721194 and by Texaco Oil Company. We thank Peter Lyttle and Debora R. Hutchinson for exceptionally thorough and helpful reviews.

REFERENCES CITED

Aleinikoff, J. N., Zartman, R. E., Rankin, D. W., Lyttle, P. T., Burton, W. C., and McDowell, R. C., 1991, New U-Pb zircon ages for rhyolite of the Catoctin and Mount Rogers Formations—More evidence for two pulses of Iapetan rifting in the central and southern Appalachians: Geological Society of America Abstracts with Programs, v. 23, p. 2.

Ando, C. J., and 11 others, 1984, Crustal profile of mountain belt: COCORP deep seismic reflection profiling in New England Appalachians and implications for architecture of convergent mountain chains: American Association of Petroleum Geologists Bulletin, v. 68, p. 819–837.

Austin, J. A., Jr., Stoffa, P. L., Phillips, J. D., Oh, J., Sawyer, D. S., Purdy, G. M., Reiter, E., and Makis, J., 1990, Crustal structure of the southeast Georgia embayment–Carolina trough: Preliminary results of a composite seismic image of a continental suture(?) and a volcanic passive margin: Geology, v. 18, p. 1023–1027.

Baird, R. A., 1989, Tectonic and geologic development of the Charlotte belt, south central Virginia Piedmont [Ph.D. thesis]: Blacksburg, Virginia Polytechnic Institute and State University, 163 p.

Baird, R. A., 1991, Stratigraphy of the northern Charlotte metamorphic belt:

Application to the Charlotte belt/Milton belt problem: Southeastern Geology, v. 32, p. 61–82.

Bobyarchick, A. R., and Glover, L., III, 1979, Deformation and metamorphism in the Hylas zone and adjacent parts of the eastern Piedmont in Virginia: Geological Society of America Bulletin, v. 90, p. 739–752.

Bond, G., 1991, Profile of the Hudson River: Lamont-Doherty Geological Observatory 1990 and 1991 Report, p. 22, 23.

Bond, G. C., Nickeson, P. A., and Kominz, M. A., 1984, Breakup of a super continent between 625 Ma and 555 Ma: New evidence and implications for continental histories: Earth and Planetary Science Letters, v. 70, p. 325–345.

Bowring, S. A., Grotzinger, J. P., Isachsen, C. E., Knoll, A. H., Pelechaty, S. M., and Kolosov, P., 1993, Calibrating rates of Early Cambrian evolution: Science, v. 261, p. 1293–1298.

Bradley, D. C., 1982, Subsidence in late Paleozoic basins in the northern Appalachians: Tectonics, v. 1, p. 107–123.

Briggs, D. F., Gilbert, M. C., and Glover, L., III, 1978, Petrology and regional significance of the Roxboro Metagranite, North Carolina: Geological Society of America Bulletin, v. 89, p. 511–521.

Brown, W. R., and Pavlides, L., 1981, Mélange terrane in the central Virginia Piedmont: Geological Society of America Abstracts with Programs, v. 13, p. 419.

Calver, J. L., compiler, 1963, Geologic map of Virginia: Virginia Division of Mineral Resources, Charlottesville; scale 1:500,000.

Conley, J. F., 1985, Geology of the southwestern Virginia Piedmont: Department of Mines, Minerals, and Energy, Division of Mineral Resources Publication 59, 33 p.

Conley, J. R., and Henika, W. S., 1973, Geology of the Snow Creek, Martinsville East, Price and Spray Quadrangles, Virginia: Virginia Division of Mineral Resources Report of Investigations 33, 71 p.

Çoruh, C., Bollinger, G. A., and Costain, J. K., 1988, Seismogenic structures in the central Virginia seismic zone: Geology, v. 16, p. 748–751.

Costain, J. K., and Çoruh, C., 1990, Tectonic setting of Triassic half-grabens in the Appalachians: Seismic data acquisition, processing, and results, *in* Tankard, A. J., and Balkwill, H. R., eds., Extensional tectonics and stratigraphy of the North Atlantic margins: American Association of Petroleum Geologists Memoir 46, p. 155–174.

Costain, J. K., Çoruh, C., and Hatcher, R. D., Jr., 1987, Geophysical signature of an inclined strike-slip duplex in the southeastern United States: Geological Society of America Abstracts with Programs, v. 19, p. 628.

Daniels, D. L., and Leo, G. W., 1985, Geologic interpretation of basement rocks of the Atlantic Coastal Plain: U.S. Geological Survey Open File Report 85-655, 45 p.

Delaney, J. R., Henika, W. S., and Christensen, J., 1966, Delineation and interpretation of the St. Stephens Church magnetic anomaly, King and Queen County, Virginia: Virginia Journal of Science, Proceedings, v. 17, p. 317.

Diebold, J. B., Stoffa, P. L., and LASE Study Group, 1988, A large aperture seismic experiment in the Baltimore Canyon trough, *in,* Sheridan, R. E., and Grow, J. A., eds., The Atlantic Continental Margin, U.S.: Boulder, Colorado, Geological Society of America, The Geology of North America, v. I-2, p. 387–398.

Drake, A. A., 1987, Pre-Taconian deformation in the Piedmont of the Patomac Valley, Penobscottian, Cadomian, or both?: Virginia Division of Mineral Resources, Contributions to Virginia Geology, v. 74, p. 1–18.

Drake, A. A., and Morgan, B. A., 1981, The Piney Branch Complex—A metamorphosed fragment of the central Appalachian ophiolite in northern Virginia: American Journal of Science, v. 281, p. 484–508.

Evans, N. H., 1991, Latest Precambrian to Ordovician metamorphism in the Virginia Blue Ridge: Origin of the contrasting Lovingston and Pedlar basement terranes: American Journal of Science, v. 291, p. 425–542.

Farrar, S. S., 1984, The Goochland granulite terrane: Remobilized Grenville basement in the eastern Virginia Piedmont, *in* Bartholomew, M. J., ed., The Grenville Event in the Appalachians and related topics: Geological Society of America Special Paper 194, p. 215–227.

Farrar, S. S., 1985, Tectonic evolution of the easternmost Piedmont, North Carolina: Geological Society of America Bulletin, v. 96, p. 362–380.

Gates, A. E., 1981, Geology of the western boundary of the Charlotte belt at Brookneal, Virginia [M.S. thesis]: Blacksburg, Virginia Polytechnic Institute and State University, p. 1–86.

Gates, A. E., and Glover, L., III, 1989, Alleghanian tectono-thermal evolution of the dextral transcurrent Hylas zone, Virginia Piedmont, USA: Journal of Structural Geology, v. 11, p. 407–419.

Gates, A. E., and Valentino, D. W., 1991, Late Proterozoic rift control on the shape of the Appalachians: The Pennsylvania reentrant: Journal of Geology, v. 99, p. 863–872.

Gates, A. E., Simpson, C., Glover, L., III, 1986, Appalachian carboniferous dextral strike slip faults: An example from Virginia: Tectonics, v. 5, p. 119–133.

Gates, A. E., 1986, Tectonic evolution of the Altavista area, southwest Virginia Piedmont [Ph.D. thesis]: Blacksburg, Virginia Polytechnic Institute and State University, 256 p.

Gates, A. E., Speer, J. A., and Pratt, T. L., 1988, The Alleghanian southern Appalachian Piedmont: A transpressional model: Tectonics, v. 7, p. 1307–1324.

Glover, L., III, 1989, Tectonics of the Virginia Blue Ridge and Piedmont, *in* Glover, L., III, Evans, N. H., Patterson, J. G., and Brown, W. C., Tectonics of the Virginia Blue Ridge and Piedmont: International Geological Congress 28th, Field Trip Guidebook T363, Washington, D.C., American Geophysical Union, p. 1–29, 52–59.

Glover, L., III and Klitgord, K., chief compilers, 1995, West Sheet, Southwestern Pennsylvania to Virginia Coastal Plain, and East Sheet,Virginia Coastal Plain to Baltimore Canyon Trough, *in* E-3 Southwestern Pennsylvania to Baltimore Canyon Trough: Geological Society of America, Centennial Continent/Ocean Transect #19, 2 sheets and text.

Glover, L., III, Sinha, A. K., Higgins, M. W., and Kirk, W. S., 1971, U-Pb dating of Carolina slate belt and Charlotte belt rocks, Virgilina district, Virginia, and North Carolina: Geological Society of America Abstracts with Programs, Southeastern Section, v. 3, p. 313.

Glover, L., III, Mose, D. G., Costain, J. K., Poland, F. B., and Reilly, J. M., 1982, Grenville basement in the eastern Piedmont of Virginia: A progress report: Geological Society of America Abstracts with Programs, Southeastern Section Meeting, v. 14, no. 1 and 2, p. 20.

Glover, L., III, Speer, J. A., Russell, G., and Farrar, S. S., 1983, Ages of regional metamorphism and ductile deformation in the central and southern Appalachians: Lithos, v. 16, p. 223–245.

Glover, L., III, Sheridan, R. E., and Wang, P., 1991, Taconic suture and the eastern extent of Laurentia in the Middle Atlantic States: EDGE project Investigations, Part I: Geological Society of America Abstracts with Programs, v. 23, p. 35.

Glover, L., III, Gates, A. E., Wang, P., and Valentino, D. W., 1992a, Anomalous heat source under the Laurentian margin during Cambrian collision with the Avalonian magmatic arc: Geological Society of America Abstracts with Programs, v. 24, p. 17.

Glover, L., III, Çoruh, C., Costain, J. K., and Bollinger, G. A. S., 1992b, Piedmont seismic reflection study: A program integrated with tectonics to probe the cause of eastern seismicity—Final Report: National Technical Information Service Report NUREG/CR-5731, 146 p.

Glover, L., III, Gates, A. E., Wang, P., and Valentino, D. W., 1994, Central Appalachian Late Proterozoic/Cambrian mélanges may be recycled Laurentian rift facies rocks formed during collision of Carolinia and Laurentia: Geological Society of America Abstracts with Programs, v. 26, p. 15.

Grow, J. A., Klitgord, K. D., and Schlee, J. S., 1988, Structure and evolution of Baltimore Canyon Trough, *in* Sheridan, R. E., and Grow, T. A., eds., Continental Margin, U.S.: Boulder, Colorado, Geological Society of America, Geology of North America, v. I-2, p. 269–290.

Hansen, H. J., and Edwards, J., Jr., 1986, The lithology and distribution of pre-Cretaceous basement rocks beneath the Maryland Coastal Plain: Maryland Geological Survey Report of Investigations 44, 27 p.

Harris, C. W., and Glover, L., III, 1988, The regional extent of the ca. 600 Ma Virgilina deformation: Implications for stratigraphic correlation in the Carolina terrane: Geological Society of America Bulletin, v. 100, p. 200–217.

Harry, D. C., and Sawyer, D. S., 1992, A dynamic model of extension in the Baltimore Canyon trough region: Tectonics, v. 11, p. 420–426.

Hatcher, R. D., Jr., 1989, Tectonic synthesis of the U.S. Appalachians, *in* Hatcher, R. D., Jr., Thomas, W. A., and Viele, G. W., eds., The Appalachian-Ouachita Orogen in the United States: Boulder, Colorado, Geological Society of America, The Geology of North America, v. F-2, p. 511–535.

Haworth, R. T., Daniels, D. L., Williams, H., and Zietz, I., 1980, Bouguer gravity anomaly map of the Appalachian orogen: Memorial University of Newfoundland Map 3a, scale 1:2,000,000.

Hess, H. H., 1962, History of ocean basins, *in* Engle, A. E. J., James, H. L., and Leonard, B. F., eds., Petrological studies: A volume in honor of A. F. Buddington: New York, Geological Society of America, p. 599–620.

Holbrook, W.S., and nine others, 1992, Deep velocity structure of rifted continental crust, U.S. Mid-Atlantic margin, from wide-angle reflection/refraction data: Geophysical Research Letters, v. 19, p. 1699–1702.

Holbrook, W. S., Purdy, G. M., Sheridan, R. E., Glover, L., III, Talwani, M., Ewing, J., and Hutcheson, D., 1994, Seismic structure of the U.S. Mid-Atlantic continental margin: Journal of Geophysical Research, v. 99, p. 17871–17891.

Hopson, C. A., 1964, The crystalline rocks of Howard and Montgomery Counties, Maryland: Baltimore, Maryland Geological Survey. (Reprinted from The Geology of Howard and Montgomery Counties, 1964.)

Horton, J. W., Jr., Blake, D. E., Wylie, A. S., and Stoddard, E. F., 1986, Metamorphosed mélange terrane in the eastern Piedmont of North Carolina: Geology, v. 14, p. 551–553.

Horton, J. W., Drake, A. A., and Rankin, D. W., 1989, Tectonostratigraphic terranes and their Paleozoic boundaries in the central and southern Appalachians: Geological Society of America Special Paper 230, p. 213–245.

Horton, J. W., Drake, A. A., Rankin, D. W., and Dallmeyer, R. D., 1991, Preliminary tectonostratigraphic terrane map of the central and southern Appalachians: U.S. Geological Survey Miscellaneous Investigations Series Map I-2163, scale 1:2,000,000.

Jachens, R. C., Simpson, R. W., Blakely, R. J., and Saltus, R. W., 1989, Isostatic residual gravity and crustal geology of the United States *in* Pakiser, L. C., and Mooney, W. D., eds., Geophysical framework of the continental United States: Geological Society of America Memoir 172, p. 405–424.

Keller, M. R., Robinson, E. S., and Glover, L., III, 1985, Seismicity, seismic reflection and geology of the central Virginia seismic zone. Part 3: Gravity: Geological Society of America Bulletin, v. 96, p. 1580–1584.

Keppie, J. D., and Dallmeyer, R. D., compilers, 1989, Tectonic map of pre-Mesozoic terranes in circum-Atlantic Phanerozoic orogens. International Geologic Correlation Programme Project 233: Geological Society of America Abstracts with Programs, v. 21, no. 6, p. 203, scale 1:5,000,000.

Kline, S. W., 1991, Provenance of arkosic metasediments in the Virginia Blue Ridge: Constraints on Appalachian suspect terrane models: American Journal of Science, v. 291, p. 189–198.

Klitgord, K. D., Hutchinson, D. R., and Schouten, H., 1988, U.S. Atlantic continental margin: Structural and tectonic framework, *in* Sheridan, R. E., and Grow, J. A., eds.; The Atlantic Continental Margin, U.S.: Boulder, Colorado, Geological Society of America, The Geology of North America, v. I-2, p. 19–55.

Lefort, J. P., and Max, M. D., 1991, Is there an Archean crust beneath Chesapeake Bay?: Tectonics, v. 10, p. 213–226

Lister, G. S., Etheridge, M. A., and Symonds, P. A., 1991, Detachment models for the formation of passive continental margins: Tectonics, v. 10, p. 1038–1064.

Manspeizer, W., ed., 1988, Triassic-Jurassic rifting: Continental breakup and the origin of the Atlantic Ocean and passive margins, Parts A and B., New York, Elsevier, 995 p.

Morgan, W. J., 1968, Rises, trenches, great faults, and crustal blocks: Journal of Geophysical Research, v. 73, p. 1959–1982.

Muller, P. D., and Chapin, D. A., 1984, Tectonic evolution of the Baltimore Gneiss anticlines, Maryland: Geological Society of America Special Paper 194, p. 127–148.

Muller, P. D., Candela, P. A., and Wylie, A. G., 1989, Liberty Complex; polygenetic mélange in the central Maryland Piedmont, *in* Horton, J. W., Jr., and Rast, N., eds., Geological Society of America Special Paper 228, p. 113–134.

Musser, D. L., 1993, Structure, stratigraphy and evolution of the Norfolk rift basin [M.S. thesis]: Newark, New Jersey, Rutgers University, 147 p.

Musser, D. L., Sheridan, R. E., and Schlische, R. W., 1992, Basin filling, subsidence and thermal history of the Norfolk rift basin, mid-Atlantic margin: Geological Society of America Abstracts with Programs, v. 24, p. 65.

Nelson, K. D., McBride, J. H., Arnow, J. A., Oliver, J. E., Brown, L. D., and Kaufman, S., 1985, New COCORP profiling in the southeastern United States. Part II: Brunswick and east coast magnetic anomalies, opening of the north-central Atlantic Ocean: Geology, v. 13, p. 718–721.

Pavlides, L., 1981, The central Virginia volcanic-plutonic belt: An island arc of Cambrian(?) age: U.S. Geological Survey Professional Paper 1231-A, p. A1–A34.

Phinney, R. A., 1986, A seismic cross section of the New England Appalachians: The orogen exposed *in* Reflection seismology: The continental crust: American Geophysical Union, Geodynamic Series Volume 14, p. 157–172.

Phinney, R. A. and Roy-Chowdhury, K., 1989, Reflection seismic studies of crustal structure in the eastern United States: Geological Society of America Memoir 172, p. 613–653.

Poland, F. B., 1976, Geology of the rocks along the James River between Sabot and Cedar point, Virginia [M.S. thesis]: Blacksburg, Virginia Polytechnic Institute and State University, 99 p.

Pratt, T. L., Çoruh, C., Costain, J. K., and Glover, Lynn, III, 1988, A geophysical study of the Earth's crust in central Virginia: Implications for Appalachian crustal structure: Journal of Geophysical Research, v. 93B, p. 6649–6667.

Quinlan, G., Beaumont, C., and Hall, J., 1993, Tectonic model for crustal seismic reflectivity patterns in compressional orogens: Geology, v. 21, p. 663–666.

Rankin, D. W., and nine others, 1989, Pre-orogenic terranes, *in* Hatcher, R. D., Thomas, W. A., Viele, G. W., eds., The Appalachian-Ouachita Orogen in the United States: Boulder, Colorado, Geological Society of America, The Geology of North America, v. F-2, p. 29–42.

Ratcliffe, N. M., Burton, W. C., D'Angelo, R. M., and Costain, J. K., 1986, Low-angle extensional faulting, reactivated mylonites, and seismic reflection geometry of the Newark basin margin in eastern Pennsylvania: Geology, v. 14, p. 766–770.

Reilly, J. M., 1980, A geologic and potential field investigation of the central Virginia Piedmont [M.S. thesis]: Blacksburg, Virginia Polytechnic Institute and State University, 111 p.

Robinson, E. S., Poland, P. V., Glover, L., III, and Speer, J. A., 1982, Some effects of regional metamorphism and geologic structure on magnetic anomalies over the Carolina slate belt near Roxboro, North Carolina [expanded abs.]: Society of Exploration Geophysicists, 52nd Annual International Meeting and Exposition, Dallas, Texas, Technical Program Abstracts and Biographies, p. 478–479.

Rogers, J. J. W., 1982, Criteria for recognizing environments of formation of volcanic suites; Application of these criteria to volcanic suites in the Carolina slate belt, *in* Bearce, D. N., Black, W. W., Kish, S. A., and Tull, J. F., eds., Tectonic studies in the Talledega and Carolina slate belts, southern Appalachian orogen: Geological Society of America Special Paper 191, p. 99–107.

Sacks, P. E., Maher, H. D., Jr., Secor, D. T., and Shervais, J. W., 1989, The Burks Mountain complex, Kiokee belt, southern Appalachian Piedmont of South Carolina and Georgia: Geological Society of America Special Paper 231, p. 75–86.

Secor, D. T., Jr., Samson, S. L., Snoke, A. W., and Palmer, A. R., 1983, Confirmation of the Carolina slate belt as an exotic terrane: Science, v. 221, p. 649–651.

Secor, D. T., Jr., Snoke, A. W., Bramlett, K. W., Costello, O. P., and Kimbrill, O. P., 1986a, Character of the Alleghanian orogeny in the southern Appalachians. Part I: Alleghanian deformation in the eastern Piedmont of South Carolina: Geological Society of America Bulletin, v. 97, p. 1319–1328.

Secor, D. T., Snoke, A. W., and Dallmeyer, R. D., 1986b, Character of the Alleghanian orogeny in the southern Appalachians. Part III: Regional tectonic relations: Geological Society of America Bulletin, v. 97, p. 1345–1353.

Secor, D. T., Jr., Murray, D. P., and Glover, L., III, 1989, Geology of the Avalonian rocks *in* Hatcher, R. D., Thomas, W. A., and Viele, G. W., eds., The Appalachian-Ouachita Orogen in the United States: Geological Society of America, The Geology of North America, v. F-2, p. 57–85.

Sheridan, R. E., 1983, Phenomena of pulsation tectonics related to the breakup of the eastern North American continental margin, *in* Sheridan, R. E., and Gradstein, F. M., eds., Initial reports of the Deep Sea Drilling Projects, Volume 76, Washington, D.C., U.S. Government Printing Office, p. 897–909.

Sheridan, R. E., Olsson, R. K., and Miller, J. J., 1991, Seismic reflection and gravity study of proposed Taconic suture under the New Jersey Coastal Plain: Implications for continental growth: Geological Society of America Bulletin, v. 103, p. 402–414.

Sheridan, R. E., and eight others, 1993, Deep seismic reflection data of EDGE U.S. and mid-Atlantic continental-margin experiment: Implications for Appalachian sutures and Mesozoic rifting and magmatic underplating: Geology, v. 21, p. 563–567.

Simpson, E. L., and Eriksson, K. A., 1989, Sedimentology of the Unicoi Formation in southern and central Virginia: Evidence for Late Proterozoic to Early Cambrian rift-to-passive margin transition: Geological Society of America Bulletin, v. 101, p. 42–54.

Sinha, A. K., Hund, E. A., and Hogan, J. P., 1989, Paleozoic accretionary history of the North American plate margin (central and southern Appalachians): Constraints from the age, origin, and distribution of granitic rocks, *in*, Hillhouse, J. W., ed., Deep structure and past kinematics of accreted terranes: International Union of Geodesy and Geophysics, American Geophysical Union, Geophysical Monograph 50, Volume 5, p. 219–238.

Southwick, D. L., 1964, Petrography of the basement gneiss beneath the Coastal Plain sequence, Island Beach State Park, New Jersey: U.S. Geological Survey Professional Paper 501-C, p. C55–C60.

Southworth, S., Burton, W. C., Froelich, A. J., Schindler, J. S., Lyttle, P. T., and Drake, A. A., 1994, Geologic maps of Loudon County, Virginia: Geological Society of America Abstracts with Programs, v. 26, p. 63.

Spencer, C., Green, A., Morel-a-l'Huissier, M., Luetgert, J., Stewart, D., Unger, J., and Phillips, J., 1989, The extension of Grenville basement beneath the northern Appalachians: Results from the Quebec-Maine seismic reflection and refraction surveys: Tectonics, v. 8, p. 677–697.

Stanley, R. S., and Hatch, N. L., Jr., 1988, The pre-Silurian geology of the Rowe-Hawley zone, *in* Hatch, N. L., ed., The bedrock geology of Massachusetts: U.S. Geological Survey Professional Paper 1366, p. A1–A39.

Stanley, R. S., and Ratcliffe, N. M., 1985, Tectonic synthesis of the Taconic orogeny in western New England: Geological Society of America Bulletin, v. 96, p. 1227–1250.

Stewart, D. B., Wright, B. E., Unger, J. D., Phillips, J. D., and Hutchinson, D. R., principal compilers, 1991, Global geoscience transect 8: Quebec–Maine–Gulf of Main transect, southeastern Canada, northeastern United States of America: U.S. Geological Survey Open File Report 91-353, 17 p.

Stewart, D. B., Wright, B. E., Unger, J. D., Phillips, J. D., and Hutchinson, D. R., 1993, Global geoscience transect 8: Quebec–Maine–Gulf of Maine transect, southeastern Canada, northeastern United States of America: U.S. Geological Survey Miscellaneous Investigations Series Map I-2329, scale 1:1,000,000.

Talwani, M., Worzel, J. L., and Landisman, M., 1959, Rapid gravity computations for two-dimensional bodies with application to the Mendocino submarine fracture zone: Journal of Geophysical Research, v. 64, p. 49–59.

Talwani, M, Ewing, J., Sheridan, R. E., Holbrook, W. S., and Glover, L., III, 1995, The EDGE experiment and the east coast magnetic anomaly, *in* Banda, E., Talwani, M., and Thorne, M., eds., Klewec NATO special volume on rifted ocean-continent boundaries: Overijs, Belgium, NATO ASI Publications Office, p. 155–181.

Thomas, W. A., Chowns, T. M., Daniels, D. L., Neathery, T. L., Glover, L., III, and Gleason, R. J., 1989, The subsurface Appalachians beneath the Atlantic and Gulf Coastal Plains, *in* Hatcher, R. D., Thomas, W. A., and Viele, G. W., eds., The Appalachian-Ouachita Orogen in the United States: Boulder, Colorado, Geological Society of America, The Geology of North America, v. F-2, p. 445–458.

Trzcienski, W. E., Jr., Rodgers, J., and Guidotti, C. V., 1992, Alternative hypotheses for the Chain Lakes "massif," Maine and Quebec: American Journal of Science, v. 292, p. 508–532.

Valentino, D. W., Gates, A. E., and Glover, L., III, 1994, Late Paleozoic transcurrent tectonic assembly of the central Appalachian Piedmont: Tectonics, v. 13, p. 110–126.

Wang, P., 1991, Geology and tectonic significance of the post-Grenville cover sequence in the eastern Blue Ridge of central Virginia [Ph.D. thesis]: Blacksburg, Virginia Polytechnic Institute and State University, 238 p.

Wang, P. and Glover, L., 1992, A tectonics test of the most commonly used geochemical discriminant diagrams and patterns: Earth Science Reviews, v. 33, p. 111–131.

Wehr, F., 1985, Stratigraphy of the Lynchburg Group and Swift Run Formation, Late Proterozoic (730–570 Ma), central Virginia: Southeastern Geology, v. 25, p. 225–239.

Wehr, F., and Glover, L., III, 1985, Stratigraphy and tectonics of the Virginia–North Carolina Blue Ridge: Evolution of a Late Proterozoic–early Paleozoic hinge zone: Geological Society of America Bulletin, v. 96, p. 285–295.

Wernicke, B., and Tilke, P. G., 1989, Extensional tectonic framework of the U.S. central Atlantic passive margin, *in* Tankard, A. J., and Balkwill, H. R., eds., Extensional tectonics and stratigraphy of the North Atlantic margin: American Association of Petroleum Geologists Memoir 46, p. 7–21.

Wilkes, G. P., Johnson, S., and Milici, R. C., 1989, Exposed and inferred early Mesozoic basins onshore and offshore, Virginia: Virginia Division of Mineral Resources Publication 94, p. .

Woodcock, N. G., and Fischer, M., 1986, Strike-slip duplexes: Journal of Structural Geology, v. 8, p. 725–735.

Wright, J. E., Sinha, A. K., and Glover, L., III, 1975, Age of zircons from the Petersburg Granite, Va.; with comments on belts of plutons in the Piedmont: American Journal of Science, v. 275, p. 848–856.

Zen E-an, Hepburn, J. C., Kidd, W. S., Robinson, P., Skehan, J. W., and Thompson, J. B., 1989, Tectonostratigraphic terranes in the northern Appalachians: Their distribution, origin, age; evidence for their existence, *in* International Geological Congress, 28th, Field Trip Guidebook T359: Washington, D.C., American Geophysical Union, 68 p.

Zietz, I., and Gilbert, F. P., 1980, Aeromagnetic map of part of the southeastern United States in color: U.S. Geological Survey Geophysical Investigations Map GP-936, scale 1:2,000,000.

Zietz, I., Calver, J. L., Johnson, S. S., and Kirby, J. R., 1977, Aeromagnetic map of Virginia: U.S. Geological Survey Geophysical Investigation Map GP-915, scale 1:500,000.

Manuscript Accepted by the Society April 30, 1996

Typset and printed in U.S.A. by Johnson Printing, Boulder, Colorado